THE

STRUCTURE AND FUNCTION

OF

ENZYMES

BIOLOGY TEACHING MONOGRAPH SERIES

CYRUS LEVINTHAL, *Editor*
Massachusetts Institute of Technology

Sidney Bernhard, *University of Oregon* THE STRUCTURE AND FUNCTION OF ENZYMES

Vernon M. Ingram, *Massachusetts Institute of Technology* THE BIOSYNTHESIS OF MACROMOLECULES

Albert L. Lehninger, *Johns Hopkins University* BIOENERGETICS

James D. Watson, *Harvard University* THE MOLECULAR BIOLOGY OF THE GENE

SIDNEY A. BERNHARD
University of Oregon

THE STRUCTURE AND FUNCTION OF ENZYMES

1968

W. A. BENJAMIN, INC. New York Amsterdam

THE STRUCTURE AND FUNCTION OF ENZYMES

Library of Congress Catalog Card Number 67–19434
Manufactured in the United States of America
2345K321098 C868

The manuscript was put into production on November 4, 1966; this volume was published on January 2, 1968

W. A. BENJAMIN, INC.
New York, New York 10016

EDITOR'S FOREWORD

GREAT CHANGES HAVE TAKEN PLACE RECENTLY IN THE TEACHING of undergraduate biology. To some extent, these changes are due, as they are in many other sciences, to the upgrading of high school science education. However, the major factor for biology would seem to be the change which has taken place during the last twenty to thirty years in the science itself. Biology, which has in the past been taught primarily as a descriptive subject, is now changing to one which is primarily analytical. One result of this change is that the undergraduate biology courses are becoming more and more dependent on physics, mathematics, and chemistry in order to enable the student to achieve a deeper understanding of his subject. Biochemistry, which until recent years has been a subject for advanced biology students and for medical students, has become an increasingly important part of the undergraduate training for all biology majors. In addition, the developments in molecular biology have not only become a part of the undergraduate biology curricula, but are increasingly necessary for a general understanding of our scientific culture.

This textbook series is designed to aid those biology teachers who are developing new programs for undergraduate biologists. We do not imagine a program lacking in the subject matter of "classical biology." Rather, we are thinking of a program in which the teaching of many of the more classical biological subjects will be firmly based on the student's background in biochemistry and molecular biology.

CYRUS LEVINTHAL

Cambridge, Massachusetts
December 1964

PREFACE

THIS BOOK HAS BEEN WRITTEN WITH THE HOPE THAT IT WILL serve as a partial text for a course in biochemistry for undergraduate students who have been exposed to introductory courses in chemical principles, organic chemistry, and cellular biology. My intent has been to emphasize both the conventionality of the chemical (electronic) mechanisms of biological reactions and the uniqueness of structure, selectivity, and reaction velocity of the catalysts (enzymes) which mediate these reactions.

The first two chapters are meant to serve as background material for a variety of students of biology and chemistry. Many who are conversant with the material in one of these chapters will have little familiarity with the other. The subject matter proper (enzymes) begins with Chapter 3. The regular covalent structure and regular configuration of polypeptides as well as the *unique* structural features of enzymes are introduced in this chapter. Deliberately, in subsequent chapters I have chosen to alternate between *structure* and *function* (catalyzed reaction mechanism). In the final chapter I have attempted to present some details of known structure-function relationships for a few "select" enzyme-substrate systems. This is a subject which is at present receiving a great deal of attention and which is consequently undergoing rapid growth and revision. For example, the three-dimensional structures of ribonuclease and chymotrypsin (two of the "select" enzymes) have both been determined to atomic resolution in the interim between the revision of galley proof and the arrival of page proof for Chapter 8. Doubtless other major changes in our understanding of specific enzyme-catalyzed mechanisms are soon forthcoming. I have attempted to restrict to the final chapter that material which is obviously subject to expansion and revision.

The three major topics relevant to enzymes that are not covered in this book are (1) the molecular details of intermediary metabolic pathways, (2) the consequent bioenergetics associated with these chemical (metabolic) conversions, and (3) the biosynthesis of proteins. Bioenergetics and biosynthesis are covered by Lehninger and by Ingram in the companion volumes in this series. Intermediary metabolism is dealt with in considerable detail in standard texts on biochemistry (see the references in Chapter 1). The subject is often discussed in sufficient detail in elementary texts and in courses on modern biology.

In writing this book I have benefited greatly by criticism from students and colleagues. Not infrequently, I have had to revise (or more accurately, *correct*) sections I had earlier considered as definitive. Impatience has undoubtedly limited the extent of my revisions. Countless colleagues and students have contributed ideas to the text, and I am indebted to all of them. A few in particular have contributed to a major extent in both thought and time: My former student, Dr. Harry Noller, took notes conscientiously in my course on enzymes, and improved on them to the extent that they could form a basis for this text. My colleague Dr. Om Malhotra read and criticized the entire manuscript, as did Prof. George Stark of Stanford University Medical School. An extensive revision of Chapter 5 was made by Prof. Stark. The entire text has been reviewed by my long standing collaborator, Prof. Freddie (Herbert) Gutfreund of Bristol University. The manuscript was prepared with the expert assistance (both in typing and in English usage) of Mrs. Susan Keizer and Miss Clara Ueland.

Sidney A. Bernhard

Eugene, Oregon
September 1967

CONTENTS

ONE ~ INTRODUCTION

1–1 HISTORICAL

THE MECHANISM OF ENZYME ACTION IS ONE OF THE IMPORtant and challenging topics confronting both chemists and biologists today. In order to gain proper perspective on the current concepts of enzyme action, it is helpful to review briefly the history of the discoveries and theories which have had a profound influence on the development of the subject.

The concept that chemical reactions in living systems occur via some special mechanism distinct from reactions studied in the chemical laboratory has been recognized for over two hundred years. It was first demonstrated that living systems could be broken up into smaller units which could not themselves be classified as "living"; yet some of these subunits could still carry out the distinctive chemical reactions of a living system. A concept current at the beginning of the nineteenth century was that these subcellular systems still contained some of the "vital force" of the living system and that this "vital force" was in some way transmitted so as to direct specific reactions. The development of more rigorous chemical concepts as to the nature of the action of these subcellular systems ("enzymes") awaited the development of a general theory of chemical catalysis. In 1835 the Swedish chemist Berzelius set forth such a theory. In the discussion of his theory, Berzelius cited a number of examples of chemical catalysis, including catalysis by enzymes. He noted the tremendous potency of an "enzyme" (diastase) compared with an inorganic catalyst (sulfuric acid) in the catalytic hydrolysis of starch. Berzelius suggested that the "vital force" previously assumed to be a part of the subliving enzyme system might, in fact, be a property of the chemical reactants themselves and that the role of the catalyst (enzyme) was to direct chemical reactions along particular paths rather than to impart to chemical compounds a special reactivity (or energy) to undergo physiological reactions. So brilliant and important was this first paper on catalysis (published in a Swedish journal) that it was translated into German by the outstanding German chemist F. Wohler in the following year (1836). Questions raised by Berzelius in this paper were to have an important consequence for research in catalysis for many years thereafter. Among these questions were: (1) Are the particular products of a chemical reaction a consequence of the nature of the catalyst? (2) Can a catalyst act on a wide variety of reactants, or is it restricted in its catalytic action to a single chemical reaction?

Enzyme research during the nineteenth century was devoted largely to the isolation of subcellular enzyme systems, and to the description of the chemical reactions which these systems catalyzed. Almost invariably, it was found that further purification of a previously investigated enzyme system resulted in a separation of the original system into fractions, each fraction exhibiting a more limited variety of catalytic functions. It became evident that the enzyme systems isolated initially were composed of many different catalysts, each catalyst enhancing the velocity of a small number of distinct chemical reactions. (The number of distinctly different biological catalysts which have been identified in the laboratory are, by now, enormous. For example, a simple bacterium contains the order of 10^4 distinctly different enzymes.)

Concurrently with the discoveries of the large number of enzyme catalysts and of their limited individual functions, extensive information was being gathered concerning catalysis by simple inorganic and organic compounds, particularly catalysis by acids, bases, heavy metals, and their salts. This information led to the generalization that these nonbiological catalysts were far less specific than enzymes in the chemical reactions which they could affect. For example, a simple molecule such as sulfuric acid was found to be catalytic for a wide variety of organic and inorganic chemical reactions.

By the middle of the nineteenth century the following generalizations concerning enzyme action could be made:

(1) The rate of an enzyme-catalyzed reaction is usually much faster than the rate of the same reaction when directed by nonbiological catalysts.

(2) Enzyme catalysts, as distinct from other catalysts studied in the laboratory, are highly specific and catalyze only one or a small number of chemical reactions.

(3) There exist a great variety of enzyme catalysts, each of which carries out a limited function. Complex biological processes involve a very large number of different chemical reactions mediated by a large number of enzyme catalysts.

The fundamental question to be answered was, "How do enzymes carry out these very rapid and highly specific catalytic functions?" The answer to this question is a primary concern of a very large group of research scientists at the present time. A wealth of relevant factual information and theory has been accumulated over the past century. This information, and the way it can be assembled towards an understanding of the mechanism (or mechanisms) of enzyme action is the subject of this book. In the remainder of the present chapter we shall consider the experimentally based generalizations about enzyme structure and mechanism which have become accepted as axiomatic, and to which we shall frequently refer in subsequent chapters.

1–2 AXIOMS

Brief statements of the axioms of enzymology are contained in this section; the justification for these axioms is herein limited to a description of experimental results which had a direct bearing at the time such conclusions were first proposed. In subsequent discussions we shall considerably amplify on additional relevant experiment and theory.

ENZYMES ARE MEMBERS OF A CLASS OF ABUNDANT BIOLOGICALLY OCCURRING MOLECULES TERMED PROTEINS

Proteins are compounds containing principally the elements carbon (~ 60% by weight), nitrogen (~ 16%), oxygen (~ 16%), and hydrogen (~ 8%). They generally contain a small amount of sulfur. All enzymes have elemental compositions typical of proteins.

PROTEINS (ENZYMES) ARE VERY LARGE (MACRO) MOLECULES

Proteins cannot permeate through many types of membranes through which smaller molecules of various types freely pass, as for example in the experiment described in Fig. 1–1.

Measurements of the physical properties of solutions which depend on the *number* of dissolved solute molecules (as for example, the osmotic pressure of a solution) indicate that the *number* of protein solute molecules per weight of protein is unusually small. The molecular weights of proteins, thus determined, are in the range of 10^4 to 10^6 g of protein per mole.

FIG. 1–1 A dialysis experiment. The large protein molecules are unable to pass through the pores of the membrane, whereas water molecules and small ions can freely permeate. Due to the high concentration of protein, the interior of the sac fills with solvent (water) until the osmotic pressure inside and outside are equal.

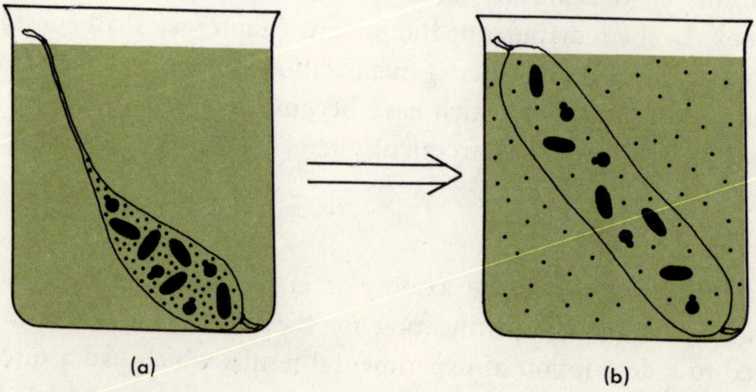

Measurements of the physical properties of solutions which depend on the *mass* of dissolved solute molecules, most notably, the velocity of sedimentation of solute in a high centrifugal field, indicate that proteins are molecules of very large mass (molecular weights of $\sim 10^4$–10^6 g).

Measurements of the physical properties of solutions which depend on the *dimensions* of the solute molecules (as for example, the scattering of monochromatic light from monodispersed protein solutions) indicate that water-soluble proteins have molecular *dimensions* which correspond to the above quoted molecular weights. The extent of the scattering of light from a monodispersed (homogeneously distributed solute) solution depends on the ratio of the dimensions of the solute molecules to the wavelength of the monochromatic light.

PROTEINS (ENZYMES) ARE CONSTRUCTED FROM SIMPLE AMINO ACID SUBUNITS LINKED TOGETHER IN LINEAR ARRAY BY COVALENT (PEPTIDE) BONDS

The repeating unit in the covalent structure of a protein is illustrated in Fig. 1–2. A single unbranched chain of amino acids linked together via *peptide* bonds is termed a *polypeptide*. As we shall demonstrate subsequently (Chapter 5), enzyme molecules may be composed of either a single polypeptide chain or a specific aggregate of polypeptides. The molecular weights

FIG. 1–2 The repetitive sequence of amino acids in polypeptide linkage.

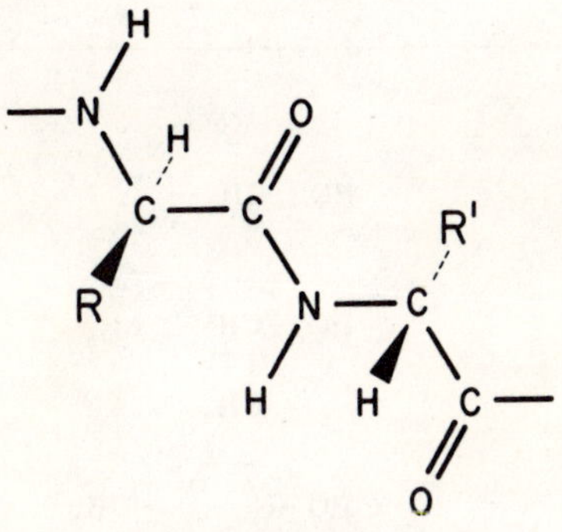

Table 1-1
Classification of R Groups

Type and Name	Structure	Comments
Hydrocarbon amino acids		Inert. Interactions due to dispersion forces
Alanine	CH_3-	
Valine	$(CH_3)_2CH-$	
Leucine	$(CH_3)_2CH-CH_2-$	
Isoleucine	$(C_2H_5)(CH_3)CH-$	
Phenylalanine	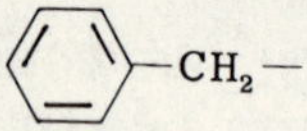	
Charged amino acids		Electrostatic interactions. Hydrogen bond acceptors (anions) and donors (cations)
Aspartate	$^{\ominus}O_2C-CH_2-$	
Glutamate	$^{\ominus}O_2C-CH_2-CH_2-$	
Lysine	$^{\oplus}H_3N-CH_2-CH_2-CH_2-CH_2-$	
Arginine	$(H_2N)_2C^{\oplus}\cdots NH-CH_2-CH_2-CH_2-$	
Hydroxylic		Hydrogen bond donors and acceptors. Dipolar interactions
Serine	$HO-CH_2-$	
Threonine	$HO-CH(CH_3)-$	
Tyrosine	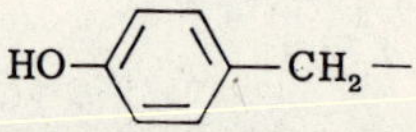	

(Continued)

Table 1-1 (Continued)
Classification of R Groups

Type and Name	Structure	Comments
Inert heteroatom		Similar to hydrocarbons. Strong dispersion forces
Methionine	$CH_3-S-CH_2-CH_2-$	
Tryptophan	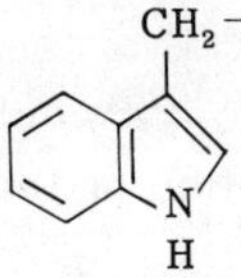	
Amide		Hydrogen bond donors and acceptors. Dipolar interactions
Asparagine	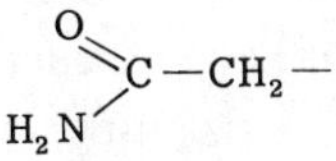	
Glutamine	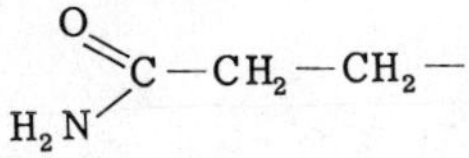	
Miscellaneous: Specialized Properties		
Glycine	$H-$	No β-carbon atom. More flexible peptide chain
Cysteine	$HS-CH_2-$	Like serine, but chemically reactive to form cystine
Cystine	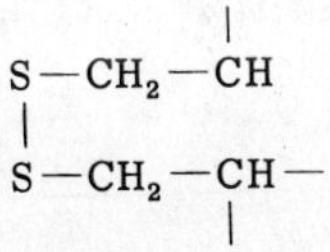	Introduces cross-linked polypeptide chains. Severely limits the flexibility of the polypeptide
Histidine	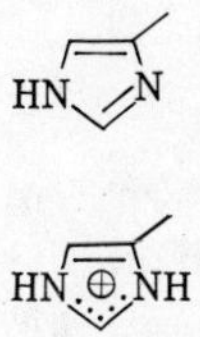	The cation and the neutral conjugate base exist in comparable amounts at pH neutrality. The cation is comparable to arginine, the neutral molecule to phenylalanine. Hydrogen bonding
Proline	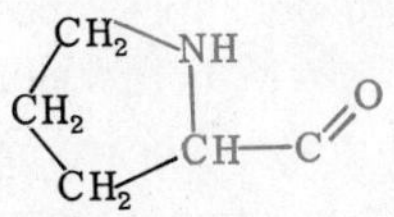	Imino acid. Causes a rigid peptide link

quoted in the above discussion vary to a large extent as a consequence of the variable number of polypeptide chains contained within different enzyme molecules.

When proteins are completely degraded by prolonged exposure to strong aqueous acid or base, small *amino acids* of known structure and of limited variety can be isolated in high yield. These constituent amino acids are summarized in Table 1–1. Note that the various amino acids differ only in the nature of the "R" groups. A classification of the twenty different amino acid constituents, according to the special features of these various "R" groups is contained in Table 1–1. When proteins are incompletely degraded (by terminating the degradative process sooner), peptides of known structure can be isolated, for example the dipeptides glycylalanine and alanylglycine (Eq. 1–1).

$$\overset{\oplus}{H_3N}-CH_2-\overset{\overset{\textstyle O}{\|}}{C}-NH-\underset{\underset{\textstyle CH_3}{|}}{CH}-CO_2^{\ominus} \qquad \overset{+}{H_3N}-\underset{\underset{\textstyle CH_3}{|}}{CH}-\overset{\overset{\textstyle O}{\|}}{C}-NH-CH_2-CO_2^{\ominus}$$

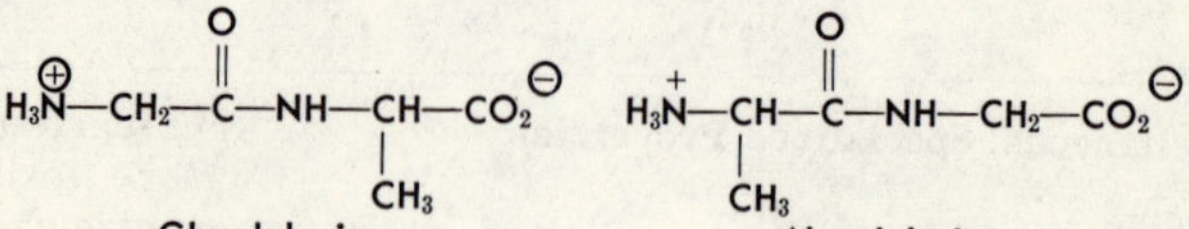

Glycylalanine   Alanylglycine   (1–1)

The acidic and basic conditions required for the degradation of proteins are similar to those required for the degradation of amides, as for example in the strong acid- and strong base-catalyzed hydrolysis of N-methyl acetamide (Eq. 1–2).

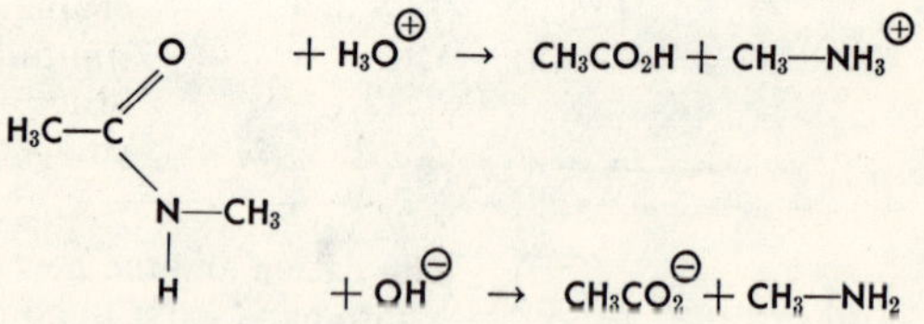

$$H_3C-\overset{\overset{\textstyle O}{\|}}{\underset{\underset{\textstyle \underset{\textstyle H}{|}}{\underset{\textstyle N-CH_3}{|}}}{C}} \quad \begin{array}{l} +\,H_3O^{\oplus} \;\rightarrow\; CH_3CO_2H + CH_3-\overset{\oplus}{NH_3} \\[20pt] +\,OH^{\ominus} \;\rightarrow\; CH_3CO_2^{\ominus} + CH_3-NH_2 \end{array} \qquad (1–2)$$

THE AMINO ACID CONSTITUENTS OF PROTEINS ARE ALL OF THE SAME GEOMETRICAL CONFIGURATION

Note that the amino acids (with the exception of glycine) can potentially be of two nonsuperimposable configurations. If an amino acid is synthesized in the laboratory from materials lacking an asymmetric tetrahedral carbon atom (as for example, in the reductive amination of pyruvic acid, illustrated in Fig. 1–3), the resultant product is invariably an exactly equal mix-

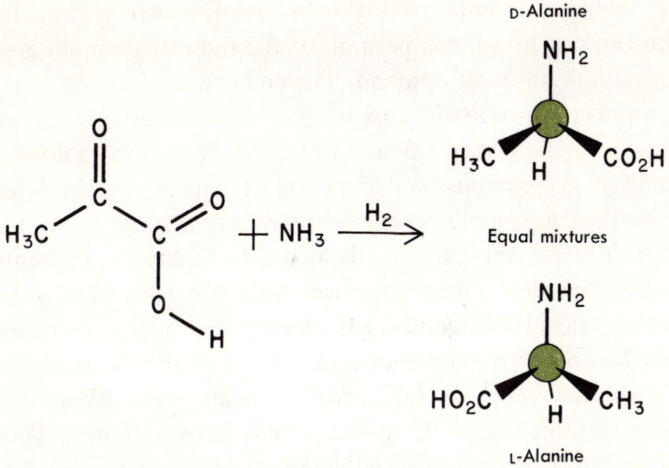

FIG. 1–3 Reductive amination of pyruvic acid.

ture of molecules of each of the two configurations. Equal (*racemic*) mixtures of DL-alanine (or any other amino acid) can be distinguished from the naturally occurring pure L isomer in two ways:

The crystalline structures of the two solids may be notably different since an equal mixture of D and L molecules cannot be arranged (packed) in the same three-dimensional array as a collection of pure L molecules. The melting points and the regular geometrical shapes may differ, being dependent on the particular *inter*molecular arrangements within the crystal.

Solutions of the pure L isomer will rotate the plane of plane-polarized light in a manner dependent on the asymmetric *intramolecular* electron density. This phenomenon is known as *optical rotation.* A solution of pure D-alanine would rotate the plane to the same extent but with an exactly opposing angle. All protein solutions exhibit this property of optical rotation. A DL mixture would lead to a cancellation of the two opposed optical rotatory effects and hence to a loss of this property.

THE PROPERTIES OF EACH PROTEIN ARE DETERMINED BY ITS PARTICULAR SEQUENCE OF AMINO ACID RESIDUES

Each distinct protein contains a unique linear sequence of the twenty kinds of amino acids in peptide linkages. Not only is the sequence of amino acids distinctive for each protein, but also the percentage composition of the individual amino acids varies from protein to protein. The amino acid compositions of a number of proteins are designated in Table 1–2. These proteins perform very different functions. From a more detailed study of the composition of peptide fragments of these proteins, the complete linear covalent sequence of amino acids has, in each case, been determined. As will be discussed in Chapter 5, this complete linear sequence could not have been determined unless the large majority of all protein molecules of one type had precisely the same amino acid sequence. A number of recognizably different (abnormal) human hemoglobins have been identified, each from a different human source. These recognizable differences in either the function or the structure of the protein have in every case been demonstrated to arise from a change of only a single amino acid residue in the polypeptide sequence.

The specific linear sequence of amino acid residues in a polypeptide not only dictates a set of unique chemical properties but also determines the three-dimensional structure of the polypeptide chain and the specific way in which polypeptide chains aggregate into more complex protein structures.

ENZYME PROTEINS HAVE ORDERLY STRUCTURES

Enzymes can be crystallized. Such crystals scatter electromagnetic radiation of wavelengths comparable to chemical bond lengths (i.e., X rays) in an orderly fashion characteristic of crystalline solids (solids in which the atomic nuclei are rigidly fixed in position in a regular crystal lattice). It has been demonstrated in a number of instances that aqueous suspensions of such enzyme crystals are catalytically active. Hence the function of enzyme can be maintained in the crystalline state. In contrast, all enzymes (like all other proteins) undergo a process called *denaturation*. This process results in the disruption of the orderly organized structure of the protein. Denaturation may be achieved by elevation of the temperature, or by subjecting the protein to particular solvents which interfere with the noncovalent intramolecular forces responsible for maintenance of the

Table 1-2
Amino Acid Composition of Some Polypeptide Chains of Diverse Function

	Beef insulin	Chicken lysozyme	Human hemoglobin (β-chain)	Tobacco mosaic virus protein	Bovine chymotrypsinogen
Alanine	3	12	15	11	22
Arginine	1	11	3	9	4
Asparagine	3	13	0	10	14
Aspartic acid	—	8	13	7	9
Cysteine (or $\frac{1}{2}$ cysteine)	6	8	2	1	10
Glycine	4	12	13	6	23
Glutamine	3	3	0	12	10
Glutamic acid	4	2	11	7	5
Histidine	2	1	9	0	2
Isoleucine	1	6	0	7	10
Leucine	6	8	18	13	19
Lysine	1	6	11	2	14
Methionine	—	2	1	1	2
Phenylalanine	3	3	8	8	6
Proline	1	2	7	8	9
Serine	3	10	5	16	28
Threonine	1	7	7	17	23
Tryptophan	—	6	2	3	8
Tyrosine	4	3	3	5	4
Valine	5	6	18	15	23
Total	51	129	146	158	245

organized structure (*denaturants*). Denatured proteins do not crystallize. Denatured enzymes are invariably catalytically inert (i.e., the enzyme function is lost). In the denaturation process, no chemical bonds are disrupted.

COROLLARY: *Intramolecular structural integrity is essential for catalytic action.*

CATALYTIC ACTION IS CARRIED OUT WITHIN A GEOMETRICALLY DISCRETE REGION OF THE ENZYME PROTEIN (THE ENZYME SITE)

Two types of experiments, carried out near the turn of the century, are relevant to this axiom. One type of experiment is concerned with the relationship between the chemical configuration of a potential reactant and the capability of the enzyme to catalyze the reaction. The other type of experiment involves the chemical kinetic method of investigation of reaction mechanism, where the dependence of the reaction rate on the concentration of reactant and catalyst is investigated. From the details of the dependence of reaction rate on these concentrations, it is often possible to draw inferences as to the mechanism by which these species interact in the course of a catalyzed reaction.

On the basis of large differences in rate of enzymic catalysis among organic molecules having very similar chemical properties, but different stereochemical configurations, it was concluded that enzymes are highly specific in the selection of *substrates* (reactants) according to a particular configuration. Emil Fischer first postulated that the high degree of stereospecificity actually observed in enzyme-catalyzed processes must reflect a stereochemical *complementary* attachment of substrate to enzyme during the course of catalysis. This stereochemical specificity is most dramatically illustrated in the specificity of particular enzymes for L or for D substrates exclusively. For example, the enzyme chymotrypsin catalyzes the hydrolysis of acetyl-L-phenylalanine methyl ester (Fig. 1–4), but is virtually inert to the corresponding D isomer, a molecule with identical chemical properties. The relative size of enzyme and substrate is of interest. Note that in this example, the substrate is approximately the size of a peptide of two amino acid residues. This size is typical of the extent of the complementary surfaces

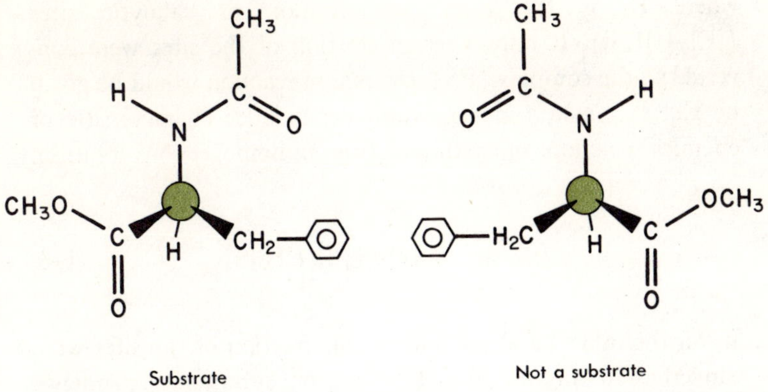

FIG. 1–4 The D and L isomers of N-acetyl-L-phenylalanine methyl ester.

between enzymes and substrates. Since the enzyme (chymotrypsin) molecule contains about 250 amino acid residues, the substrate represents about 1% of the total enzyme-substrate mass. In later sections it will be demonstrated that there is only one catalytic site per chymotrypsin molecule.

Fischer's postulate was utilized first qualitatively, and later quantitatively, by chemical kineticists in the correlation of the effect of substrate concentration on the rate of enzyme-catalyzed reactions. Many enzyme-catalyzed reactions can be described by the formal reaction pathway illustrated in Eq. 1–3.

$$\begin{array}{l} E + S \rightleftharpoons ES \text{ (enzyme-substrate complex)} \\ \qquad\quad \downarrow k_P \\ \qquad\quad E + \text{Products} \end{array} \tag{1–3}$$

If the concentration of substrate is very much greater than the concentration of catalytic enzyme sites, or if the rate of decomposition of the *enzyme-substrate complex* (ES) is much slower than the rate of dissociation of the complex to enzyme and substrate, the rate of appearance of product or disappearance of substrate is given by Eq. 1–4

$$-\frac{d\,[S]}{dt} = \frac{d\,[P]}{dt} = \frac{k_P\,E_0}{1 + \dfrac{[E]}{[ES]}} \tag{1–4}$$

where E_0 is the total concentration of catalytic sites ([E] + [ES]). If only a trivial fraction of the sites were converted to the complex (ES), the rate of reaction would be given by Eq. 1–5, a rate-concentration dependence characteristic of chemical reactions of small molecules in homogeneous solution.

$$-\frac{d[S]}{dt} = k_P[ES] \simeq k'E_0[S] \tag{1–5}$$

If, on the other hand, an appreciable fraction of the sites were converted to the complex, the rate-concentration dependence would be as given in Eq. 1–6.

$$-\frac{d[S]}{dt} = \frac{k_P E_0 [S]}{K + [S]} \qquad \text{where } K = \frac{[E][S]}{[ES]} \tag{1–6}$$

Equation 1–6 (which is derived explicitly in Chapter 4) is commonly known as the *Michaelis-Menten equation*. It is applicable to many enzyme catalyzed reactions; under appropriate conditions of concentration and environment, Eq. 1–6 can be demonstrated to be applicable to nearly all enzyme catalyzed reactions. Note that a plot of reaction velocity as a function of substrate concentration, under the condition $[S] > E_0$, will have the curvature shown in Fig. 1–5, and that at sufficiently high concentrations of substrate ($[S] > K$), the reaction velocity becomes independent of the substrate concentration. Such *zero-order* dependence of reaction velocity on substrate concentration is easily demonstrable in most enzyme-specific substrate systems. Concentrations of substrate in the range 10^{-3}–$10^{-5}M$ are typically sufficient for nearly complete "saturation" of the enzyme sites (when $[S] > K$, $[ES] \sim E_0$). This saturation phenomenon, a characteristic of enzyme-substrate systems, is a reflection of the strong and specific interaction of the enzyme protein with the substrate. It was intuitively obvious very early, that such specific and strong interactions could not occur repeatedly in many different regions along a single polypeptide chain, particularly since different enzymes (different polypep-

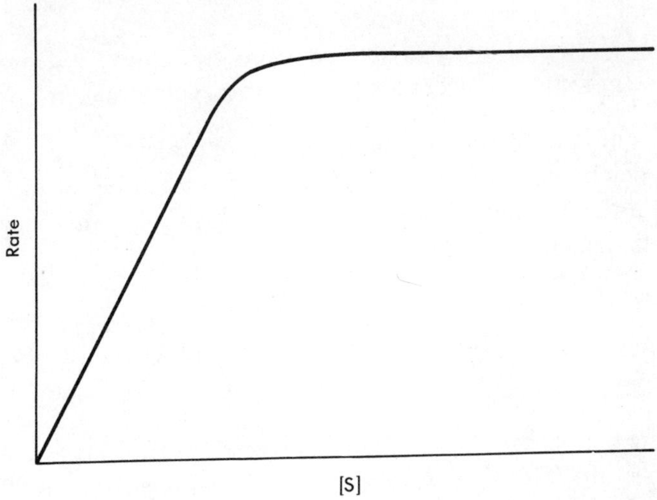

FIG. 1–5 The dependence of reaction rate on substrate concentration, for an enzyme-substrate system obeying Michaelis-Menten kinetics.

tides) are highly specific for very different stereochemical configurations. Hence, there is strong implication that the number of enzyme sites are limited and restricted to particular regions of the large polypeptide chain.

Note that in Eq. 1–6 the number of enzyme sites (E_0) cannot be determined independent of the *specific rate* (k_P) of chemical reaction. The rate of reaction at limiting high substrate concentration is $k_P E_0$. The number of enzyme sites per enzyme molecule was not determined for any enzyme until long after the first derivations of Eq. 1–6. However, it was apparent that the number of sites per protein molecule was small in specific instances, from the observation that various enzymes which require *cofactors* (nonprotein small molecules essential for particular enzymic catalyses) can function optimally at very low concentrations of cofactor of the order of the enzyme protein concentration, and from the fact that various "poisons" (i.e., *irreversible inhibitors*) of enzyme activity are effective in abolishing catalytic activity at concentration levels stoichiometrically comparable to that of the enzyme protein molecules.

THE RATES OF CONVERSION OF ENZYME-SUBSTRATE COMPLEXES TO PRODUCTS ARE EXTREMELY RAPID

In instances where the concentrations of actual sites (E_0) can be determined, the specific rate constant, k_P, can be calculated from the limiting velocity of reaction at high substrate concentration, namely

$$-\left(\frac{d[S]}{dt}\right)_{[S]\to\infty} \to k_P E_0$$

Note that k_P has only the dimensions of $(\text{time})^{-1}$. This parameter is often referred to as the *turnover number*, that is, the number of substrate molecules converted to product per enzyme site per unit time. Some typical known turnover numbers are listed in Table 1–3. Note both the enormous velocities and their relatively small variability.

Table 1-3
Maximum Turnover Numbers of Enzymes

	Turnover number, sec^{-1}
Acetylcholinesterase	10^4
Chymotrypsin	10^2 (to 10^3)
Trypsin	10^2 (to 10^3)
Ribonuclease	10^2
Papain	10
Carboxypeptidase	10^2
Myosin ATPase	10^2
Kinases	10^3
Urease	10^4
Fumarase	10^3
Aldolase	10^2
Enolase	10^2
Carbonic anhydrase	10^5
Transaminases	10^3
Dehydrogenases	10^3
Catalase	10^7
Peroxidase	10

1–3 ENZYME ACTION: PRESENT, PAST, AND FUTURE PROBLEMS

It is noteworthy that the "vital force" hypothesis of enzyme action remained prevalent throughout the nineteenth century. The uniqueness of biological reaction processes could not be intuitively rationalized on the basis of the existent chemical and physical theory during that period. Proponents of the "vital force" theory included the eminent microbiologist and biochemist, Louis Pasteur. The proponents of *vitalism* emphasized the uniqueness of biological processes, a fact which might otherwise have been overlooked by physical scientists of that time. Contemporary physical and chemical theory does, however, present further insight into the nature of *biological specificity*. For this reason a review of atomic and molecular forces of interaction, pertinent both to enzyme-substrate complex formation and to catalyzed reaction mechanisms, is presented in the following chapter.

The problems facing the contemporary student of enzyme action may be summarized as follows:

What are the conformations and the detailed chemical bond configurations of enzyme sites?

Why do these particular structures lead to enormously rapid and stereospecific chemical reactions?

In what way is the mechanism of interaction between a small substrate molecule and a specific enzyme site affected by the much larger remaining structure of the enzyme (protein) molecule?

REFERENCES

Mahler, H. R., and Cordes, E. H., *Biological Chemistry*, Harper and Row, New York, 1966. A comprehensive modern treatise on molecular biochemistry, including the mechanism of enzyme action and the structure of proteins. Each chapter contains an excellent list of references to the recent literature.

Fruton, J. S., and Simmonds, S., *General Biochemistry*, 2nd ed.,

Wiley, New York, 1963. An extremely well-written and comprehensive text on the chemistry of cellular components of known chemical structure. Chapters 2 and 3 deal with the properties of proteins and their amino acid constituents.

Segal, H. L., "The Development of Enzyme Kinetics," in *The Enzymes*, 2nd ed., Vol. 1, Academic Press, New York, 1959. A fascinating and scholarly historical account of the development of the concept of enzyme sites and complexes.

Anfinsen, C. B., *The Molecular Basis of Evolution*, Wiley, New York, 1960 (available in paperback). A clear description of protein chemistry, including chemical aspects of structure, is contained in early sections.

Watson, J. D., *Molecular Biology of the Gene*, Benjamin, New York, 1965. This contemporary masterpiece deals with all aspects of modern biology which can now be rationally interpreted at the molecular level. Chapter 2 lays to rest the "vital force" hypothesis and presents a biological rationale for the writing of my own text.

TWO — THE PHYSICAL FORCES OF INTERACTION BETWEEN ATOMS AND MOLECULES

IN ORDER TO UNDERSTAND THE VERY HIGH AFFINITY OF ANY enzyme site for a substrate and how such a specific interaction leads ultimately to rapid catalysis, we shall review briefly the nature of the forces of interaction between atoms and between molecules. The following discussion is not intended to be comprehensive, but merely to serve as an outline of those forces which are known to play an important role in specific enzyme-substrate interaction. Such forces can be divided (somewhat arbitrarily) into two classes. These two classes are distinguished by the interatomic distances at which they become of significance to the total intermolecular interaction.

We may choose to look at the problem of enzyme-substrate interaction in terms of the two-phase system illustrated in Fig. 2–1; an *aqueous solution phase,* in which the substrate molecules are on the average separated one from another by a very large distance (in atomic units), and an *enzyme site phase,* in

0
10
20
angstroms

which the substrate molecule is close (in atomic units) to some of the atoms of the enzyme protein.

In the aqueous solution phase, interactions will be restricted to substrate-solvent interactions and solvent-solvent interactions. Usually, substrate-solvent interactions will fall into only one of the two classes of intermolecular interactions mentioned above. We shall term this class *physical interactions*. These substrate-solvent interactions will occur at separation distances between the two molecules which are greater than the average chemical bond length between covalently adjacent atoms of the substrate, but at distances much smaller than the average distance between one substrate molecule and another in dilute solution.

In the enzyme site phase, interaction between substrate and enzyme will also involve these physical forces. Some interactions involving a collection of atoms of the enzyme and the substrate must occur at distances of the order of a chemical bond length, in order for subsequent chemical reaction to take place. Such "short range" interactions we shall term *chemical interactions*. Let us consider the origins of the physical interactions first.

2–1 ELECTROSTATIC FORCES

Perhaps the most familiar physical force of interaction is that which exists between any two charged particles. The energy of interaction (E) between two point charges as a function of their distance of separation is given by Coulomb's law (Eq. 2–1),

$$E = \frac{Z_A Z_B \epsilon^2}{r_{AB}} \tag{2–1}$$

FIG. 2–1 A schematic view of the relative dimensions of enzyme, enzyme site, substrate, and solvent (water). Neighboring water molecules are separated by distances of approximately 2 Å; the other dimensions are essentially to this scale.

where r_{AB} is the distance between the two charges, Z is the charge number, and ϵ is one unit of electronic charge. Many biological substrates, particularly those involved in the biosynthesis of complex molecules from structurally simpler metabolites, are charged molecules (ions). Likewise, proteins (enzymes) contain a number of charged amino acid constituents (Fig. 2–2).

These complex organic ions as well as the simpler inorganic constituents of enzyme-substrate systems (for example, HPO_4^{2-}, Fe^{3+}, Co^{3+}, Mg^{2+}, $H_2P_2O_7^{2-}$, NH_4^+) are obviously not point charges. Moreover, Coulomb's law is strictly applicable only in a vacuum, that is, only where there are no polarizable solvent molecules to dissipate the potential energy between the charges. It is therefore not particularly surprising that Eq. 2–1 is not an exact expression for the energy of interaction between ions in organic and biological systems. However, in dilute solution, Coulomb's law gives a reasonable approximation to the actual energetic situation when account is taken of the dissipation of part of the electrical energy between charges by the polarizable solvent molecules (see Fig. 2–3). In dilute solution we can *approximate* the electrical energy between charges by Eq. 2–2

$$\Delta E_{\mathrm{el}} = \frac{Z_A Z_B \epsilon^2}{D\, r_{AB}} \qquad (2\text{–}2)$$

where ϵ is the electronic charge and Z_A and Z_B are the algebraic charge numbers of the ions. In Eq. 2–2, D is known as the *dielectric constant* of the solvent. This constant is an independently measurable property of the bulk solvent, arising from the polarizability of the solvent molecules. We can expect Eq. 2–2 to be valid whenever the number of solvent molecules between the two interacting charges is large enough so that the properties of this *microregion* of solvent can be expected to be similar to that of the bulk solvent. As the two interacting charges come closer together and the number of solvent molecules separating

FIG. 2–2 A few examples of charged residues in proteins and of charged groups in substrates.

Charged residues in protein

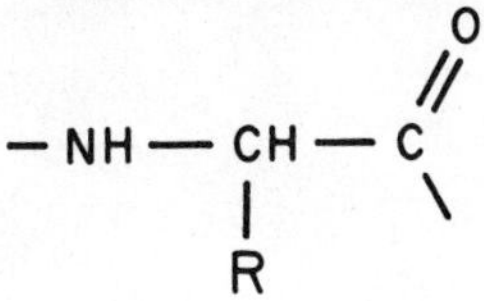

R = $-(CH_2)_4\overset{\oplus}{N}H_3$ (Lysine)

$-(CH_2)_3NH\overset{\oplus}{C}(NH_2)_2$ (Arginine)

$-CH_2CO_2^{\ominus}$ (Aspartate)

$-CH_2CH_2CO_2^{\ominus}$ (Glutamate)

Charged groups in substrates

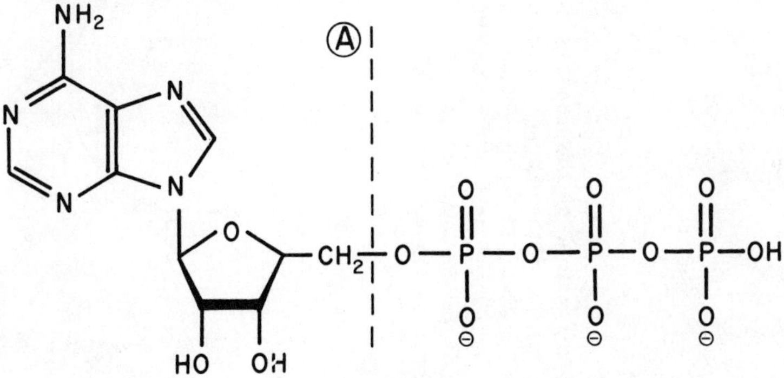

Adenosine-5′-triphosphate (ATP)

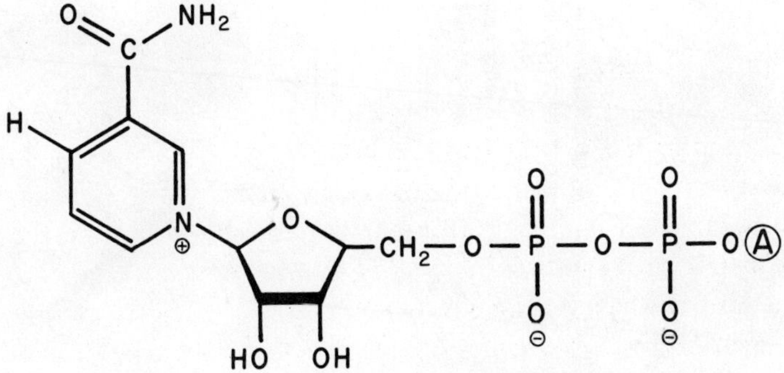

Nicotinamide adenine dinucleotide (oxidized form)

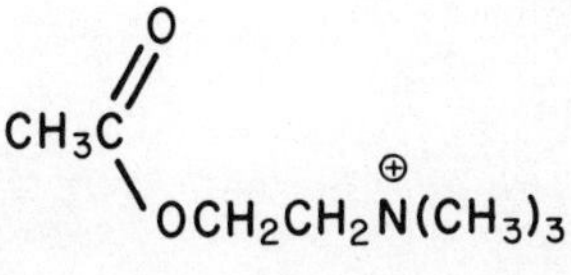

Acetylcholine

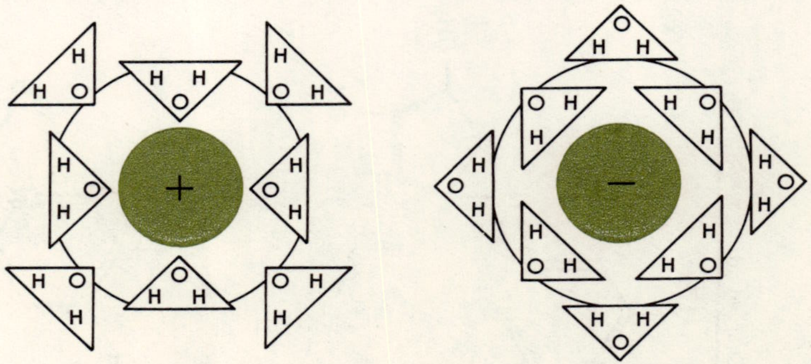

FIG. 2–3 A qualitative picture of the nature of the aqueous solvent environment separating two ions, as a function of the distance between the charges. As the ions come closer together, the intervening solvent becomes more completely oriented and hence less polarizable. As a result, the *effective dielectric constant* decreases and the charge interaction energy becomes larger than would be anticipated if the polar water molecules were disoriented to their usual extent.

the charges becomes smaller, the properties of this particular group of solvent molecules begin to deviate markedly from that of the bulk of the solvent. Under such conditions the modified expression of Coulomb's law (Eq. 2–2) becomes invalid.

Biological processes occur in aqueous solvents. The dielectric constant (a unitless parameter) is nearly 80 for water; that is to say, for a given distance of separation the electrical interaction energy between two ions in dilute aqueous solution will be about $1/80$ the energy which would obtain in a vacuum (as in Eq. 2–1). Water is a highly polarizable molecule; its dielectric constant is a reflection of this polarizability. Ions interact very strongly with the aqueous solvent, thereby reducing the magnitude of the interaction energy with other ions in the vicinity. As two ions are brought closer together, the region of solvent between the two interacting ions becomes less polarizable (and hence D becomes smaller) than that of the bulk of the solvent. With decreasing distance of separation between two ions, the electrostatic energy of interaction increases more rapidly with the inverse of the distance of separation ($1/r$) than would be anticipated on the basis of Eq. 2–2 (which predicts a linear relationship between ΔE_{el} and the inverse of the distance of separation). Two examples will suffice to illustrate these effects. In Table 2–1 are listed the ionization constants for the dissociation of acetic acid in dilute solutions of varying dielectric constant. Acetic acid is a neutral molecule; the products of its

Table 2-1
Dissociation Constants of Acetic Acid (A) and Water (W) in Water-Dioxane Mixtures at 25°[a]

% Dioxane $\left(\frac{V_{dioxane}}{V_{total}}\right) \times 100$	D	pK_A $-\log\left(\frac{a_{H^+} a_{Ac^-}}{a_{HAc}}\right)$	pK_W $-\log(a_{H^+} a_{OH^-})$
0	79	4.75	14.0
20	61	5.41	14.6
45	38	6.33	15.8
76	18	8.35	17.8
82	9.5	10.16	—

[a]The symbol a is the thermodynamic activity.

dissociation (acetate ion and a proton) are oppositely charged. In dilute solution, the tendency for these oppositely charged ions to separate will be dependent on the bulk dielectric constant of the solvent medium. The change in electrostatic free energy of dissociation in changing from one solvent medium to another is given according to Eq. 2–2 by the relationship

$$\frac{E_2}{E_1} = \frac{D_1}{D_2}$$

This difference will be reflected in a difference in the ionization constant of the acid, according to the equilibrium thermodynamic relationship in Eq. (2–3)

$$\Delta G_{el} = E_2 - E_1 = RT \ln \left(\frac{K_2}{K_1}\right) \tag{2–3}$$

where K is the dissociation constant of the acid in the appropriate solvent.

$$HA \overset{K}{\rightleftharpoons} H^{\oplus} + A^{\ominus}$$

Hence, the effect of lowering the dielectric constant of the solvent will be to inhibit the dissociation of the acid. The data in Table 2–1 bear out this prediction. Likewise in our model system (Fig. 2–1), if the substrate is negatively charged and the "enzyme phase" is positively charged (or vice versa), the electrostatic energy contribution to the formation of enzyme-substrate complex will depend on the dielectric constant of the enzyme site solvent relative to that of water. Although a precise value cannot be assigned to this parameter, it will usually be smaller than the dielectric constant of pure water and greater than unity (the value of D in a vacuum); typically, D is estimated to be in the range of 10–40. Thus, the tendency towards association for two ions of opposite charge will be greater in the "enzyme phase" than in aqueous medium.

The data in Table 2–2 are illustrative of what occurs as the

Table 2-2

$pK_a (H_2N{-}R{-}\overset{\oplus}{N}H_3) - pK_a(H_3\overset{\oplus}{N}{-}R{-}\overset{\oplus}{N}H_3)$
as a Function of the Distance (d) between Charged Centers[a]

R	d, Å	ΔpK_a (H_2O)	ΔpK_a (80% C_2H_5OH)
$-(CH_2)_2-$	4.2	2.35	2.76
$-(CH_2)_3-$	5.4	1.43	1.77
$-(CH_2)_4-$	6.6	0.90	1.25
$-(CH_2)_5-$	7.8	0.68	1.05
$-(CH_2)_8-$	11.6	0.27	0.95

[a]Corrected for the statistical probabilities of dissociation and association of the various species.

two interacting charges are brought closer together. In the example cited, we observe the effect of electrostatic repulsion between charges on the ionization constant of an acid. In this example, when the acid ionizes the electrostatic repulsion between the two like charges is removed.

$$H_3\overset{\oplus}{N}{-}(CH_2)_n{-}\overset{\oplus}{N}H_3 \rightleftharpoons H_2\overset{0}{N}{-}(CH_2)_n{-}\overset{\oplus}{N}H_3 + \overset{\oplus}{H}$$

To a good approximation, the methylene ($-CH_2-$) groups that separate the two charges can be considered as "insulators." Consequently, there will be no *covalent* perturbations of one of the charged residues on the dissociation energy of the other. Since there is free rotation about all of the single bonds in the molecule, it is likely that the molecules will be fully extended so as to avoid, as much as possible, the electrostatic repulsions between the charged ends. From the known covalent bond lengths and angles, we can calculate the distance of separation between the two charges in every example in Table 2–2. It is therefore possible to calculate the electrostatic free energy of interaction between the charges in each molecule as follows: When one of the charged groups dissociates in a dilute solution, the intramolecular electrostatic repulsion is removed. We may therefore compare the electrostatic free energy calculated from Eq. 2–3 with that actually observed. The best way of making this comparison is to refer the electrostatic free energy to a standard state in which the charges are separated by a nearly infinite distance. As can be seen from the results listed in the table, the electrostatic free energy of interaction increases with decreasing distance between charges at a greater rate than that predicted from Eq. 2–2, once the two charges come into proximity (about 10 Å or less).

If, in the enzyme site phase, enzyme and substrate are charged, molecular geometrical factors governing the distance to which these two charges can approach will determine the energy to be derived from the interaction more critically than is predicted by Eq. 2–2. Such geometrical factors may contribute to the observed molecular specificity in enzyme-substrate interactions.

We might anticipate on the basis of the above discussion that two oppositely charged species will tend to approach each other to within chemical bonding distance. This is not the case. As the two oppositely charged atoms or molecules come closer together, their filled electronic orbitals begin to overlap. This will result in very strong interelectronic repulsions, which at some point will counterbalance the attractive interionic electrostatic interaction. The distance of separation between the two interacting ions at which there is maximal stability is known as the *ionic contact distance*.

2–2 LONDON DISPERSION FORCES

For *any* pair of atoms, including for example, two noble gas atoms, there exists a force of attraction, the origins of which could only have been predicted subsequent to the development of modern quantum theory. Let us consider any atom: the electronic charge distribution about the nucleus is on the average spherically symmetric (i.e., there is no permanent dipole moment). At any *instant of time*, the distribution of electronic charge will, however, be localized in some specific asymmetric geometry dependent on the distribution of electrons into the various electronic orbitals (Fig. 2–4). At this specific instant, there will be an induction of complementary electronic charge distribution in the neighboring atoms. This induced dipolar interaction will lead to a net attraction between any pair of neighboring atoms. However, when the two atoms come very close together, the repulsion between the electron clouds of the two atoms will become very strong and eventually counterbalance these induced dipolar attractions. The distance of separation at which attraction between a pair of atoms is maximal is known as the *van der Waals contact distance;* it is a characteristic distance for the particular atoms involved in the interaction. Each atom has a characteristic *van der Waals radius*, the van der Waals contact distance being the sum of the radii for a

FIG. 2–4 Dispersion forces between atoms.
(a) Promotion of electrons into higher energy orbitals at any instant of time results in an "instantaneous" dipole moment, the magnitude of which depends on the geometry of the electronic arrangement (the *polarizability*, α). The number of atoms so perturbed will depend on the activation energy for the promotion of the electron to an excited state, and should be proportional to the *first ionization potential* (I_A) of the atom ($A \rightarrow A^+ + \epsilon^-$), if only a small fraction of the total atoms are so excited. The instantaneous dipoles will tend to arrange themselves so as to yield the most favorable electrostatic interaction.
(b) An approximate equation first derived by London gives the net mutually induced attractive energy (W) between two atoms (*A and B*),

$$E_{\text{attractive}} \propto \frac{\alpha_A \alpha_B}{r^6_{AB}} \frac{I_A I_B}{(I_A + I_B)}$$

where α is the polarizability of the electrons of each atom. For the most common atoms in biological systems (C, H, O, N) α is roughly proportional to the atomic number, and, for the second row atoms, I is nearly constant.

When the electronic orbitals of two atoms come into very close proximity, the repulsions between electrons oppose this induced dipolar attraction. Equation 2–4 gives the net energy due to attractive and repulsive contributions, according to a theoretical treatment proposed by Lennard-Jones. The energy of interaction as a function of distance between a pair of atoms is plotted accordingly.

particular pair. To a good approximation the interaction between otherwise inert atoms can be expressed by the potential energy function given in Eq. (2–4). The attractive and repulsive forces involved in this interaction are known collectively as *London dispersion forces,*

$$W_{AB} = -\frac{A_0}{r_{AB}^6} + \frac{B_0}{r_{AB}^{12}} \tag{2–4}$$

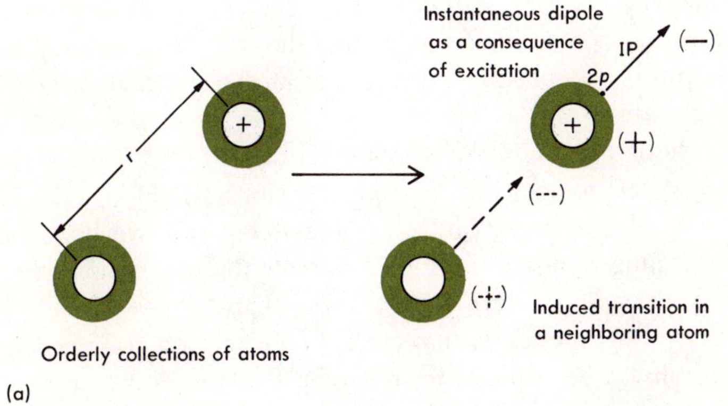

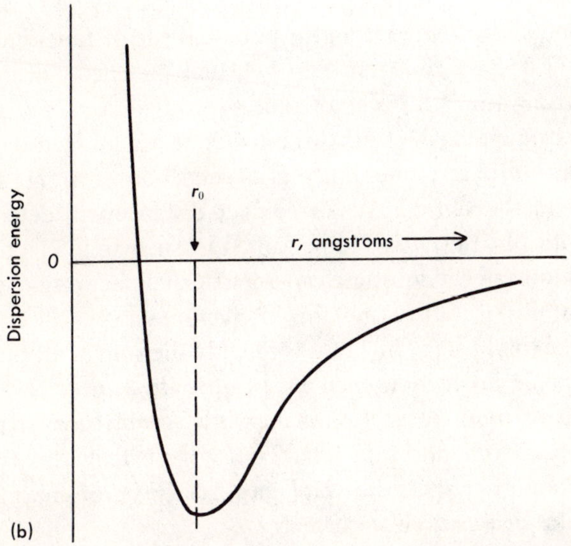

Table 2-3
Van der Waals Contact Radii of Atoms

Atom	r, Å
H	1.2
CH (tetrahedral as in methane)	2.0
CH (planar trigonal as in benzene)	1.8
N	1.5
NH (as in NH_3)	2.1
O	1.4
S	1.85
P	1.9

where r_{AB} is the distance separating the two atomic centers and A_0 and B_0 are constants characteristic of the electronic structure of the atoms.

The position of the energy minimum in Fig. 2–4 is the van der Waals contact distance. Typical contact radii $(r_{AA}/2)$ for atoms of interest in the biological systems with which we will be dealing are listed in Table 2–3. Note that the sums of these contact radii are very much longer than the covalent bond lengths. In order to form a covalent bond, a good deal of energy will always be required to overcome the interelectronic repulsions which exist in the pathway between the van der Waals contact distance and the much shorter chemical bond distance. It should also be noted from the potential energy function given above (Eq. 2–4) and from Fig. 2–4, that the energy of interaction arising from this type of process is highly sensitive to the distance of approach of the two interacting atoms. Hence, in the enzyme site phase, the ability of a collection of atoms to approach to the van der Waals contact distance will determine, very critically, the energy to be gained by such an interaction. If many atoms of the substrate can approach contact distance with many atoms of the enzyme, this interaction energy will be large, since the total energy involved is roughly the sum of all the pairwise atomic interaction energies. *In order to achieve this situation, there must be a "complementary" geometrical structure between enzyme and substrate.* To a large extent, this type of interaction gives rise to the *biological specificity* of enzymes for substrates of a particular geometry.

2–3 DIPOLAR INTERACTIONS

Intermediate between the interionic and London forces of interaction discussed above are the intermolecular forces which arise from the interaction between polar molecules; for example, interactions between an ion and a dipolar molecule, between dipoles and induced dipoles, ions and induced dipoles, dipoles and dipoles, and so on. The dependence of the energy of interaction on the separation distance will be intermediate between that of the purely ionic ($E \propto r_{AB}^{-1}$) and the purely induced dipole type interaction ($E \propto r_{AB}^{-6}$). Although these intermediate type physical interactions are doubtless of importance in many situations involving enzyme substrate interactions, they will vary in relative importance from situation to situation, and in most instances their net contribution to the total interaction energy will be small in comparison to the contribution from the sum of the two previously discussed forces.

2–4 THE HYDROGEN BOND

For inert liquids, a measure of the London dispersion energy is the heat of vaporization, since in these cases the process of liquefaction is due essentially to dispersion force interactions. The heats of vaporization of inert hydrocarbons (Table 2–4)

Table 2-4
Heats of Vaporization (kcal/mole) at the Normal Boiling Point[a]

Compound	MW	ΔH_v	BP (°C)	ΔH_v/MW	$\Delta H_v - \Delta H_{dispersion}$
CH_4	16	2.0	−162	0.125	
C_2H_6	30	3.5	− 89	0.12	
n-C_3H_8	44	4.5	− 42	0.10	
n-C_4H_{10}	58	5.4	− 0.5	0.09	
Benzene	78	7.6	80	0.10	
NH_3	17	5.6	− 33	0.33	3.5
CH_3OH	32	8.4	65	0.26	4.7
C_2H_5OH	46	9.4	78	0.20	4.7
H_2O	18	8.7	100	0.48	6.5

[a] $\Delta H_{dispersion} = \Delta H_v$ of a hydrocarbon of the same molecular weight as interpolated from the upper values in this table.

increase at a nearly constant rate, with increase in the number of carbon atoms per molecule.

From this data on liquid hydrocarbons, we might attempt to make some predictions about the heats of vaporization and boiling points for some other common liquids on the assumption that *these liquids depend on dispersion forces exclusively* for their integrity. On this basis we would expect water to be a gas even at the temperature of liquid air; we would anticipate methanol and ethanol to be gases at room temperature. That these substances are not gaseous at room temperature is a reflection of the fact that other forces of interaction are operative. In ethanol, methanol, water, ammonia, and other liquids of unusually high boiling points (and high enthalpies of vaporization), the stability of the liquid structure arises primarily from an interaction which may be classified as intermediate between "physical" and "chemical." The average distance between oxygen atoms in liquid water is less than the van der Waals contact distance,[1] but greater than the chemical bonding distance. This phenomenon arises due to the formation of *hydrogen bonds* between the oxygen atoms.

Information regarding the types of hydrogen bonds which

Table 2-5
Typical Hydrogen Bond Lengths

Type of bond	Typical compounds	Separation distance, Å X—H.....Y ←d→
O—H-----O	Water	3.0
O—H-----O	Primary alcohols	2.7
O· H-----O	Carboxylic acids	2.6
N—H-----N	Ammonia, amines	3.1
N—H-----O	Urea, amides, peptides	2.9

[1] The average distance between water molecules can be calculated from the density of water. At 4°C, 16 ml of liquid water contains one mole or 6.02×10^{23} water molecules. The average volume per molecule is 2.66×10^{-23} ml or 26.6 Å^3. Hence the average separation of water molecules is $(26.6)^{1/3} \times 10^{-8}$ cm or 2.98 Å.

occur is summarized in Table 2–5. In hydrogen bond formation, the nature of the "donor" atom and the "acceptor" atom must be distinguished. In enzyme-substrate systems, only oxygen, nitrogen, (and possibly sulfur) atoms take part in hydrogen bonding. Residues of the type —OH and —NH can act as good hydrogen bond donors. Residues such as carbonyl oxygens, hydroxylic oxygens, and to a more limited extent nitrogen bases such as imidazole can act as acceptors. Typical energies for such hydrogen bonds are in the range of 3–6 kcal/mole. Although these absolute bond energies appear to be quite substantial, we must be constantly aware that our particular interests lie in an aqueous system; a hydrogen bond formed between a substrate molecule and a region of the enzyme may yield a substantial energy but may still be counterbalanced by the potential of both the donor and the acceptor molecules to alternatively form hydrogen bonds with water. To make a better estimate of the net energy gained in hydrogen bonding, we must consider a process such as the following: A substrate molecule in the aqueous phase must, to an extent predictable from its geometry, disrupt part of the hydrogen bonding which exists in liquid water. In doing so, it may (dependent upon chemical structure and geometry) break one or more hydrogen bonds in the liquid water structure. The removal of this substrate molecule from the aqueous phase into the enzyme-site phase will restore the maximal hydrogen bonding potential of pure water. In the enzyme-site phase, formation of a hydrogen bond between substrate and enzyme may release water molecules from the site into the aqueous phase (where hydrogen bonding of water is nearly maximal). The *net energy* derived from this process will be the algebraic sum of each of the hydrogen bonds made or broken multiplied by the energy involved in each individual process. This net energy will, on the average, be substantially smaller than the energy derived from the formation of hydrogen bonds between enzyme and substrate viewed as an isolated process in the absence of water. It will, however, tend to be a favorable interaction due to the high efficiency of hydrogen bond formation in liquid water.

2–5 WATER STRUCTURE AND THE STRUCTURE OF PROTEINS

The hydrogen bonded structure of water is a very stable one, and the interaction of water with ions is a very strong interaction due to the high polarizability of water. When an enzyme is dissolved in water, the liquid water structure is necessarily broken up at the protein-water interface and some hydrogen bonds between neighboring water molecules are lost. An energetically favorable situation for a protein dissolved in water would be one in which hydrogen bond donors and acceptors on the protein molecules replace at least some of the hydrogen bonds lost by the breakup of the liquid water structure at the interface. Another energetically favorable situation would be one in which charges on the protein molecule were situated near this interface so as to interact strongly with the solvent. These interactions, which are a consequence of the highly polar structure of water, play an important role in governing the conformation of a complex enzyme protein molecule.

As we shall see (in Chapter 3), the specific amino acid sequence and the nature of the repeating polypeptide "backbone" play a major role in the determination of protein structure in solution. That the peptide "backbone" and the side chain structures are of consequence to protein conformation, exclusive of the influence of solvent, is inherent from two essential features of the chemical bond, namely, the high degree of restriction in the allowable bond lengths, and in the allowable bond angles. The way in which these restrictions govern the three dimensional conformation of a complex protein molecule is best illustrated by the original work of Pauling, Corey, and their collaborators on the nature of the chemical bonds in peptides, the results of which led to the prediction of the "α-helical" structure of polypeptides, discussed in the following chapter.

REFERENCES

Pauling, L., *The Nature of the Chemical Bond*, 3rd ed., Cornell University Press, Ithaca, New York, 1960. The forces of interaction between atoms and molecules are described with commendable insight and clarity and with transmitted enthusiasm in this great work. Much emphasis is placed on the hydrogen bond.

Edsall, J. T., and Wyman, J., *Biophysical Chemistry*, Academic Press, New York, 1958. A long well-written advanced text which deals primarily with the solution thermodynamic properties of polyelectrolytes and proteins. It contains an excellent early chapter on "Water and Its Biological Significance."

Watson, J. D., *Molecular Biology of the Gene*, Benjamin, New York, 1965. Chapter Four of this superbly written book contains an informed biologist's view of the role of weak forces of interaction in the determination of biological structure and function.

Pimentel, G. C., and McClellan, A. L., *The Hydrogen Bond*, Freeman, San Francisco, 1960. The definitive volume on the variety, geometry, and strength of hydrogen bonds, and the physical methods by which they have been examined.

Pauling, L., and Pressman, D., "The Serological Properties of Simple Substances, IX," *J. Amer. Chem. Soc.*, 67, 1003 (1945). This paper deals with the physical-chemical nature of the antibody-antigen interaction. It contains an excellent description of the potential physical forces which may be operative in a system entirely analogous to an enzyme-substrate system, and how the strengths of such interactions may be estimated. Theoretical calculations and experimental results are compared.

THREE ~ CHEMICAL BONDS AND THE STRUCTURE OF PROTEINS

IN ORDER TO INVESTIGATE THE CONFORMATION (THE PARTICULAR three-dimensional arrangement) of a large polypeptide or protein, a prior knowledge of the stereochemical arrangement of atoms (the atomic *configuration*) of smaller molecular constituents (amino acids and peptides) is first required. The techniques of X-ray crystallography can yield detailed quantitative information concerning the relevant *intra*molecular bond lengths and bond angles. The geometrical details of the regular arrangement of molecules within a crystal (the conformation) is *necessarily* determined in the course of the investigation of the intramolecular parameters. These latter conformational parameters are pertinent to the physical interactions which may occur among the various molecular residues within a complex polypeptide. They inform us, for example, of the distance of closest approach of two nonbonded atoms (the van der Waals contact distance).

3–1 CHEMICAL BONDING AND THE CONFIGURATION OF PEPTIDES

With the problem of polypeptide conformation in mind, Pauling, Corey, and their collaborators undertook an investigation of the three-dimensional structures of some relevant simple organic molecule crystals. Some of the molecules investigated are illustrated in Fig. 3–1.

A complete three-dimensional structure analysis of even such simple organic molecules leads to a wealth of useful information. A partial summary of results for one such molecule, the

FIG. 3–1 Some typical organic molecules whose molecular structure and arrangement in crystals provide information about characteristic features of polypeptides. Notice among these molecules the presence of peptide bonds, tetrahedral α-carbon atoms, terminal amino and carboxyl groups, and some of the typical side chain groups present in polypeptides and proteins.

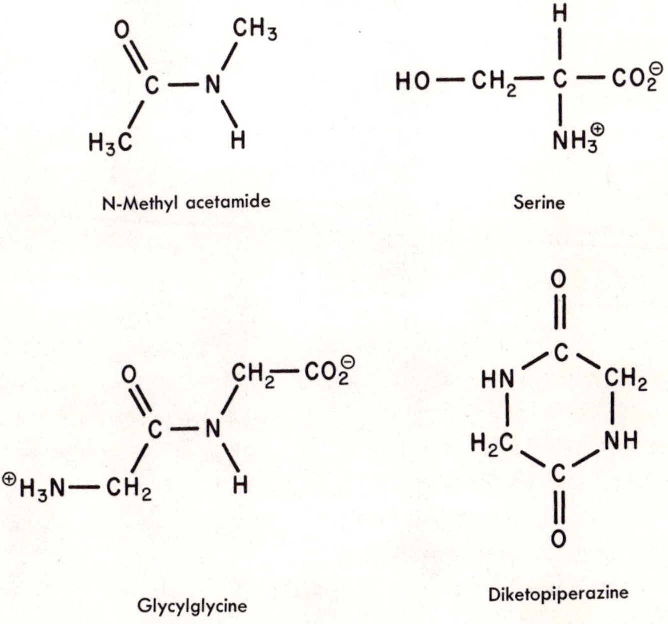

amino acid L-threonine, is illustrated in Fig. 3–2, and some quantitative parameters are listed in Table 3–1.

FIG. 3–2 The atomic coordinates, in three dimensions, of a group of L-threonine molecules sufficient to describe the arrangement of all atoms in the molecular crystal. (a) The atomic coordinates of each L-threonine molecule. (b) The actual electron density maps on the three crystallographic axes. (c) A representative view of the inter- and intramolecular details derived from the electrons density maps. (d) A three-dimensional representation of the arrangement of molecules in the crystal. Notice carefully that although the relative coordinates of atoms within a molecule are identical in all of the threo-

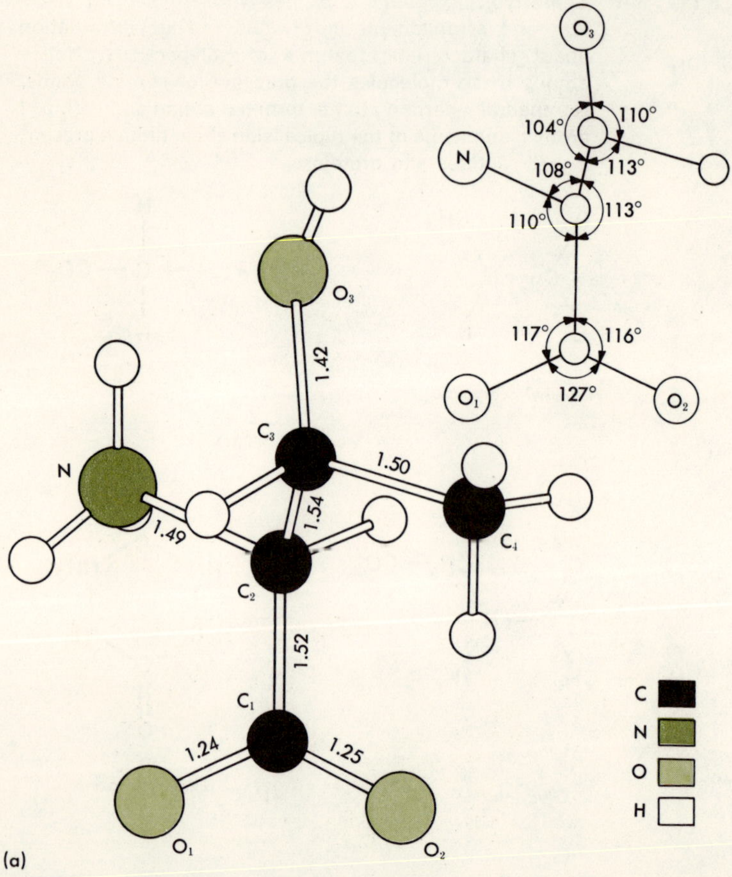

nine molecules shown, there are several different *relative orientations* of molecules in the crystal. In order to describe the relative atomic coordinates between any pair of atoms in adjacent molecules, it is necessary to consider a collection of four such molecules. This smallest unit of crystal space, characteristic of the entire crystalline array, is known as the *unit cell.* The dimensions given in Fig. 3–2c are the unit cell dimensions. This unit cell contains four molecules of L-threonine. [From D. P. Shoemaker *et al., J. Am. Chem. Soc.* **72,** 2343 (1950).]

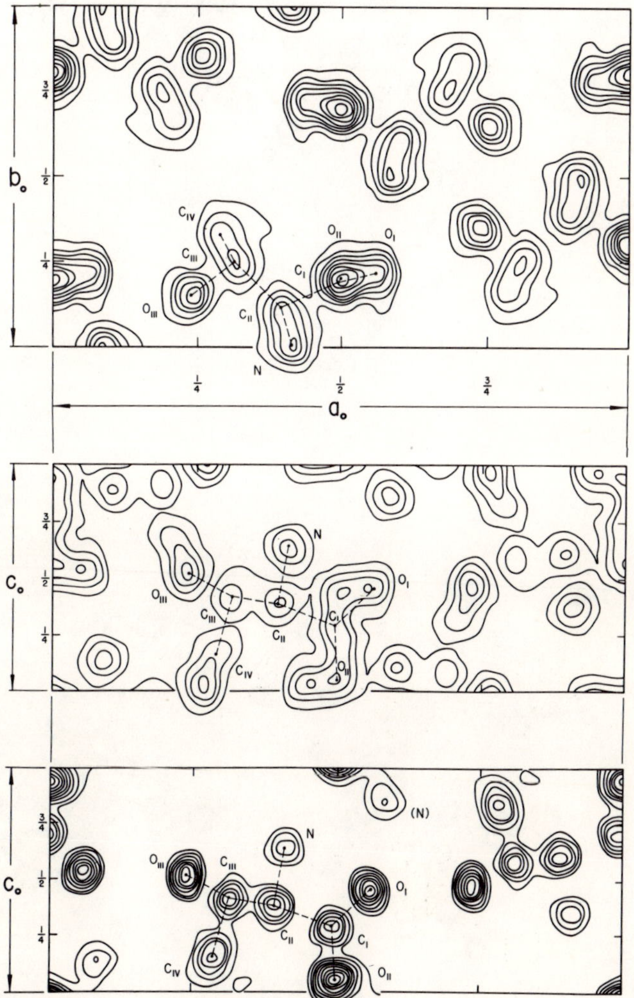

(b)

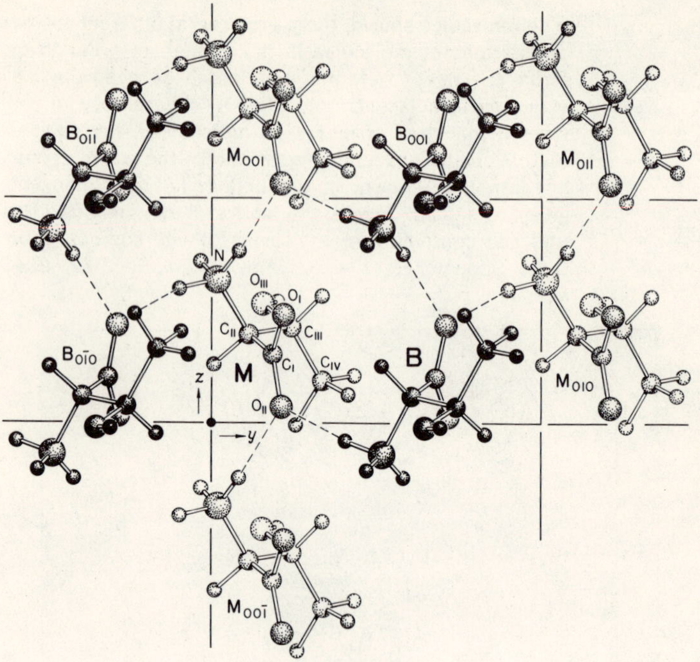

FIG. 3–2c

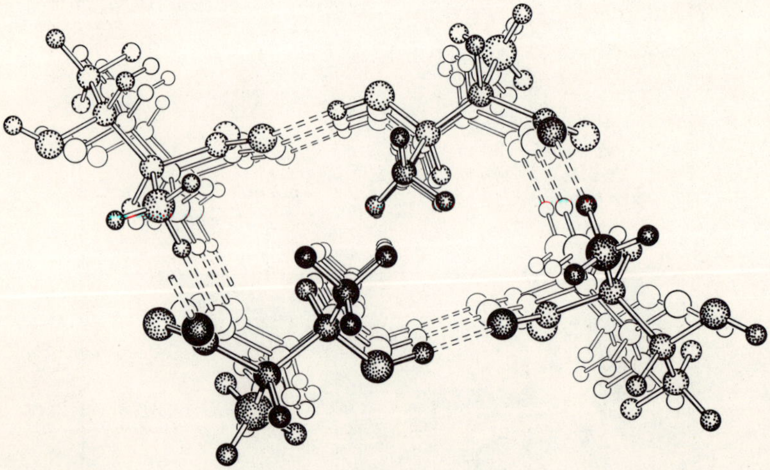

FIG. 3–2d

Table 3-1

Intramolecular and Intermolecular Interatomic Distances and Bond Angles for the Threonine Molecule[a,b]

Distance, Å		Angle, deg	
C_I-C_{II}	1.517	$O_I-C_I-O_{II}$	126.9
$C_{II}-C_{III}$	1.542	$O_I-C_I-C_{II}$	117.0
$C_{III}-C_{IV}$	1.505	$O_{II}-C_I-C_{II}$	116.1
$C_{II}-N$	1.490	$C_I-C_{II}-C_{III}$	113.4
C_I-O_I	1.236	$C_I-C_{II}-N$	110.4
C_I-O_{II}	1.253	$C_{II}-C_{III}-C_{IV}$	112.5
$C_{III}-O_{III}$	1.424	$C_{II}-C_{III}-O_{III}$	104.1
		$N-C_{II}-C_{III}$	108.0
$N \ldots O_I$	2.672	$O_{III}-C_{III}-C_{IV}$	110.5
$N \ldots O_{III}$	2.678		
$C_I \ldots C_{IV}$	3.084		
$O_{II} \ldots C_{IV}$	3.106		

Intermolecular Interatomic Distances and Angles

From atom (X) on molecule M	Atom (Y)	on molecule (1)	Equiv. contact to molecule[c] (2)	Distance X ... Y, Å	Angle CX...Y	Angle X...YC
		A. Hydrogen bond contacts				
N	O_{II}	M_{001}	$M_{00\bar{1}}$	2.90	116°	128°
O_{III}	O_I	$A_{\bar{1}00}$	A	2.66	120°	136°
N	O_{II}	$B_{0\bar{1}0}$	B	2.80	98°	122°
N	O_{III}	$C_{0\bar{1}0}$	$C_{0\bar{1}\bar{1}}$	3.10	132°	120°
		B. Other distances[d]				
N	C_{II}	M_{001}	$M_{00\bar{1}}$	3.92		
C_{III}	C_{IV}	M_{001}	$M_{00\bar{1}}$	3.94		
N	C_{IV}	M_{001}	$M_{00\bar{1}}$	3.73		
O_I	O_{II}	M_{001}	$M_{00\bar{1}}$	3.14		
O_I	C_{IV}	A	$A_{\bar{1}00}$	3.67		
C_{IV}	O_I	B	$B_{0\bar{1}0}$	3.69		
C_I	C_{II}	B	$B_{\bar{0}10}$	3.71		
O_I	C_{II}	B	$B_{\bar{0}10}$	3.59		
O_I	O_{II}	B	$B_{0\bar{1}0}$	3.92		
O_{II}	O_I	B	$B_{0\bar{1}0}$	4.00		
N	O_I	$B_{0\bar{1}1}$	B_{001}	3.08	161°	134°
C_{IV}	C_{IV}	C	$C_{00\bar{1}}$	3.79		
C_{III}	C_{IV}	C	$C_{00\bar{1}}$	3.81		
O_{III}	C_{II}	$C_{0\bar{1}0}$	$C_{0\bar{1}\bar{1}}$	3.28		

(continued)

Table 3–1 (*continued*)

From atom (X) on molecule M	Atom (Y)	on molecule (1)	Equiv. contact to molecule[c] (2)	Distance X...Y Å	Angle CX...Y	Angle X...YC
B. Other distances[d]						
O_{III}	C_{IV}	$C_{0\bar{1}0}$	$C_{0\bar{1}\bar{1}}$	3.74		
O_{III}	C_{IV}	$C_{0\bar{1}0}$	$C_{0\bar{1}\bar{1}}$	3.83		
N	C_{IV}	$C_{0\bar{1}0}$	$C_{0\bar{1}\bar{1}}$	3.87		

[a]Data from Shoemaker *et al.*. *J. Am. Chem. Soc.*, **72**, 2328 (1950).

[b]The shape of the threonine molecule, as it exists in the crystal, is given in Fig. 3-2.

[c]In every case where the contact is from X on *M* to Y on molecule (1), the equivalent contact is from Y on *M* to X on molecule (2).

[d]All distances of 4 Å or less, excepting those listed in Part A and those involving hydrogen atoms, are given here.

Following a massive accumulation of data (such as are illustrated in Table 3–1) on the chemical bond lengths and bond angles, and on the distances between adjacent nonbonded atoms in the crystals, the following firm conclusions were arrived at:

(1) *In every peptide investigated all the atoms comprising the linkage lie in one plane* (Fig. 3–3). That all of these atoms are mutually coplanar is a consequence of the "partial double-bond character" of the carbonyl carbon to nitrogen bond.

(2) *The two α-carbon atoms at each peptide bond are oriented in a trans configuration.* The conformity to the trans configuration arises from the fact that in the cis configuration, the distance between C-1 and C-2 (Fig. 3–3) is 2.8 Å. In the trans configuration the distance between C-1 and carbonyl oxygen is again about 2.8 Å. The van der Waals contact distance between two carbon atoms is 4.0 Å; that between carbon and

oxygen is about 3.4 Å. Thus, repulsive forces will be minimized in the trans configuration.

(3) *In the crystals, the maximum hydrogen bonding potential is actually realized.* Each peptide bond contains one potential hydrogen bond donor (N—H) and one potential acceptor (carbonyl oxygen). In all amide and peptide crystals examined, the shortest distance between nonbonded peptide nitrogen and oxygen atoms are 2.9 ± 0.1 Å. This distance is far shorter than the van der Waals contact distance. About each peptide bond in the crystal there are complementary nonbonded (carbonyl) oxygen and (N—H) nitrogen atoms in precisely this close proximity. Moreover, the N—H- - -O hydrogen bond is linear or nearly linear (Fig. 3–4).

Making use of these three firm conclusions derived from crystal structure analysis, and the accumulated information on bond lengths and bond angles, Pauling and Corey proceeded to investigate the allowable configurations for a polypeptide chain on the assumption that these conclusions regarding the structure of small molecules were applicable to noncrystalline polypeptides. In addition, they took cognizance of the important role of dispersion forces in the stabilization of conformation. It is an axiom in molecular crystallography, and in molecular "model building," that the most stable structure will always be a "closely-packed" structure; that is, one in which nonbonded atoms are surrounded by other nonbonded atoms at distances equal (as far as is possible) to the sums of the van der Waals contact radii.

In attempting to utilize the maximum hydrogen bonding potential of the polypeptide, two types of structural models must be considered:

(1) Structures involving *inter*chain hydrogen bonds: If each polypeptide chain hydrogen bonds to more than one other

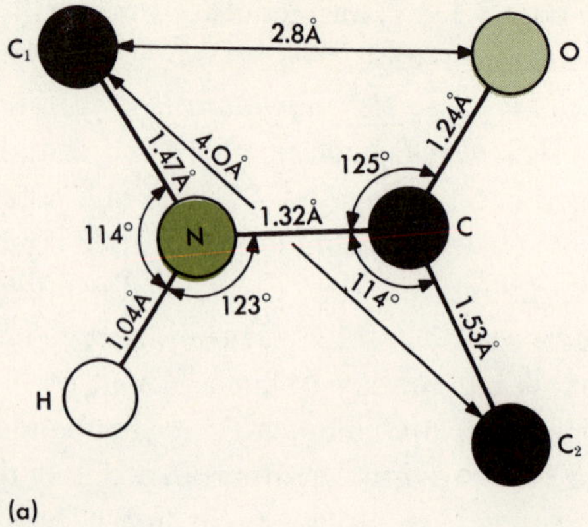

FIG. 3-3 The geometry of the peptide bond (color code as in Fig. 3-2a). (a) Coordinates of the peptide bond. (b) Spatial relationship between adjacent peptide units.

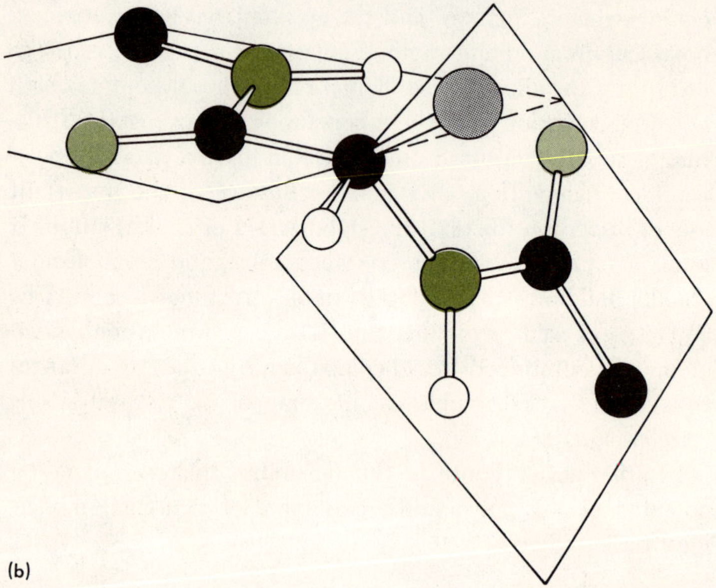

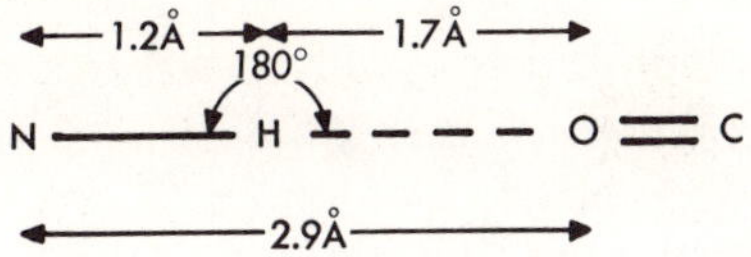

FIG. 3–4 The geometry of a hydrogen bond between peptide units.

FIG. 3–5 Two potential helical configurations of a polypeptide composed of L-amino acid residues. (a) A "right-handed" α-helix containing approximately 3.7 residues per turn of the helix. (b) A "left-handed" α-helix, again containing approximately 3.7 residues per turn.

chain, the resultant structure will be either a macromolecular sheet or a three-dimensional crosslinked macromolecular structure.

(2) Structures in which the carbonyl oxygen of one amino acid residue hydrogen bonds with the NH group of another residue in the same chain: Such an *intra*molecular structure might be realized by twisting the two freely rotating bonds so that a carbonyl and an NH pair are stereochemically adjacent and in the proper configuration for hydrogen bonding. As can be seen from the restrictions of planarity and trans configuration (Fig. 3–3), the only peptide bonds in which free rotations are allowable are the C–1 carbonyl carbon and C–2 nitrogen bonds. An examination of molecular models clearly demonstrates that in a polypeptide chain, hydrogen bonding is impossible between a carbonyl oxygen and an NH from covalently adjacent amino acid residues. Thus, if *intra*molecular hydrogen bonding were to occur, it would have to occur between residues separated by greater than one covalent unit. In order to realize *all* of the hydrogen bonding possibilities via *intramolecular interactions*, the polypeptide chain will necessarily have to conform to a particular *repetitive* structure. A repetitive structure within a linear polymeric chain gives rise to a *helix*. Two different polypeptide helices are shown in Fig. 3–5.

3–2 THE α-HELIX

The two helical configurations in Fig. 3–5 are both in accord with the conclusions reached from structural studies of related small molecules. Of the two helices cited, Pauling and Corey postulated that the right-handed α-helix is more stable on the grounds that dispersion force interactions are optimal for this structure.

In principle, it is possible to obtain information as to the existence of α-helices in naturally occurring proteins by an examination of the X-ray diffraction pattern photographs from crystalline samples. In practice, such "examinations" are extremely complicated, mainly because of the quantity of experimental data which results as a consequence of the large number of different, but rigid interatomic arrangements within the protein crystal. As we shall see later, it is impossible to ascertain the structure of a particular region of the protein without determining the entire three-dimensional structure of the crystal. An experimental investigation as to the existence of the α-helix could be carried out more readily with a polypeptide in which the number of *different* atomic arrangements were greatly reduced. This would necessarily be the case for a synthetic polypeptide in which all of the R groups (amino acid side chains) were identical. Two such polypeptides were known at the time the α-helix was postulated, namely poly-γ-benzyl-L-glutamate (PBLG) and poly-L-alanine. These synthetic polymers can be isolated as fibers—an important structural feature in simplifing the interpretation of the X-ray diffraction pattern. It is inherent in the nature of fibers that there exists a rigid, orderly structure in *one dimension* (in the direction of the fiber axis), and lesser order elsewhere. In a polypeptide fiber the direction of the fiber axis is presumably the direction of the majority of the helix axis, if the fiber is indeed constructed from a helical polypeptide. Were such fibers to consist of long α-helical polypeptides, the observed diffraction pattern would reflect three strong features: the regular, repetitive distance between identical atoms along the fiber axis (i.e., the helical axis), the distance between identical atoms separated by a single peptide bond, and the distance between identical atoms on adjacent "turns" of the helix. These distances and their projections along the helix axis are shown in Fig. 3–6. The projections along the helix axis are related, in a *reciprocal* manner, to strong features in the anticipated X-ray diffraction pattern according to a theory proposed by Cochran, Crick, and Vand. The actual observed diffraction pattern obtained from PBLG and from poly-L-alanine offer strong support for the existence of α-helical structure. Particularly noteworthy in this regard is the occurrence of a 1.5 Å spacing, characteristic of the projection on the helix axis of the shortest repeat distance shown in Fig. 3–6. The fiber X-ray diffraction pattern gives us information about rigidly oriented and regularly repeated atomic configurations within the fiber. Since not all of the atoms are rigidly oriented in the fiber, it is impossible to deduce from the diffraction pattern

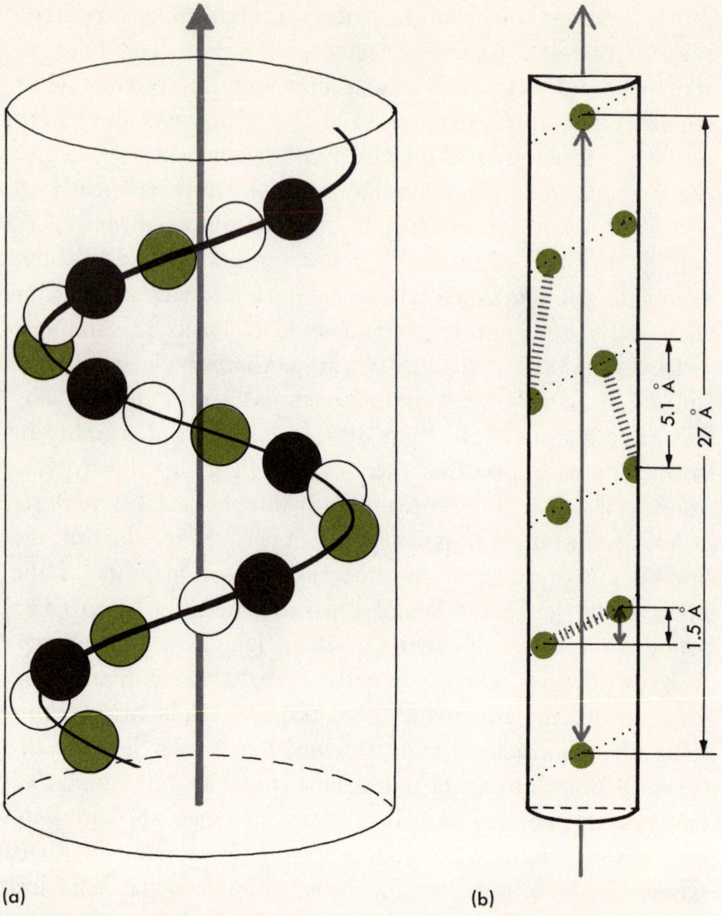

FIG. 3–6 Topological model of a helical polymer. The characteristic vector dimensions of the helical array.

the complete atomic configuration within the molecule. Hence, it is not possible to examine the fibers with regard to the details of α-helix structure illustrated in Fig. 3–5. In order to obtain this level of structural information it is necessary to investigate a system in which all of the atomic coordinates of the system are rigidly specified, that is, a crystal.

X-RAY DIFFRACTION AND CRYSTAL STRUCTURE

Let us consider an actual X-ray diffraction pattern, for example one of the many photographs that can be obtained by the scattering of an X-ray beam from a crystal of the modified enzyme protein diisopropylphosphoryl (DIP)-trypsin (Fig. 3–7). Notice the regular distribution of intensity maxima: The *symmetry* of the pattern is a reflection of the *symmetry* of the crystal, that is, the precise manner in which molecules are packed in regular repeating units to form the crystal. The spacing of the maxima are related to the dimensions of the repeat unit or *unit cell* of the crystal in a *reciprocal* manner: The larger the dimensions of the unit cell (the smallest volume element of the crystal from which the entire crystal can be built up by *translations* only), the closer are the intensity maxima on the film. Both the specific symmetry in the arrangement of unit cells and the *dimensions* of the unit cell are readily obtainable from the information contained in three diffraction photographs (taken with the crystal differently oriented in each). The inner diffraction maxima (those closest the center) and therefore, according to the reciprocal relationship, those related to the largest repeat distances, are sufficient to ascertain both the symmetry and the dimensions.

The *intensity* of the spots is a function of the distribution of atoms within the unit cell, and of the electron density of the particular atoms. The unit cell symmetry of the L-threonine crystal utilized for the structure analysis (Table 3–1) is shown in detail in Fig. 3–2. The diffraction pattern illustrated in Fig. 3–7 has the same characteristic symmetry as does the L-threonine crystal (space group $P2_12_12_1$). The actual separation of the spots, however, will be different in the two crystals, due to the differences in unit cell dimensions. The larger unit cell (DIP-trypsin) will give rise to much closer spacings. Moreover, the diffraction pattern of L-threonine differs from that of DIP-trypsin in the relative intensity of particular spots. The relative intensity of each of the spots can be quantitatively predicted from a knowledge of the configuration of all of the atoms in the molecule, and of the unit cell symmetry. The relationship among the unit cell symmetry, the atomic configuration, and the intensity of the diffraction maxima is illustrated in Figs. 3–8 and 3–9. Notice that the configuration of the atoms and the symmetry of the molecular packing in *real space* permits a quantitative prediction of the *intensity* and

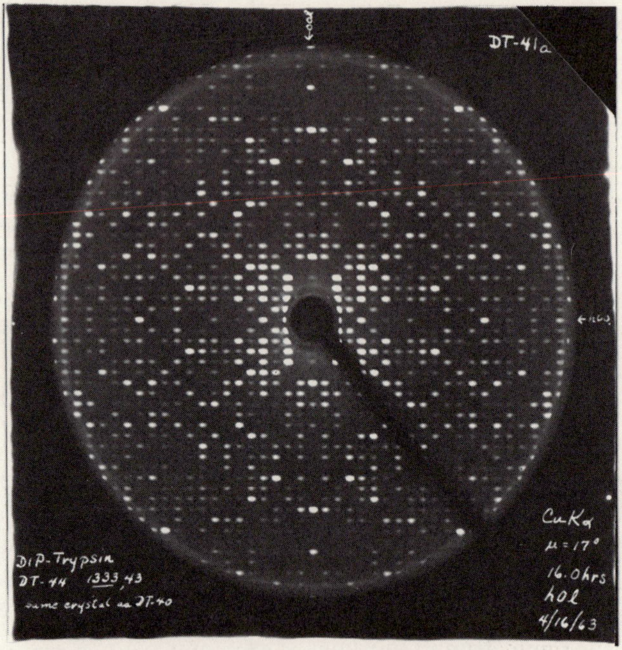
00l
DT-41a
h00
DIP-Trypsin
DT-44 1333, 43
same crystal as DT-40
CuKα
μ = 17°
16.0 hrs
h0l
4/16/63

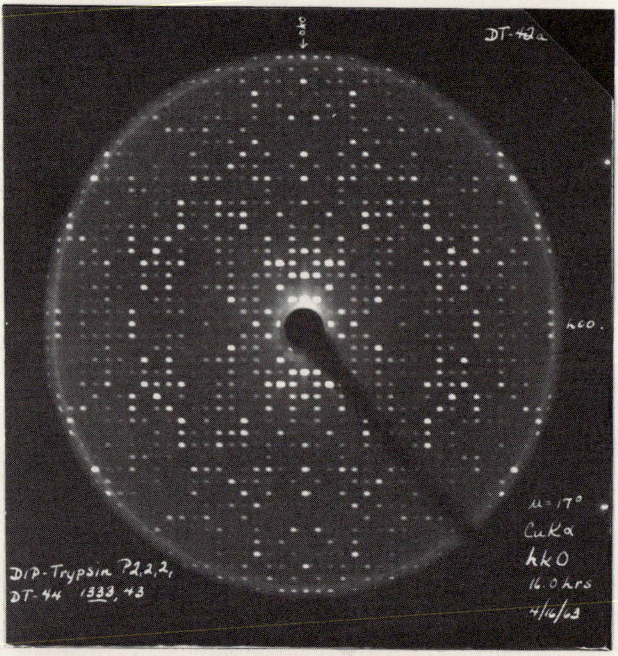
0k0
DT-42a
h00.
μ = 17°
CuKα
hk0
16.0 hrs
4/16/63
DIP-Trypsin $P2_12_12_1$
DT-44 1333, 43

the *phase* of each spot in the *transformed space* of the diffraction photograph. The converse would equally be true: Given the (relative) intensity and the phase for each of the diffraction maxima, the coordinates of atoms and molecules in real space could be predicted. Unfortunately, only the intensities can actually be measured from the X-ray diffraction photograph; hence the atomic coordinates are not directly soluble. Given a crystal of a small molecule, and having a knowledge of its unit cell symmetry and dimensions and its density, a clever structural chemist might make some informed guess as to what kind of atomic and molecular configurations could fit within the unit cell dimensions. Having made such a guess, he could thereby predict the phases for all of the atoms in each of the reflections, and in this way calculate the relative intensities of each of the maxima. Only a good guess would lead to any correspondence between the predicted and observed relative intensities. Once there is some correspondence between prediction and experiment, more refined phase determinations can be carried out by a series of successive approximations until the best possible match of predicted and observed intensities is obtained. This type of refinement is now carried out utilizing high-speed digital computors. Nevertheless, an organic molecule with a molecular weight of several hundred represents the practical limits for which the correct phases, and hence the atomic coordinates, can be found by such guessing procedures. With larger molecules (proteins are much larger molecules) it is necessary to solve the phase problem for each of the maxima

FIG. 3–7 Two x-ray diffraction photographs of a crystal of diisopropylphosphoryl-trypsin. The symmetry of the reflections is characteristic of a particular unit cell arrangement (space group—$P2_12_12_1$). The *space group* can be identified from a set of x-ray diffraction photographs of the crystal taken with the beam perpendicular to each of the three crystallographic axes. This particular space group symmetry is also found in L-threonine crystals. Notice the variations in relative intensity of different reflections. This particular quantitative variation in relative intensity is characteristic of the atomic coordinates within the unit cell of DIP-trypsin. (Courtesy of Prof. R. Corey.)

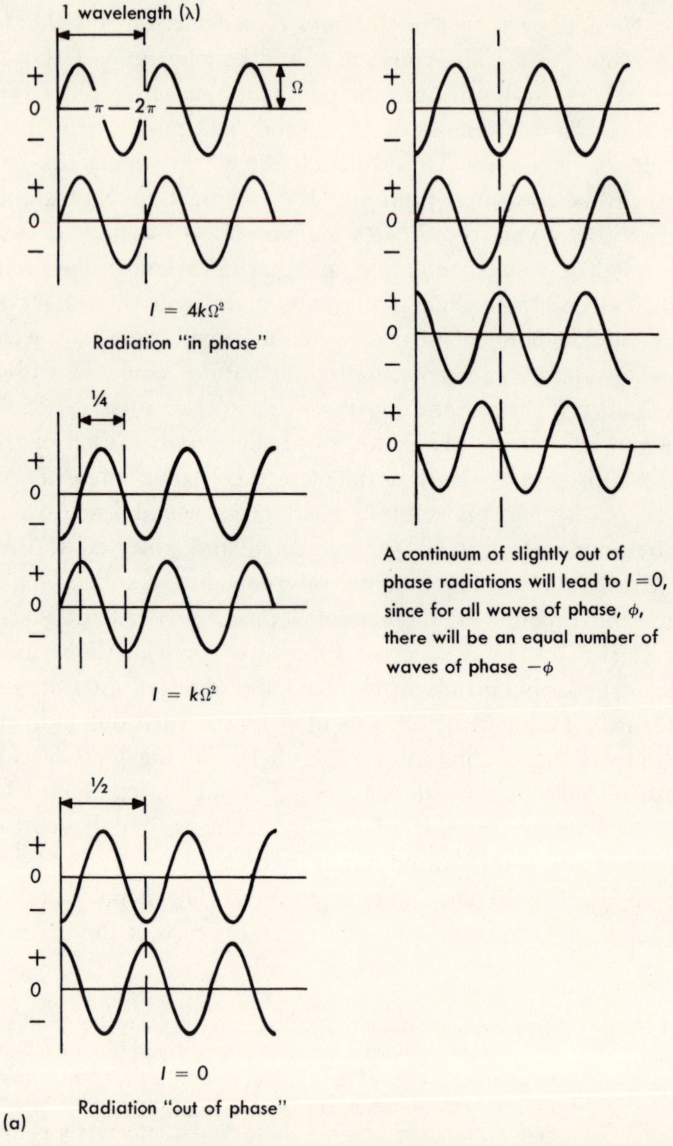

FIG. 3–8 X-ray diffraction. (a) Characteristics of electromagnetic radiation; wavelength, amplitude, and phase. (b) Conditions leading to a diffraction maximum: The scattered waves must be in phase, that is, the difference in phase, $d_2 - d_1$, must be an integral multiple (n) of the wavelength (λ). The symbol Ω is the amplitude of the wave and I is the resultant intensity of a group of waves propagated in the same direction.

Two-dimensional array of unit cells

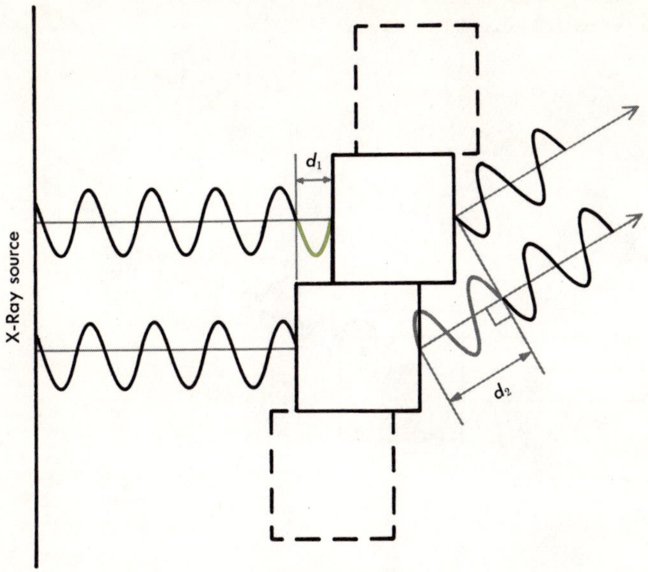

Condition for a diffraction maximum

$$d_2 - d_1 = n\lambda$$

FIG. 3–8b

before the structure determination can be carried out. The importance of practical methods for the solution of the phase problem for a large molecule has recently been recognized by the award of three Nobel prizes.

3–3 CHAIN REGULARITY AND THE STRUCTURE OF PROTEINS IN SOLUTION

Once the α-helical structure had been established in the abovementioned fibers, it became of interest to determine to what extent helical structures are present in other synthetic polypeptides and in proteins, and to what extent such structures maintain their integrity in the aqueous environment where most enzymes function.

The three-dimensional structures of protein molecules in solution cannot be determined by X-ray diffraction techniques. If

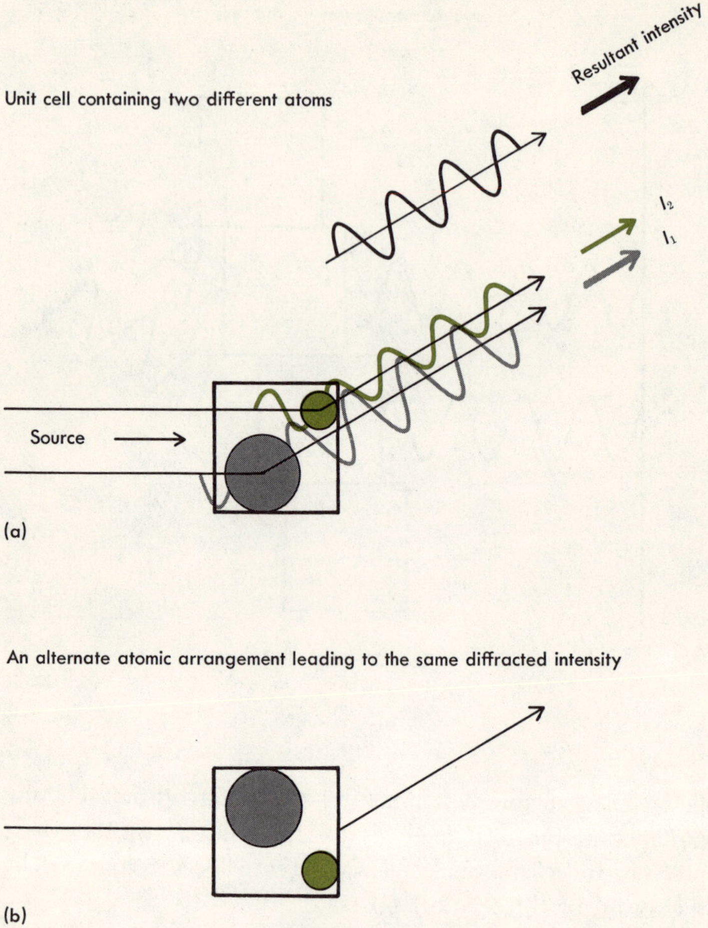

FIG. 3–9 Intensity and phase. The effect of interatomic configuration on the intensity of scatter in the direction of a diffraction maximum.

the α-helix were to remain intact in solution, the regular features, in particular the regular geometric arrangement of carbonyl and NH groups, and the regular hydrogen bonding patterns, would lead to some predictable and distinctive physical properties. Methods of searching for regular structural features in protein solutions involve techniques such as infrared spectroscopy, ultraviolet spectroscopy, and optical rotatory dispersion.

The *infrared absorption spectrum* (i.e., the vibration-rotation spectrum) of a polypeptide will be altered if it is in an α-helical configuration, as a consequence of the altered characteristics of the peptide N—H stretching and bending frequencies due to hydrogen bonding.

The *ultraviolet spectrum* of the peptide bond (i.e., the electronic transition spectrum) will change due to the regularly oriented and electronically altered configurations about each carbonyl group accompanying the hydrogen bonded arrangement in the helix. Similar effects are noted in the spectra of simple amides, when these molecules are arranged in an orderly hydrogen bonded array in crystals.

Since *optical rotation* derives from the difference between the interaction of an asymmetric polar environment with the ground electronic state and an excited electronic state of a *chromophore* (an absorber of electromagnetic radiation), the above-mentioned changes in the ultraviolet absorption spectra should also give rise to differences in optical rotation between helical and nonhelical protein molecules. It should be noted that a right-handed α-helix composed of L-amino acid residues is itself an asymmetric structure (Fig. 3–5). Hence, there should be a contribution to the total optical rotation from the asymmetric helix, as well as from the individual asymmetric centers of each amino acid residue.

By measurements of such physical properties, it has been possible to demonstrate that the above-mentioned synthetic polypeptides can exist in an α-helical configuration in solution as well as in fibers. Typical experimental data are illustrated in Fig. 3–10. Moreover, by appropriate change of physical conditions (such as change of temperature or change of solvent), the α-helical configuration can be destroyed. When a high molecular weight synthetic polypeptide such as PBLG is heated in a solvent in which the α-helical conformation is stable at room temperature, a transition from a helical to a nonhelical (and presumably random) structure occurs rather abruptly (over a narrow range of temperature). Such phenomena can be followed by measuring a physical property which is dependent upon macromolecular conformation such as optical rotation or viscosity, as illustrated in Fig. 3–11. The general shape of this figure resembles that obtained for the melting of a crystalline solid, and is characteristic of *cooperative phenomena*. That is to say, the breakup of one region of helical structure aids in the disruption of the entire helix. A cooperative process would be predicted in the breakup of an α-helix, since the carbonyl oxygen and peptide nitrogen from a single amino acid residue of the polypeptide are hydrogen bonded to conjugate NH donors and oxygen acceptor atoms in two different regions of the molecule (i.e., the oxygen to NH donor bonds point downward, whereas the NH to oxygen acceptor bonds point upward, in the model of Fig. 3–5). A region of approximately seven covalently linked amino acid residues separates the two extreme donor and acceptor atoms in formation of a pair of hydrogen bonds with any *single* —CONH— residue in the middle of the chain. Disruption of one hydrogen bond would still leave the entire helix intact. Disruption of three adjacent hydrogen bonds would break up one turn of the helix, which would in turn help in the disruption of helical structures stabilized by residues preceding and following this nonhelical region. Cooperative transitions of structure in solution are common in proteins. Consider for example the denaturation of egg albumin, a phenomenon which occurs abruptly at a temperature near the boiling point of water.

FIG. 3–10 Some examples of physical properties of polypeptides and proteins characteristic of their hydrogen-bonded helical structure. (a) The far ultraviolet spectrum of poly-L-glutamic acid in the helical configuration (pH 4.9) and in the random coil (pH 8.0). (b) The specific optical rotation in the ultraviolet for ferrihemoglobin and ferrimyoglobin at pH 6.6 and 25° C. The two proteins contain essentially the same percentage of α-helical structure. The upper curve, which exhibits no trough at 230 mμ is for the denatured and presumably randomly coiled structure. [Figure 3–10a redrawn from data of Tinoco, Simpson, and Halpern in *Polyamino Acids, Polypeptides and Proteins* (M. A. Stahman, ed.), University of Wisconsin Press, 1962; Fig. 3–10b redrawn from S. Beychok and E. R. Blout, *J. Mol. Biol.* **3**, 769 (1961). and data of B. Littman and J. Schellman.]

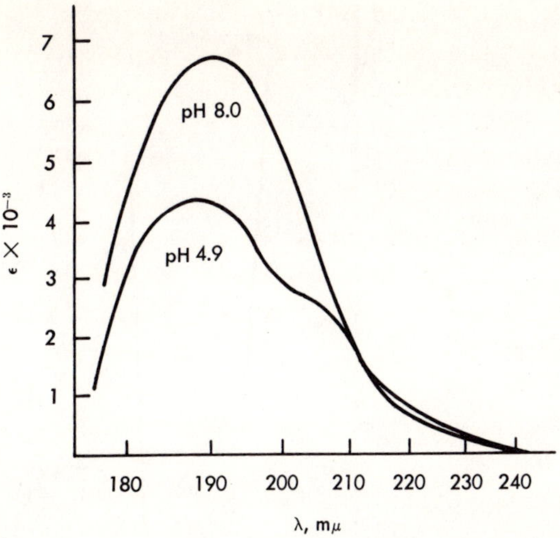
7
6
5
4
3
2
1
$\epsilon \times 10^{-3}$
pH 8.0
pH 4.9
180
190
200
210
220
230
240
λ, mμ

(a)

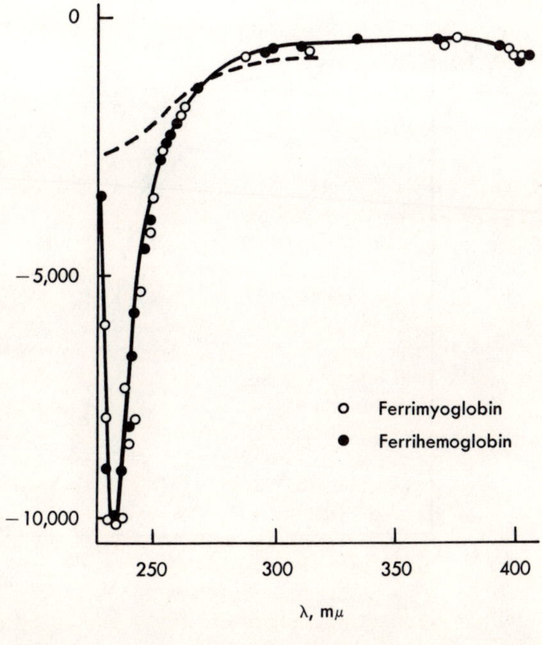
0
(α)
−5,000
−10,000
Ferrimyoglobin
Ferrihemoglobin
250
300
350
400
λ, mμ

(b)

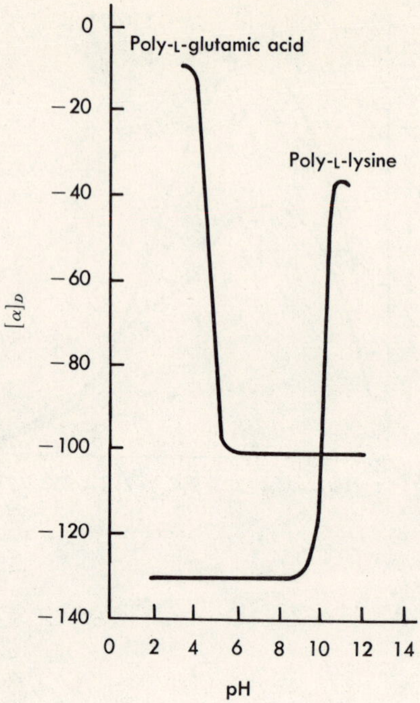

Poly-L-glutamic acid
Poly-L-lysine
$[\alpha]_D$
0
−20
−40
−60
−80
−100
−120
−140
0 2 4 6 8 10 12 14
pH

(a)

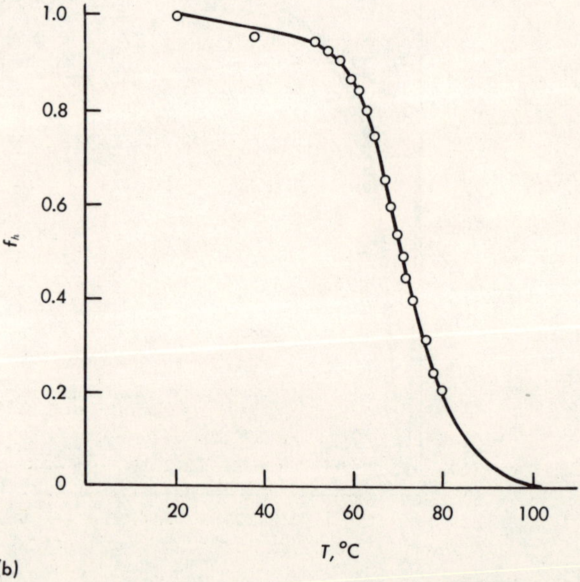

f_h
1.0
0.8
0.6
0.4
0.2
0
20 40 60 80 100
T, °C

(b)

Following the demonstration of the existence of α-helical structures in synthetic polypeptide fibers, and with the availability of methods for the detection of structure in solution, a good deal of effort has been devoted to the study of different synthetic polypeptides in order to ascertain whether the α-helix is common to all proteins, or if not, to what extent the structure of any particular polypeptide is governed by the nature of its particular side chain residue. By examination of molecular models, it is clear that certain types of residues cannot exist in the α-helical configuration. For instance, a polypeptide composed of a racemic mixture of D- and L-amino acid residues is not compatible with the α-helix since a particular R group will lie either above or below a plane passing through the α-carbon atom perpendicular to the helical axis (Fig. 3–5), depending on whether the residue is D or L. Strong repulsions will arise wherever a D residue coincides with an L residue one helical repeat unit above or below it in the direction of the helical axis. This prediction has been established by the demonstration that the polypeptide poly-DL-alanine will under no circumstances form an α-helix, whereas poly-L-alanine is completely helical, both in solvents which do not disrupt interpeptide hydrogen bonds and in the fiber.

The amino acid proline, when it occurs in peptide linkage, demands a fixed angle between the planes of the two peptide residues on either side of the α-carbon atom as is shown in Fig. 3–12. This fixed (dihedral) angle between the two planes differs from that demanded in the α-helical arrangement. Thus, proline must terminate the α-helical conformation, whenever it occurs within a polypeptide chain, due to chemical bonding restrictions on the rotation about the N—C-2 bond (Fig. 3–3).

FIG. 3–11 Helix coil transitions. (a) The pH dependence of the specific rotation of poly-L-glutamic acid and poly-L-lysine. Notice the sharp transitions from coil to helix. (b) The helix coil transition as a function of temperature, for sperm whale myoglobin. The fraction helix (f_h) is calculated from data on the optical rotation of the protein solution as a function of temperature.

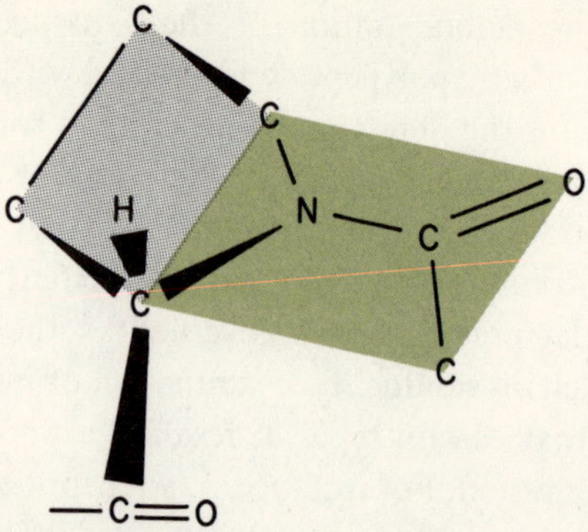

FIG. 3–12 The configuration of an L-proline residue in peptide linkage.

Most amino acids contain a β-CH_2 group. When the β-carbon positions in Fig. 3–5 are all CH_2R, there is no interference with the geometrical structure of the helix. However, when one of the hydrogens in the β-CH_2 is replaced by substituents with larger van der Waals radii, as for example by carbon atoms in the amino acids isoleucine and valine, considerable repulsion will arise due to the short contacts between the second β-carbon substituent and neighboring atoms on the α-helical backbone. This will tend to lower the stability of the helix.

Structural studies with a number of synthetic polypeptides each containing only one type of amino acid residue are summarized in Table 3–2. From this table it is clear that some residues are compatible with (and in fact stabilize) the α-helix, whereas others tend to disrupt the helix. An α-helix may not be a stable structure for two alternative reasons: (1) The side chain interactions may be such as to prevent the formation of a stable helix due to repulsions of one type or another which are concomitant with helix formation (as in poly-L-isoleucine, poly-L-proline, poly-L-glutamate$^-$, and poly-L-aspartate$^-$, and poly-L-lysine$^+$), or (2) there may be a more stable alternative structure. In poly-L-serine, hydrogen bonding involving the serine hydroxyl group gives rise to a structure more stable than the α-helix (since the total hydrogen bonding potential of the serine hydroxyls cannot be realized in the α-helix). Polyglycine lacks

Conformational Classification of α-Amino Acids[a]

α-Helix forming	α-Amino acid side chains (R) $-NH-CH(R)-C(=O)-$	Non α-helix forming	α-Amino acid side chains (R) $-NH-CH(R)-C(=O)-$
Alanine	$-CH_3$	For steric reasons	
Glutamic acid (neutral)	$-CH_2-CH_2-C(=O)-OH$	Valine	$-CH(CH_3)-CH_3$
Leucine	$-CH_2-CH(CH_3)-CH_3$	Isoleucine	$-CH(CH_3)-CH_2-CH_3$
Lysine (neutral)	$-CH_2-CH_2-CH_2-CH_2-NH_2$	Threonine[b]	$-CH(CH_3)-OH$
Methionine	$-CH_2-CH_2-S-CH_2$	For other reasons	
Phenylalanine	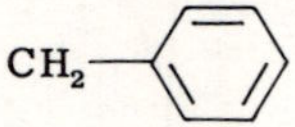	Glycine	$-H$
Tyrosine	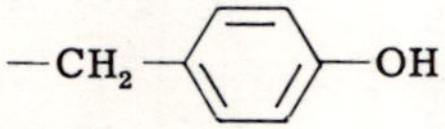	Serine	$-CH_2-OH$
		Proline	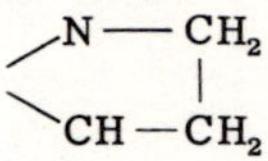
		Glutamate (anion)[c]	
		Lysine (cation)[c]	
		Asparate (anion)	

[a]Modified form from a table presented by E. R. Blaut in *Synthetic Polypeptides* (see References, this chapter).
[b]Hydrogen bonding may also contribute to the helix instability.
[c]At pH's such that the charge per residue <0.3, these residues can be "helix-forming."

the restrictions which a β-carbon atom places on the configuration of the peptide chain, and as a result, tends to form "sheet-like structures" in which different polypeptide molecules are held together via *intermolecular* hydrogen bonds.

Most noteworthy, in light of previous discussion, are the data on poly-L-glutamate$^-$ and poly-L-lysine$^+$. When these polymers are studied in aqueous solution at a pH such that most, but not all, of the carboxyl or amino groups are uncharged (pH $\sim$ 4 for polyglutamate, pH $\sim$ 10 for polylysine), they remain soluble in *aqueous* solution and form stable α-helices. Near pH neutrality, both poly-L-glutamate$^-$ and poly-L-lysine$^+$ are virtually completely charged (they contain one negative and one positive charge per monomeric residue respectively). In the α-helix these charges would be packed closely together (note the close spatial relationships between R groups which are either three or four covalent monomeric units apart). In a more extended conformation such electrostatic repulsions would be reduced. For this reason, neither of these (completely charged) polymers has a helical conformation in aqueous solution at neutral pH. In the absence of such strong and specific repulsive forces (weakly acidic solutions of polyglutamate and weakly basic solutions of polylysine), the α-helices are stable even in aqueous solution where the solvent molecules are themselves strong hydrogen bond donors and acceptors. This experimental finding tends to minimize or negate the argument that the α-helical conformation proposed by Pauling and Corey for solid (fibrous) polypeptides would not maintain its integrity in aqueous solution.

From studies of the conformation of the various synthetic polypeptides (such as are summarized in Table 3–2) we can make a rough estimate of the extent of helical configuration in a protein, if the amino acid composition of the protein is known. We can also make an estimate of the protein conformation by comparing the *magnitude* of a property arising from helical structure in a protein with that observed for a fully helical synthetic polypeptide. The conformation of a protein is precisely defined by the specific amino acid sequence of the polypeptide chain. Specific regions of the polypeptide will assume specific structures, depending on both the covalent sequence of that region and on the *intra-* or *intermolecular* interactions between that region and other regions of the protein.

Further predictions of protein structure will rely heavily on the accumulated direct information on the three-dimensional structure of proteins of known sequence. The three-dimensional structure of a specific polypeptide sequence (a protein) has now been determined to atomic resolution for the proteins *myoglobin* and *hemoglobin,* and the enzymes *lysozyme, ribonuclease, carboxypeptidase, chymotrypsin,* and *papain.* Kendrew and his collaborators, utilizing X-ray diffraction techniques for the study of crystalline myoglobin, were the first to determine a complete three-dimensional structure. A schematic model of the structure of myoglobin is illustrated in Fig. 3–13. A functionally related protein, hemoglobin, has been examined in

FIG. 3–13 An outline of the conformation of the myoglobin peptide backbone.

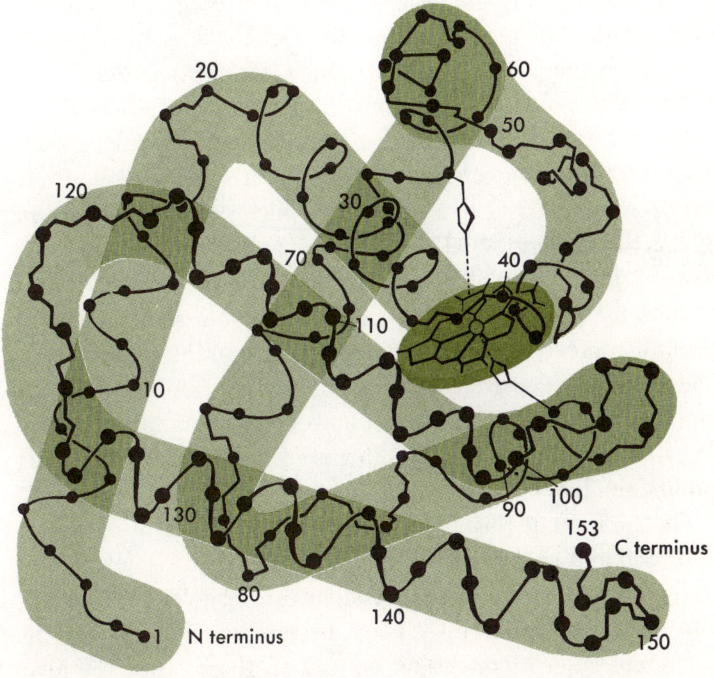

great detail crystallographically by Perutz and his co-workers. The greater complexity of this protein has until recently restricted the resolution of the structure to distances somewhat greater than chemical bond distances (see discussion below).

3–4

AMINO ACID SEQUENCE AND THE STRUCTURE AND FUNCTION OF PROTEINS

The fraction of the total amino acid residues in myoglobin which are found in an α-helical conformation is strikingly close to that predicted from physical measurements (particularly optical rotatory dispersion measurements) in aqueous solution. Although the molecule is nearly 70% α-helical, regions of nonhelical structure appear in many parts of the sequence; that is to say, the total length of any particular helical segment never represents a major fraction of the total amino acid sequence. One might be led to assume that the various bends and twists in the molecule result from the highly specific sequence of amino acid residues in myoglobin molecules. Surprisingly, the general shape of the polypeptide components of the structurally more complex hemoglobin molecule, which consists of four "myoglobin-type" polypeptide chains of two different kinds (see Fig. 3–14), are each strikingly similar to the myoglobin polypeptide. The three-dimensional structure of the hemoglobin molecule has been determined to a resolution which permits comparison of the three-dimensional shape of the polypeptide chains with that of myoglobin. The two different polypeptide constituents of hemoglobin and the single myoglobin polypeptide all contain approximately the same number of amino acid residues. They differ extensively, one from another, in their precise amino acid sequence. There are, however, some occasional striking similarities (for example, in the location of helix-breaking proline residues).

On the basis of this very limited survey of proteins of known sequence and structure, it would appear that many changes in the sequence of amino acid residues can be tolerated without any obviously significant change in the conformation of the macromolecule. A conclusion derived on the basis of the myo-

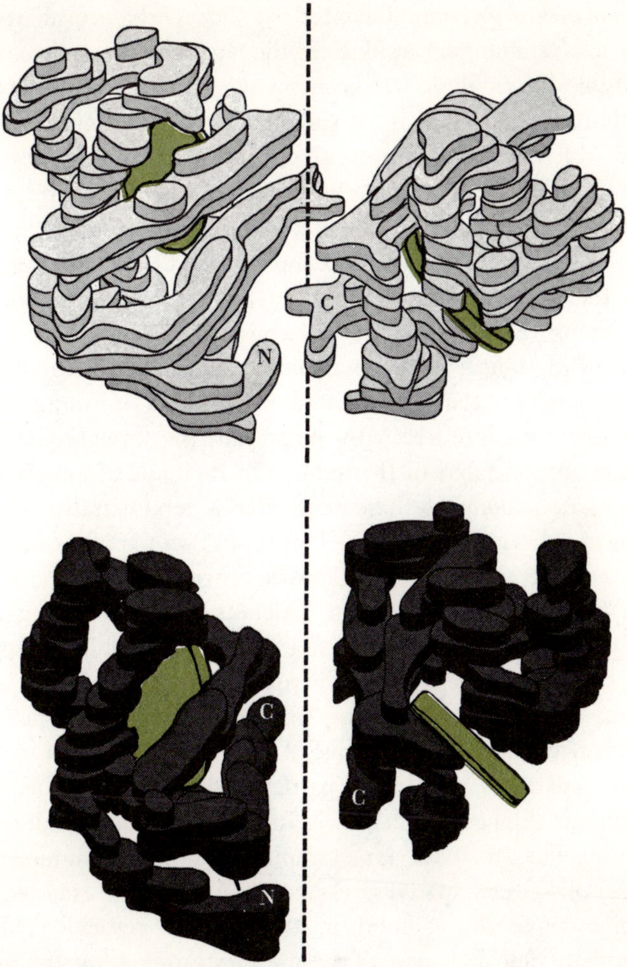

FIG. 3–14 The conformation of the polypeptide chains in hemoglobin. (From "The Structure of the Hemoglobin Molecule" by M. F. Perutz. Copyright © 1964 by Scientific American, Inc. All rights reserved.)

globin structure is that although nearly all potential hydrogen bonds are actually realized, the major interactions governing the three-dimensional conformation (aside from covalent bonding) are dispersion force interactions.

The total sequence of amino acid residues is known for only a limited number of enzymes at this time (a number of complete sequences are given in Chapter 8). The partial covalent sequence of amino acid residues in the region of the enzyme responsible for catalysis has been investigated more extensively. With many enzymes, it is possible to "label" a particular region —*the active center*—by means of a specific chemical reagent. The general principle underlying the method is described in Sections 4–2, 4–3, and in Chapter 5. The "labeled" protein can then be partially degraded to shorter polypeptide fragments, and the covalent amino acid sequence of peptide fragments containing the label can be determined. The amino acid sequences of some enzyme fragments containing a part of the active catalytic center are listed in Table 3–3. Amino acid residues which interfere with, or prevent the formation of, an α-helix are indicated in this table. On the basis of amino acid composition, none of these sequences is representative of regions of helical conformation. This result is not at all surprising. The enzyme site must have a specific geometrical conformation for the specific function which it carries out. It is unlikely that such a highly specialized function could be specified by a regular repetitive structure such as the α-helix. One would expect that a catalytic site would have a well-defined conformation, but a unique one. A unique conformation might be anticipated to arise from a particular sequence of amino acid residues in the peptide chain within this region. A large number of proteolytic enzymes, for example, contain the common site sequence -glycyl-aspartyl-seryl-glycyl-. A striking example of specific sequence is found in the enzyme glyceraldehyde-3-phosphate dehydrogenase. Two enzymes carrying out the same catalytic reaction have been isolated, one from yeast, and the other from rabbit muscle. Although these two enzymes differ in total amino acid composition, they share a common sequence of at least eighteen amino acid residues in the region of the catalytic site. This result is all the more striking in light of the fact that variations in amino acid sequence are often found when a protein of highly specific function is isolated from a number of closely related species. In the protein hormone insulin, for example, a polypeptide chain (of nearly the same length as the above-mentioned eighteen residue common se-

Table 3-3

Amino Acid Sequence in the Covalent Vicinity of the Active Site for a Variety of Enzymes[a,b]

	S ... S
Chymotrypsin	CYS* · MET · GLY* · ASP* · SER* · GLY* · GLY* · PRO* · LEU · VAL* · CYS*
Trypsin	CYS* · GLN · GLY* · ASP* · SER* · GLY* · GLY* · PRO* · VAL* · VAL* · CYS*
	S ... S
Glyceraldehyde-3-phosphate dehydrogenase	ILEU* · VAL* · SER* · ASN · ALA · SER* · CYS · THR* · THR* · ASN · CYS
Yeast alcohol dehydrogenase	VAL* · ALA · THR* · GLY* · ILEU* · CYS · ARG* · SER* · ASP* · ASP* · HIS · ALA
Alkaline phosphatase	LYS* · PRO* · ASP* · TYR · VAL* · THR* · ASP* · SER* · ALA · ALA · SER* · ALA
Phosphoglycomutase	GLY* · VAL* · THR* · ALA · SER* · HIS · ASP* · GLY* · GLU* · SER* · ALA · GLY*
Acetoacetate decarboxylase	GLU* · LEU · SER* · ALA · TYR · PRO* · LYS* · LYS* · LEU
3-Glycerophosphate aldolase	GLY* · THR* · LEU · LEU · LYS* · ASN · PRO* · MET · VAL* · THR* · PRO* · GLY*
Glutamate-aspartate transaminase	GLU* · (ALA,ASP*,GLY*,ILEU*, LYS*)[c] · GLY* · SER* · ASP* · PHE
Trypsin	HIS · PHE · CYS* · GLY* · GLY* · SER* · LEU · ILEU* · ASN · SER* — GLN
	· CYS* · HIS · ALA · ALA · SER* · VAL* · VAL* · TRY — GLN

[a] Residues which are known to function catalytically are underlined.
[b] Residues which tend to disrupt helical structures (Table 3-2) are starred.
[c] The internal order of the residues has not been determined.

quence fragment) is found to contain variations in amino acid sequence depending upon whether the protein is derived from beef, hog, or sheep. Many differences in amino acid sequence are found among the individual chains of hemoglobins, fetal hemoglobins, and myoglobin. The common sequence of (at least) eighteen amino acids in the two (glyceraldehyde-3-phosphate dehydrogenase) enzymes must reflect a highly specific chemical function for this region of sequence—a sequence in which there is no possibility for change without impairment of function.

REFERENCES

Pauling, L., *The Nature of the Chemical Bond*, 3rd ed., Cornell University Press, Ithaca, New York (1960). The classic text on the structure of molecules and the forces of interaction between atoms and molecules.

Pauling, L., Corey, R. B., and Branson, H. R., "The Structure of Proteins: Two Hydrogen-Bonded Helical Configurations of the Polypeptide Chain," *Proc. Nat. Acad. Sci.*, **37**, 205 (1951). The definitive paper on the α-helix.

Perutz, M. F., *Proteins and Nucleic Acids: Structure and Function*, Elsevier, Amsterdam, 1962. The first half of this book gives an excellent description of the structures of proteins and nucleic acids, and the methods by which these structures were determined.

Crick, F. H. C., and Kendrew, J. C., "X-Ray Diffraction and Protein Structure," *Advan. Protein Chem.*, **12**, 133 (1957). An informative and often entertaining nonmathematical account of how X-ray diffraction techniques are utilized for protein structure determinations, and in particular, how one goes about solving the "phase problem."

Cochran, W., Crick, F. H. C., and Vand, V., "The Structure of Synthetic Polypeptides. I. The Transform of Atoms on a Helix," *Acta Cryst.*, **5**, 581 (1952). For the mathematically inclined, an elegant derivation of the transform in *reciprocal* space for a helical molecule. This paper is of considerable historical interest: Although the theory was developed to analyze protein fiber structures for α-helices, it was utilized by one of the authors very shortly thereafter in proposing a structure for DNA fibers [Watson, J. D., and Crick, F. H. C., *Nature*, **171**, 737 (1953)].

Perutz, M. F., "The Hemoglobin Molecule," *Scientific American*,

November, 1964. This article is, in essence, the author's Nobel prize lecture. It describes how the "phase problem" was actually solved and what is now known about the conformation of the hemoglobin molecule.

Kendrew, J. C., "Myoglobin and the Structure of Proteins," *Science*, **139**, 1259 (1963). Adapted from the Nobel prize lecture, this nonmathematical article describes the methodology utilized in the first determination of protein structure to atomic resolution, and the implications in regard to the general problem of protein structure.

In papers by Bluhm, M. M., Bodo, G., Dintzis, H. M., and Kendrew, J. C. [*Proc. Royal Soc., London*, **A246**, 369 (1958)] and by Bodo, G., Dintzis, H. M., Kendrew, J. C., and Wyckoff, H. W. [*Proc. Royal Soc., London*, **A253**, 70 (1959)], the important details of analysis of the crystallographic data are commendably described such that they are intelligible, with sufficient effort, by students having a modicum of mathematical background.

Shoemaker, D. P., Donohue, J., Schoemaker, V., and Corey, R. B., "The Structure of L-Threonine," *J. Am. Chem. Soc.*, **72**, 2328 (1950). Probably the most illuminating example of the X-ray crystallographic method of molecular structure determination, and of what can thereby be learned about *inter-* and *intramolecular* forces of interaction.

Kauzmann, W., "Some Factors in the Interpretation of Protein Denaturation," *Advan. Protein Chem.*, **14**, 1 (1959). An excellent account of the various forces of interaction governing native protein structure and of the way these interactions can be examined and differentiated by denaturation studies.

Kopple, K. D., *Peptides and Amino Acids*, Benjamin, New York, 1966. An elementary text on amino acids, the chemical synthesis of peptides and polypeptides, and their structures and conformations. Chapter 2 (Amino Acids) and Chapter 5 (Conformation of Peptide Chains) contain excellent background material for this chapter.

Polyamino Acids, Polypeptides and Proteins (Stahman, M. A., editor), University of Wisconsin Press, 1962. A multi-authored symposium volume dealing primarily with physical chemical methods for examining polypeptide structure. Parts III and IV (pp. 111–279) are particularly relevant to the discussion contained in this chapter.

FOUR ~ ENZYME KINETICS AND FORMAL MODELS OF CATALYSIS

BEFORE CONSIDERING MOLECULAR MECHANISMS OF ENZYME action, we shall review the more formal studies of the chemical reaction kinetics of enzyme-substrate systems, on which many of the fundamental mechanisms are based. The mathematical formalism of enzyme kinetics can readily become so complex as to make analytical comparison between theory and experiment a difficult, if not impossible, process. The complexity of the mathematical equations of enzyme kinetics can often be reduced by making certain limiting assumptions about the system; these assumptions may or may not be of general validity. Conversely, plausible mechanisms of catalysis may be overlooked by avoiding the complex mathematical equations which such mechanisms may demand. In this chapter we shall summarize a number of the more common formal models of enzyme catalysis, and inhibition of catalysis, and we shall indicate the bases for the applicability of these models. Mathematical analysis of models is essential for the correlation of proposed molecular mechanism with the kinetic behavior actually observed in the experimental system. Our primary goal in undertaking a formal chemical kinetic study of an enzyme-substrate system is to make use of the derived information towards the understanding of molecular mechanism.

4–1 EQUILIBRIUM AND THE STEADY STATE

Let us return to the simple model of enzyme substrate interaction and reaction discussed in the introduction.

$$E + S \underset{k_{-S}}{\overset{k_S}{\rightleftharpoons}} ES \xrightarrow{k_P} E + P \qquad (4\text{–}1)$$

The model of Eq. 4–1 is sufficient to account formally for the behavior of many enzyme-substrate systems particularly during the early phase of reaction, before much product has accumulated. This model will lead to a mathematically tractable rate expression under the condition that the total concentration of enzyme, E_0, is very much less than the initial concentration of substrate, S_0, an experimental condition which can usually be realized. The experimentally measurable concentration variable will, in general, be either the substrate or the product concentration ([S] or [P]). If the above-mentioned condition ($S_0 >> E_0$) is applicable, the low concentration of ES will not be of significance relative to changes in [S] or [P], and the rate of reaction will be given by Eq. 4–2, since $[P] = S_0 - [S] - [ES] \simeq S_0 - [S]$.

$$\frac{d[P]}{dt} = \frac{-d[S]}{dt} = k_P[ES] \qquad (4\text{–}2)$$

The substitution of the rate expression for S, defined by the assumed model (Eq. 4–1), into Eq. 4–2 yields Eq. 4–3.

$$-\frac{d[S]}{dt} = k_S[E][S] - k_{-S}[ES] = k_P[ES] \qquad (4\text{–}3)$$

The concentration of intermediate enzyme-substrate complex is therefore given by Eq. 4–4.

$$[ES] = \frac{E_0[S]}{\{(k_{-S} + k_P)/k_S\} + [S]} \qquad (4\text{–}4)$$

where E_0, the total concentration of enzyme sites, is the sum of [ES] and [E]. By substitution of Eq. 4–4 into Eq. 4–2, we obtain the expression (Eq. 4–5) commonly referred to as the *Michaelis-Menten* or *Henri* equation

$$\frac{d[\mathrm{P}]}{dt} = v = \frac{k_{\mathrm{P}}\mathrm{E}_0[\mathrm{S}]}{K_M + [\mathrm{S}]} \tag{4–5}$$

where $K_M = (k_{-\mathrm{S}} + k_{\mathrm{P}})/k_{\mathrm{S}}$ is the *Michaelis constant.*

Equation 4–5 is applicable to any enzyme-substrate system which follows a catalytic pathway corresponding to the formal model of Eq. 4–1, provided that the total enzyme concentration is small enough to satisfy the condition of Eq. 4–2.

Notice that the rate of change of [ES] with time is given by

$$\frac{d[\mathrm{ES}]}{dt} = k_{\mathrm{S}}[\mathrm{E}][\mathrm{S}] - (k_{-\mathrm{S}} + k_{\mathrm{P}})[\mathrm{ES}]$$

and therefore from Eq. 4–3

$$\frac{d[\mathrm{ES}]}{dt} = 0$$

Since $E_0 = [E] + [ES]$, it follows that the same condition applies for $d[E/dt]$. This condition is commonly known as the *stationary* or *steady-state* condition. When the concentration ranges of initial substrate and measurable product greatly exceed the total concentration of enzyme (E_0), the steady-state assumption (in regard to enzyme species) is valid. As we shall see, much of enzyme kinetics is dependent on the validity of this assumption.

Equation 4–4 gives the *steady state* concentration of the species ES. If the *equilibration* of substrate between the solvent and the enzyme site is rapid compared to the reaction rate of ES ($k_{-\mathrm{S}} > k_{\mathrm{P}}$), the *equilibrium* concentration of ES will obtain during the course of reaction. This concentration of ES is given by Eq. 4–6.

$$[\mathrm{ES}]_{\mathrm{eq}} = \frac{[\mathrm{S}]}{(k_{-\mathrm{S}}/k_{\mathrm{S}}) + [\mathrm{S}]} \tag{4–6}$$

Note that Eq. 4–4 is a valid expression provided only that certain *experimentally controllable* conditions of concentrations are maintained. Equation 4–6, however, will pertain only when specific relationships among the *invariant* kinetic parameters *happen* to exist. The condition $k_{-S} \gg k_P$ reduces Eq. 4–4 to Eq. 4–6. Satisfying this latter condition will depend on the nature of the complex which we have formally represented by ES. If, for example, the formation of complex involves only weak interactions such as might arise from the complementary dispersion forces discussed previously (Section 2–2), it is likely that the rate of dissociation of the complex will be rapid, since the breaking of the complex will involve only the disruption of these weak (low energy) interactions. The breakdown of the complex to yield products will involve the making and breaking of chemical bonds. Since chemical processes involve the close approach of atoms through a region of high interelectronic repulsion, they will usually be characterized by high *activation energies* and hence slower rates. In this situation, Eqs. 4–4 and 4–6 will be equivalent since $k_{-S} \gg k_P$. The formation of the complex ES may, however, involve chemical bond formation between enzyme and substrate. In this latter instance, the parameter k_{-S} may be smaller than or of the order of k_P and hence Eq. 4–4 but not Eq. 4–6 may pertain. A more detailed knowledge of the enzyme-substrate system will help to distinguish between the *steady state* (Eq. 4–4) and the *equilibrium* (Eq. 4–6) assumptions.

In the simplest model (Eq. 4–1), the mathematical form of the kinetic equations is the same whether the steady state or the equilibrium assumption is made. The physical interpretation of the constant K depends, however, on which of these assumptions is valid.

4–2 REVERSIBLE AND IRREVERSIBLE REACTIONS

For simplicity, let us begin the kinetic discussion with model reactions which proceed *almost* stoichiometrically to completion. Such reactions may be termed "irreversible reactions" although in reality every reaction is, to some extent, reversible.

The essential reversibility of chemical processes is formalized in the *law of microscopic reversibility* which states that "every chemical reaction pathway is necessarily a pathway for the reverse reaction."

Microscopic reversibility is a necessary assumption for the derivation of the laws of chemical equilibrium from kinetic arguments. Consider, as an illustration, the model reaction system of Eq. 4–1. Let us assume that the concentration of the enzyme catalyst is extremely small compared to that of the substrate. The rates of formation of product and disappearance of substrate are

$$\frac{-d[\mathrm{S}]}{dt} = k_{\mathrm{S}}[\mathrm{E}][\mathrm{S}] - k_{-\mathrm{S}}[\mathrm{ES}]$$

$$\frac{d[\mathrm{P}]}{dt} = k_{\mathrm{P}}[\mathrm{ES}]$$

By definition, at equilibrium, the concentrations of substrate and product no longer change; the *net* rate of each of the two processes given above is equal to zero. Hence, according to the model of Eq. 4–1,

$$[\mathrm{ES}] \text{ at equilibrium} = 0$$

The concentration of substrate (at equilibrium) would therefore be zero as well! This absurdity arises because the model pathway described by Eq. 4–1 is, strictly interpreted, a violation of the law of microscopic reversibility. If the formation of product occurs via the pathway $\mathrm{ES} \rightarrow \mathrm{E} + \mathrm{P}$, the formation of ES from E + P must also be a reaction pathway. The correct model *must* include the reverse pathway

$$\mathrm{E} + \mathrm{P} \xrightarrow{k_{-\mathrm{P}}} \mathrm{ES}$$

Applying microscopic reversibility, the rate of formation of product is given by Eq. 4–7.

$$\frac{d[\mathrm{P}]}{dt} = k_{\mathrm{P}}[\mathrm{ES}] - k_{-\mathrm{P}}[\mathrm{P}][\mathrm{E}] = 0 \text{ (at equilibrium)} \qquad (4\text{–}7)$$

Substituting for the finite (although possibly very low) concentration of ES, we have

$$-\frac{d[\mathrm{S}]}{dt}\ (\text{at equilibrium}) = 0 = k_{\mathrm{S}}[\mathrm{E}][\mathrm{S}] - \frac{k_{-\mathrm{S}}k_{-\mathrm{P}}}{k_{\mathrm{P}}}[\mathrm{E}][\mathrm{P}]$$

Hence,

$$\frac{[\mathrm{P}]_{\mathrm{eq}}}{[\mathrm{S}]_{\mathrm{eq}}} = \frac{k_{\mathrm{S}}k_{\mathrm{P}}}{k_{-\mathrm{S}}k_{-\mathrm{P}}} = K_{\mathrm{eq}} \tag{4–8}$$

which is the familiar expression for chemical equilibrium. Note that if $[\mathrm{P}]_{\mathrm{eq}} \gg [\mathrm{S}]_{\mathrm{eq}}$, it follows from Eq. 4–8 that at all times $k_{\mathrm{S}}k_{\mathrm{P}} \gg k_{-\mathrm{S}}k_{-\mathrm{P}}$ but not necessarily that $k_{\mathrm{P}} \gg k_{-\mathrm{P}}$ [P].

The catalyst (E) can only enhance the *rate* at which chemical equilibrium is established. A more effective catalyst, E′, which might for example increase k_{P}, would also increase $k_{-\mathrm{P}}$ by a corresponding factor, as is demanded by microscopic reversibility. The ratio, $k_{\mathrm{P}}/k_{-\mathrm{P}}$, and hence K_{eq}, would remain unchanged. The same reaction will eventually take place even in the absence of catalyst.

$$\mathrm{S} \underset{k_{-1}}{\overset{k_1}{\rightleftharpoons}} \mathrm{P} \tag{4–9}$$

Although the rate at which the uncatalyzed reaction (Eq. 4–9) approaches equilibrium may be exceedingly slow compared with the catalyzed pathway, equilibrium will eventually be established. In the absence of catalyst the equilibrium conditions are

$$\frac{[\mathrm{P}]_{\mathrm{eq}}}{[\mathrm{S}]_{\mathrm{eq}}} = \frac{k_1}{k_{-1}}$$

If, in the catalyzed reaction, the enzyme concentration is very small compared to both $[\mathrm{S}]_{\mathrm{eq}}$ and $[\mathrm{P}]_{\mathrm{eq}}$, and (or) if the complex, ES, is unstable such that it never achieves concentrations comparable to either [S] or [P], the enzyme (or ES complex) concentration will not measurably perturb the equilibrium conditions. The two pathways of reactions will be related by a thermodynamic identity (Eq. 4–10).

$$\frac{k_1}{k_{-1}} \equiv \frac{k_S k_P}{k_{-S} k_{-P}} \tag{4–10}$$

Although the two specific rate constants in the uncatalyzed pathway (Eq. 4–9) may differ from, and be exceedingly smaller than, any of the specific rate constants of the catalyzed pathway, certain interrelationships, as are given in Eq. 4–10, must obtain. Since the *specific rate constants* are invariant to the concentration of reactants and products, the interrelationships we have derived from equilibrium considerations (Eq. 4–10) must equally well pertain throughout the entire course of reaction.

An essential feature of nearly every enzyme is the specific and restricted catalytic reaction pathways which it opens. Consider, for example, a system which at equilibrium and in the absence of enzyme consists only of reactants and products (Eq. 4–9). The enzyme catalyst may open an entirely new pathway of reaction, for example (Eq. 4–11), via the rapid formation of an intermediate (B).

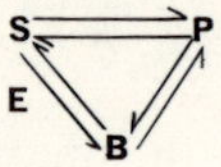

(4–11)

The enzyme will necessarily catalyze the reverse reaction (B → S) as well. If the catalyst is restricted in its action to these two steps, and if the enzymic pathway is far more rapid than any of the uncatalyzed pathways (S ↔ P, B ↔ P), the ultimate equilibrium distribution of products will be reached at a rate dependent upon the nonenzymic pathway B ↔ P. The intermediate B may be even less reactive than S, and, in this event, equilibrium will be reached more slowly due to the presence of the enzyme. In such a situation, an intermediate can achieve a long-lived steady state concentration. In biological systems, such situations are the rule rather than the exception. In the isolated representation of Eq. 4–11, the intermediate is a "dead-end" trap for substrate molecules. In biological systems such intermediates are utilized in the synthesis of other molecules;

synthesis being directed by other enzyme catalysts. It should be noted that the rate of approach to equilibrium can be slowed down by enzyme only if the rate of formation of the intermediate B is rapid compared to the rate of the reverse reaction. Otherwise the uncatalyzed pathway S → P would not be perturbed by the small drain of substrate molecules into the enzymic pathway. In the situation we have assumed (the enzymic pathways being much more rapid than the nonenzymic pathways), S and B are in virtual equilibrium after a short period. The rerouting of chemical reaction via the intermediate B can hence be achieved only if B is thermodynamically more stable than the original substrate. This implies that the formation of B is energetically favorable, that is, that there is an available energy source for driving the reaction in this direction.

"IRREVERSIBLE" REACTIONS

We may define a reaction as virtually "irreversible" if the reverse reaction in no way perturbs the concentration of substrate or enzyme-substrate intermediates during the course of kinetic measurements. The simplest model of an irreversible enzyme-catalyzed reaction is the Henri or Michaelis-Menten model (Eq. 4–1). The model is valid when there is only one substrate. This is rarely the case in enzymic reactions. Often, however, one of the two (or more) substrates is present in great excess. In such situations, the concentration of the reactant present in excess is only trivially changed during the course of reaction and hence the kinetics can be formally analyzed as if this excess reactant were not present. Hydrolytic reactions are typical examples. Moreover, hydrolytic reactions often lead to products which dissociate at the buffered pH of the biological system. Such dissociations can lead to a quasi-irreversible pathway; for example, in the hydrolysis of esters, the carboxylic acid product is virtually completely ionized at pH neutrality.

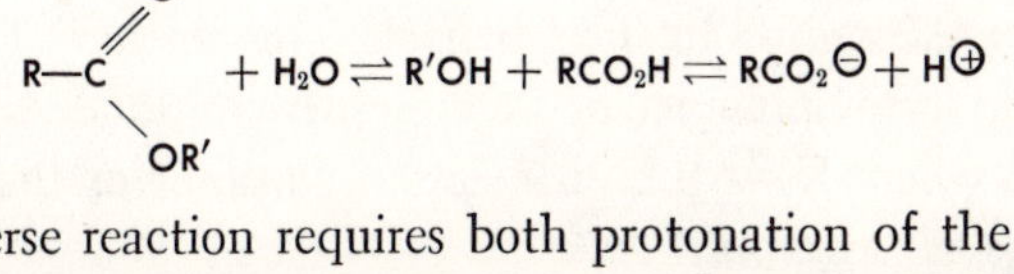

The reverse reaction requires both protonation of the product and the presence of the alcohol. Since the protonation process requires considerable energy at neutrality and since the concentration of the water reactant is far in excess of the alcohol product, the reverse reaction is very slow relative to the forward reaction. Hydrolytic reactions catalyzed by enzymes usually obey the predictions of the Michaelis-Menten model. Isomerization reactions involving a single substrate and product are often in accord with the model when the product is thermodynamically far more stable than the reactant.

Reactions which at equilibrium yield comparable concentrations of reactants and products may still be treated as "irreversible" *during the early stages,* provided that such reactions are initiated in the absence of products. As products accumulate, the reverse reactions must be taken into account. Owing to a special feature of enzyme catalysis, the accumulation of product very often complicates enzyme kinetic analysis even in quasi-irreversible systems. This complication, *inhibition by product,* when properly analyzed, can lead to important details concerning the nature of the enzyme site, as is discussed in the following section.

4–3 INHIBITION OF CATALYSIS OF IRREVERSIBLE REACTIONS

Irreversible single substrate reactions are the simplest systems for enzyme kinetic study. Much information concerning the nature of the enzyme site in such systems can be derived from studies of the inhibition of enzyme catalysis by other (nonreactive) molecules. Two models for the inhibition of specific enzymic reactions can be envisaged. These are illustrated in Fig. 4–1. In the first model of inhibition, the presence of the inhibitor prevents the access of substrate to the site. Three enzyme site components are allowed, namely ES, E, and EI. Inhibition of this type, provided that EI is dissociable, is termed *competitive inhibition*. According to the Michaelis-Menten model, the rate of reaction in the presence of a *competitive inhibitor* is given by Eq. 4–12 where K_I is the dissociation constant of the enzyme inhibitor (EI) complex.

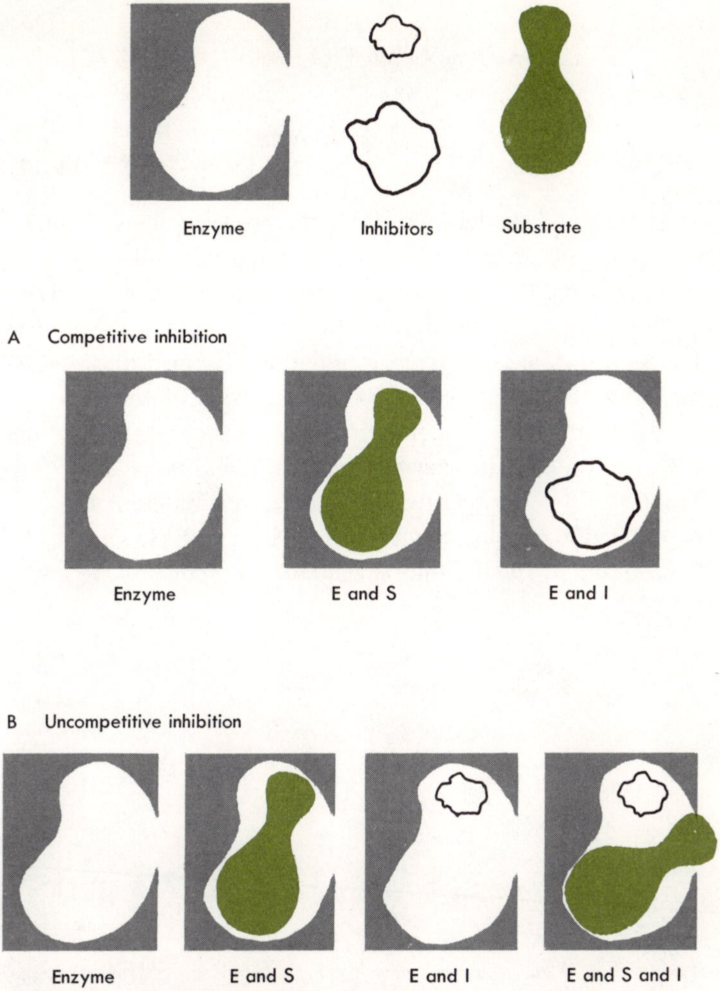

FIG. 4–1 Models for the reversible inhibition of enzyme-substrate reactions.

Competitive Inhibition:

$$E_0 = [E] + [ES] + [EI]$$

$$E + S \rightleftharpoons ES$$
$$E + I \rightleftharpoons EI$$

$$ES \xrightarrow{k_P} E + P$$

$$[\mathrm{ES}] = \frac{\mathrm{E}_0}{\{1 + (k_{-\mathrm{S}}/k_{\mathrm{S}}[\mathrm{S}]) + (k_{\mathrm{P}}/k_{\mathrm{S}}[\mathrm{S}])\}(1 + [\mathrm{I}]/K_{\mathrm{I}})}$$

$$v = \frac{k_{\mathrm{P}}\mathrm{E}_0}{(1 + K_{\mathrm{M}}/[\mathrm{S}])(1 + [\mathrm{I}]/K_{\mathrm{I}})} \tag{4–12}$$

In the second model (Fig. 4–1), the presence of inhibitor at the enzyme site does not necessarily exclude the substrate. On the other hand, the inhibitor prevents chemical catalytic action by the enzyme. Provided that the enzyme-inhibitor complex (EI) is dissociable, this type of inhibition is termed *uncompetitive inhibition*. The special case of inhibition of this type, in which the presence of the inhibitor has no effect on the binding of the substrate to the enzyme site, is known as *noncompetitive inhibition*. In uncompetitive inhibition, four different enzyme site species are possible, namely, ES, E, EI, and ESI. The steady state conditions for the four enzyme site components are given by Eq. 4–13.

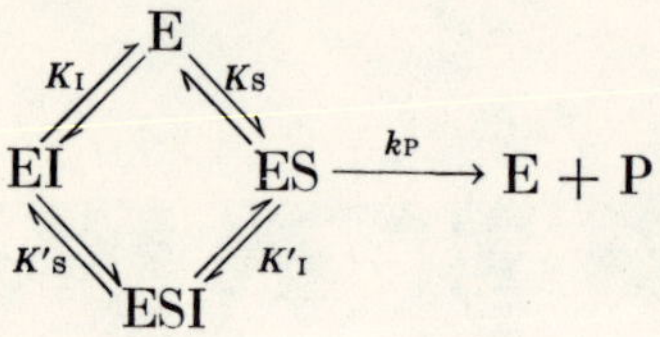

$$K_J = k_{-J}/k_J$$

$$\frac{-d[\mathrm{E}]}{dt} = 0 = (k_{\mathrm{S}}[\mathrm{S}] + k_{\mathrm{I}}[\mathrm{I}])[\mathrm{E}] - k_{-\mathrm{S}}[\mathrm{ES}] - k_{-\mathrm{I}}[\mathrm{EI}] - k_{\mathrm{P}}[\mathrm{ES}]$$

$$\frac{-d[\mathrm{ES}]}{dt} = 0 = (k_{-\mathrm{S}} + k_{\mathrm{P}} + k'_{\mathrm{I}}[\mathrm{I}])[\mathrm{ES}] - k_{\mathrm{S}}[\mathrm{S}][\mathrm{E}] - k'_{-\mathrm{I}}[\mathrm{ES}]$$

$$\frac{-d[\mathrm{EI}]}{dt} = 0 = (k_{-\mathrm{I}} + k'_{\mathrm{S}}[\mathrm{S}])[\mathrm{EI}] - k'_{-\mathrm{S}}[\mathrm{ESI}] - k_{\mathrm{I}}[\mathrm{E}][\mathrm{I}]$$

$$\frac{-d[\mathrm{ESI}]}{dt} = 0 = (k'_{-\mathrm{I}} + k'_{-\mathrm{S}})[\mathrm{ESI}] - k'_{\mathrm{S}}[\mathrm{S}][\mathrm{EI}] - k'_{\mathrm{I}}[\mathrm{I}][\mathrm{ES}] \tag{4–13}$$

In order to derive the rate equation, we must solve the four simultaneous algebraic equations in terms of the measurable variables, [S], [I], and total enzyme concentration (E_0) and some combination of the kinetic constants. The algebraic solution is simplified by substitution of the condition for the conservation of mass (Eq. 4–14).

$$E_0 = [E] + [ES] + [EI] + [ESI] \qquad (4\text{–}14)$$

In this way we can eliminate one of the enzyme species concentration variables. Nevertheless, the resultant expression for ES is extremely complicated. It is given in Eq. 4–15, where $r = k_P/k_{-S}$ and $K_M = (k_{-S} + k_P)/k_S$.

$$v = k_P[ES] =$$

$$\frac{\dfrac{k_P E_0[S]}{K_M}\left[1 + r\dfrac{k_{-I}k_{-S}k'_I[I]}{k_{-I}k'_{-S}\{k'_I[I] + k_{-S}(1+r)\} + k'_{-I}k_{-S}(1+r)(k'_S[S] + k_{-I})}\right]}{1 + \dfrac{[S]}{K_M} + \dfrac{[I]}{K_I} + \dfrac{[S][I]}{K_M K'_I}\left[1 + r\dfrac{k'_{-S}\{(k_{-I} - k'_{-I})k_{-S}(1+r) + k_{-I}k'_{-I}(1 + [I]/K'_I)\}}{k_{-I}k'_{-S}\{k'_I[I] + k_{-S}(1+r)\} + k'_{-I}k_{-S}(1+r)\{k'_S[S] + k_{-I}\}}\right]} \qquad (4\text{–}15)$$

This equation, as it stands, has no practical quantitative application other than to illustrate how rapidly the algebraic solution becomes unmanageable with increasing complexity of the formal kinetic model, and to demonstrate that v^{-1} need not necessarily be linear in either $[S]^{-1}$ or [I]. Careful inspection of Eq. 4–15 (and Eq. 4–16 below) reveals that [*ES*] *can increase no more rapidly than does* [S] *nor can it decrease more rapidly than* [*I*] *increases*. There exist, however, enzyme-substrate-inhibitor (or activator) systems which violate this generalization. A substantial number of enzymes of importance in the regulation of metabolic pathways fall into this latter category. The formal model of catalysis *must* differ from the schemes of both Eq. 4–13 and Eq. 4–12 in these instances. In a later section we shall discuss in detail, examples of enzyme-substrate-modifier reaction rates which exhibit such seemingly anomalous concentration dependencies. The apparent anomalies arise from the implicit assumption in all the reaction rate equations thus far derived, that *the enzyme sites behave independently of one another*. This assumption is valid wherever there is one site per enzyme molecule. When an enzyme molecule contains more than one site, the presence of bound substrate or modifier at one site *may* alter the binding or catalytic properties at another site within the same molecule. Those multisited enzymes which exhibit "cooperative" interactions have been termed *allosteric enzymes*.

Let us now return to the discussion of nonallosteric systems. Two assumptions greatly simplify the form of the resultant rate expression (Eq. 4–15):

(1) If the binding of the substrate to the site is unaffected by the presence of bound inhibitor and vice versa ($K_S = K'_S$, $K_I = K'_I$), it is reasonable to assume that $k_S = k'_S$, $k_{-I} = k'_{-I}$, and so on. In this noncompetitive situation the concentration of ES is given by Eq. 4–16.

$$v = k_P[\mathrm{ES}] = \frac{\dfrac{k_P E_0[S]}{K_M}\left[1 + r\dfrac{k_I[I]}{k_I[I] + (k_S[S] + k_{-S} + k_{-I})(1+r)}\right]}{1 + \dfrac{[S]}{K_M} + \dfrac{[I]}{K_I} + \dfrac{[S][I]}{K_M K_I}\left[1 + r\dfrac{k_{-I} + k_I[I]}{k_I[I] + (k_S[S] + k_{-S} + k_{-I})(1+r)}\right]} \tag{4–16}$$

The expression for ES even with this assumption is still quite cumbersome.

(2) If the specific rate of chemical reaction of ES (k_P) is slow compared to the rates of disappearance of ES to E and ESI ($r \simeq 0$), the concentrations of all enzyme species are defined by their equilibrium dissociation constants, and the rate expression (Eq. 4–17) is greatly simplified. If the inhibition is noncompetitive with substrate, Eq. 4–17 reduces to the rate expression commonly known as the *noncompetitive inhibition* rate equation (Eq. 4–18). It should be noted that Eq. 4 18 contains the implicit assumption that the intermediates equilibrate. Hence, Eq. 4–18 is valid only under the conditions assumed in the derivation of Eq. 4–6 ($K_S = K_M$).

Quasi-equilibrium ($r = 0$)

$$v = \frac{k_P E_0}{1 + \dfrac{K_S}{[S]}\left(1 + \dfrac{[I]}{K_I}\right) + \dfrac{[I]}{K'_I}} \tag{4–17}$$

Quasi-equilibrium, noncompetitive inhibition ($r = 0$, $K_I = K'_I$)

$$v = \frac{k_P E_0}{\left(1 + \frac{K_S}{[S]}\right)\left(1 + \frac{[I]}{K_I}\right)} \qquad (4\text{–}18)$$

In the presence of a noncompetitive inhibitor, the question of whether the intermediates (ES, ESI) equilibrate or are present at steady state concentrations can be investigated by detailed study of the dependence of the reaction rate on the concentrations of substrate and noncompetitive inhibitor. Note particularly, that the variation of v with [I] differs significantly depending on the value of r.

We might expect that molecules which "look like" the substrate molecule, but which lack the reactive center of the substrate, will function as competitive inhibitors of the enzyme. For a reaction which is virtually irreversible, a product of reaction often has these features. Indeed, in systems which follow the Michaelis-Menten model, product inhibition can usually be accounted for by Eq. 4–12. Often, one structural feature of the substrate molecule is primarily responsible for the binding to the enzyme site. In such cases molecules which contain this single structural feature may function as effective competitive inhibitors, as is illustrated in Fig. 4–2a, where an aromatic ring compound, lacking all other features of the substrate, competitively inhibits the enzyme-substrate reaction. Competitive inhibition of glyceraldehyde-3-phosphate dehydrogenase (GPD) reactions is illustrated in Fig. 4–2b.

A small molecule which can in some way reversibly block a *catalytic center* within the site, but still permit the substrate molecule to enter the site, will function as an "uncompetitive" inhibitor. If the energy of binding of substrate to enzyme is due primarily to interactions between catalytically inert amino acid side chains of the enzyme and the nonreactive backbone of the substrate, and if the inhibitor is a very small molecule lacking this backbone, noncompetitive inhibition may occur. A very small molecule which interacts reversibly with many catalytic

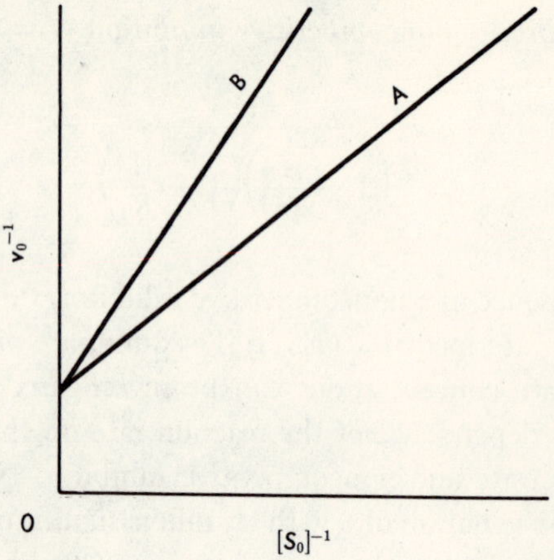

FIG. 4–2a The reversible competitive inhibition of the hydrolysis of acetyl-L-tryptophan amide catalyzed by α-chymotrypsin. Curve A is for the hydrolysis in the absence of indole and curve B, the hydrolysis in the presence of 10^{-3} *M* indole.

enzyme centers in this way is the proton. Most enzymes appear to require a constituent weak base and (or) a conjugate weak acid for catalytic action. The association of hydrogen ion with such weak bases and (or) the dissociation of a conjugate acid will result in loss of catalytic activity. Due to the small molecular volume, this proton may not interfere with the ability of the enzyme to bind substrate. Noncompetitive inhibition by hydrogen ion is illustrated in Fig. 4–3.

Irreversible inhibition occurs when a stoichiometric reaction of enzyme with inhibitor leads to inactivation of the enzyme (Fig. 4–4). Irreversible inhibition is to be distinguished from processes which involve the destruction of the catalytic site, either by denaturation (destruction of the specific polypeptide conformation), or by chemical degradation of the enzyme. An irreversible inhibitor reacts with, and thereby destroys, a catalytic group of the enzyme. An irreversible inhibitor may or may not contain structural elements of the substrate. Heavy metal ions, because of their ability to complex firmly with bases or *nucleophiles* (functional groups which are good electron pair donors) are common irreversible inhibitors. Mercuric com-

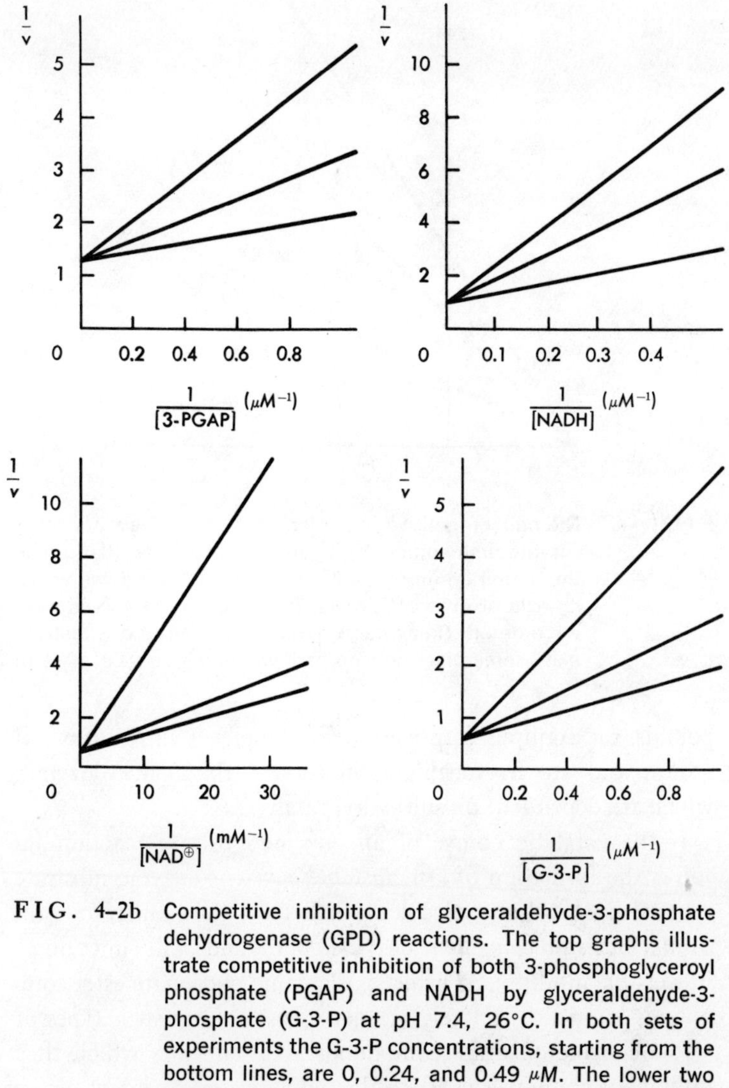

FIG. 4–2b Competitive inhibition of glyceraldehyde-3-phosphate dehydrogenase (GPD) reactions. The top graphs illustrate competitive inhibition of both 3-phosphoglyceroyl phosphate (PGAP) and NADH by glyceraldehyde-3-phosphate (G-3-P) at pH 7.4, 26°C. In both sets of experiments the G-3-P concentrations, starting from the bottom lines, are 0, 0.24, and 0.49 μM. The lower two graphs illustrate competitive inhibition of both NAD^+ and G-3-P by NADH at pH 7.4, 26°C. In both sets of experiments (lower frame) the NADH concentrations, starting from the bottom lines, are 0, 0.94, and 4.7 μM. (See pages 114–116 for the reaction components.) [Figures from C. S. Furfine and S. F. Velick, *J. Biol. Chem.* **240,** 844 (1965).]

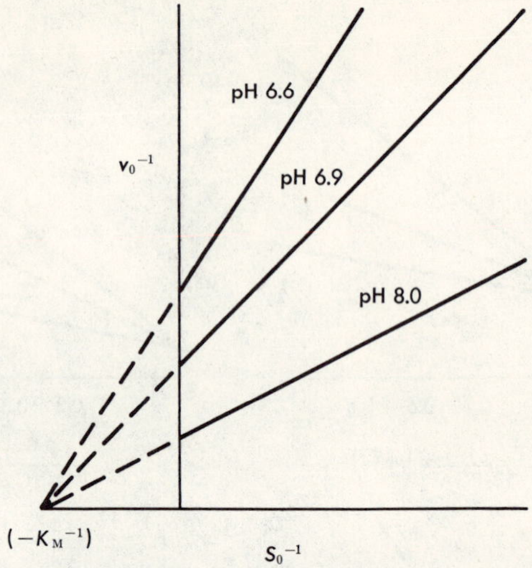

FIG. 4–3 The noncompetitive inhibition of the hydrolysis of acetyl-L-tryptophan amide by hydrogen ion. Note that unike the inhibition in Fig. 4–2, the curves intersect the velocity axis at different points. In this case the curves have a common (imaginary) intersection on the substrate axis, indicating noncompetitive inhibition (Eq. 4–18).

pounds, for example, form very tight complexes with sulfhydryl groups and are irreversible inhibitors of the many enzymes which are dependent on sulfhydryl catalysis.

If the catalytic course of an enzyme-substrate reaction involves the formation of a *metastable covalent* enzyme-substrate intermediate, other molecules which have the ability to form similar but *stable covalent* enzyme compounds will function as irreversible inhibitors. The class of organic phosphate ester compounds known collectively as "nerve gases" form such types of stable enzyme-substrate compounds. The enzymes which they inhibit all contain a catalytically active serine hydroxyl group and the enzyme-inhibitor compounds formed are all stable serine phosphate esters. Inhibition by such compounds is illustrated in Fig. 4–4a.

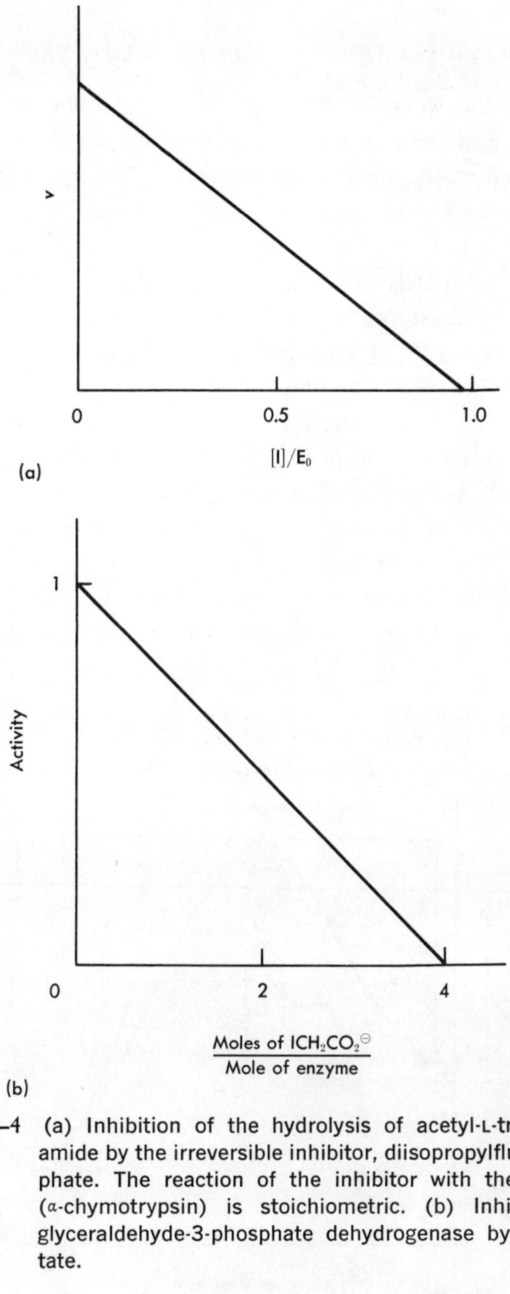

FIG. 4–4 (a) Inhibition of the hydrolysis of acetyl-L-tryptophan amide by the irreversible inhibitor, diisopropylfluorophosphate. The reaction of the inhibitor with the enzyme (α-chymotrypsin) is stoichiometric. (b) Inhibition of glyceraldehyde-3-phosphate dehydrogenase by iodoacetate.

4-4 ENZYME-SUBSTRATE INTERMEDIATES

Although the Michaelis-Menten model can formally account for a substantial fraction of the observed dependencies of reaction velocities on substrate concentrations, analyses of enzymic reaction rates according to this model necessarily avoid some of the most interesting molecular details of the mechanism of catalysis. Ultimately, we are concerned with the mechanism by which the transformation $ES \rightarrow E + P$ occurs. In order to probe further into this mechanism a more detailed chemical analysis of the system is called for. Over the past few years it has been demonstrated that for many enzyme-substrate systems, chemical intermediates in the pathway, $ES \rightarrow P$ exist. The occurrence of intermediates in a reaction pathway is illustrated schematically in the energy diagram of Fig. 4–5. The discrete minima in the energy diagram imply the existence of intermediates. In a limited number of cases, such intermediates have actually been "trapped" and identified, as for example, in spe-

FIG. 4–5 Energy diagrams for a reaction proceeding via a single step (upper curve), and via a number of metastable intermediates (lower curve).

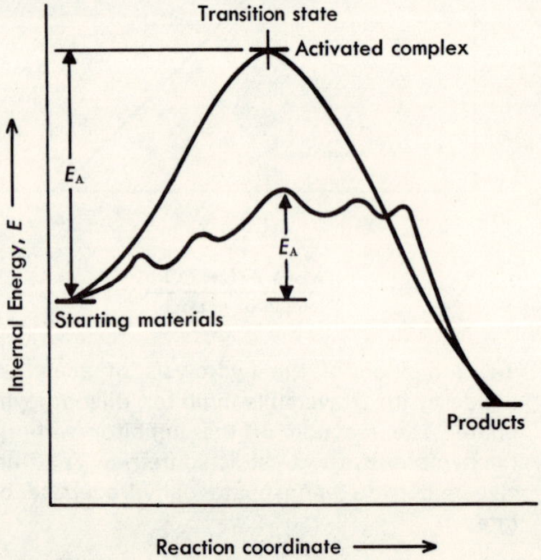

cific enzyme-catalyzed acyl and phosphoryl transfer reactions (Eq. 4–19).

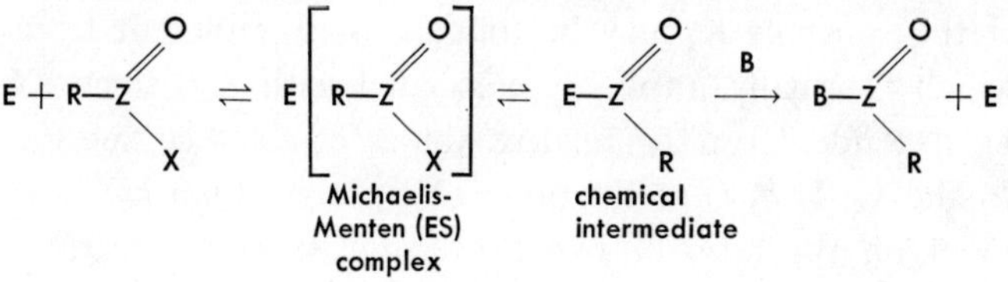

(4–19)

In a virtually irreversible pathway, the kinetics of such a process can be formalized in models such as Eq. 4–20.

$$\mathrm{E} + \mathrm{S} \overset{K_S}{\rightleftharpoons} \mathrm{ES} \overset{k_1}{\longrightarrow} \underset{\mathrm{P_1}}{\underset{+}{\mathrm{ES'}}} \overset{k_2}{\longrightarrow} \mathrm{E} + \mathrm{P_2} \qquad (4\text{–}20)$$

Invoking the steady state hypothesis, the concentration of ES′ and hence the rate equation is given by Eq. 4–21.

$$\begin{aligned} v &= k_2[\mathrm{ES'}] \\ &= \frac{k_1\mathrm{E_0}}{1 + \dfrac{K_\mathrm{S}}{[\mathrm{S}]} + \dfrac{k_1}{k_\mathrm{S}[\mathrm{S}]} + \dfrac{k_1}{k_2}} \\ &= \frac{\dfrac{k_1k_2\mathrm{E_0}}{k_1 + k_2}}{1 + \dfrac{k_2}{k_1 + k_2}\left\{\dfrac{K_\mathrm{S}}{[\mathrm{S}]} + \dfrac{k_1}{k_\mathrm{S}[\mathrm{S}]}\right\}} \end{aligned} \qquad (4\text{–}21)$$

Note that Eq. 4–21 can be rewritten in the form

$$v = \frac{k_0\mathrm{E_0}}{1 + K'/[\mathrm{S}]}$$

where

$$k_0 = \frac{k_1k_2}{k_1 + k_2}$$

$$K' = \frac{k_2}{k_2 + k_1}\left(K_\mathrm{S} + \frac{k_1}{k_\mathrm{S}}\right)$$

This equation is identical in form to that of Eq. 4–5; hence a system which follows this seemingly more complex pathway would still obey the Michaelis-Menten model. The interpretation of the constant K' may be fundamentally different from that which is usually implicitly associated with K_M, since K' may be dependent on the relative values of the two specific rates k_I and k_2. If k_1 is of the order of, or larger than k_2, K' is dependent on the ratio of two irreversible specific rates and hence, interpretations of the constant K' in terms of specific complementary binding forces between substrate and enzyme are no longer valid. The ratio [ES]/[ES′] becomes increasingly smaller as k_1 increases in magnitude relative to k_2. The constant K' is a measure of the extent of "bound" substrate without reference to the nature of the binding, that is, without distinction between the "physically bound" complementary enzyme-substrate complex and the covalent enzyme-substrate compound.

If a series of homologous substrates and a corresponding series of homologous competitive inhibitors are compared (as for instance, substrates and products in an irreversible reaction), the relative values of K' for the substrates versus K_I for the inhibitory products may be indicative of the extent to which enzyme-substrate compounds are being formed in the enzyme-substrate reaction.

4–5 REACTIONS INVOLVING MORE THAN ONE SUBSTRATE

In the previous section it was shown that two different model pathways were mathematically indistinguishable by steady state kinetic methods, and that mechanistic details beyond the Michaelis-Menten model could be inferred only by intuitive chemical reasoning or by chemical analysis. When an enzyme-catalyzed reaction involves two or more substrates whose concentrations can be varied independently, greater mechanistic detail can be inferred from kinetic experiments. Consider the reaction

$$A + B + E \rightarrow \cdots \rightarrow \text{Products} + E$$

In an irreversible pathway, we may distinguish between two types of potential reaction sequences (Eqs. 4–22, 4–23).

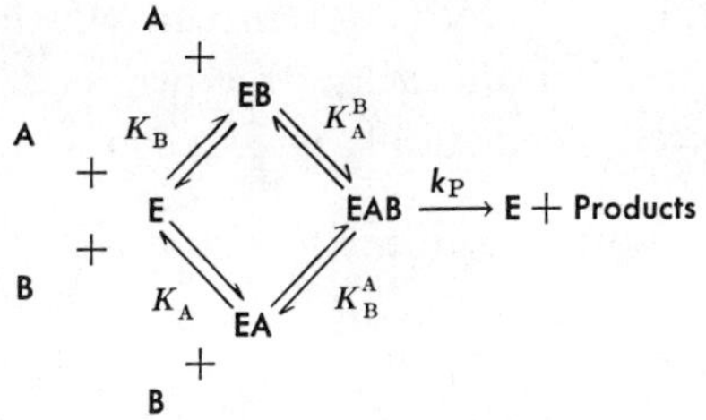

Random pathway (4–22)

$$\mathrm{E} + \mathrm{A} \underset{}{\overset{K_A}{\rightleftharpoons}} \mathrm{EA} \xrightarrow{k_1} \underset{\substack{+\\ \mathrm{P_1}}}{\overset{\substack{\mathrm{B}\\ +}}{\mathrm{EA'}}} \overset{K_B}{\rightleftharpoons} \mathrm{EA'B} \xrightarrow{k_2} \mathrm{E} + \mathrm{P_2}$$

Ordered Pathway (4–23)

In the first pathway either reactant A or reactant B can freely associate and dissociate prior to a reaction involving the triple complex (EAB). In the second model (analogous to the formation of discrete chemical intermediates in the above single substrate discussion), the pathway is *ordered*. Formation of products in this latter case involves the prior formation of an enzyme-substrate compound with one of the reactants. The rate equations governing the two models are quite different. For simplicity of mathematical manipulation, let us assume that the dissociation of all enzyme-substrate complexes are rapid compared to all irreversible chemical reactions. At a given concentration of reactant and enzyme, the concentrations of all enzyme-substrate complexes (but not enzyme-substrate compounds) are then defined by the equilibrium dissociation constants. Under this assumption, the rate equations in the two situations are given by Eqs. 4–24 and 4–25. If the enzyme has a measurable affinity for both reactants, "saturation phenomena" can be measured, that is, the rate of reaction can be studied under conditions where the enzyme is saturated with respect to one of the two reactants. Combining the information derived from such experiments with a study of the constant K' for one of the reactants, under various conditions of concentration of the other reactant, the chemical intermediate pathway can be readily distinguished provided that the enzyme-substrate compounds attain concentrations which are comparable or large compared to the concentration of enzyme-substrate complexes. The expected results from each of the two models (Eqs. 4–22, 4–23) are plotted schematically in Fig. 4–6.

Nonobligatory order as in the model of Eq. 4–22

$$v = \frac{k_P E_0}{1 + \frac{K_A^B}{[A]} + \frac{K_B^A}{[B]} + \frac{K_B K_A^B}{[A][B]}} \qquad (4\text{–}24)$$

Note that $K_A K_B^A = K_B K_A^B$

If the binding of A *and* B *is completely independent* ($K_A^B = K_A$, $K_B^A = K_B$),

$$v = \frac{k_P E_0}{\left(1 + \frac{K_A}{[A]}\right)\left(1 + \frac{K_B}{[B]}\right)}$$

Obligatory order as in the model of Eq. 4–23

$$v = \frac{k_2 E_0}{\left(1 + \frac{K_B}{[B]}\right) + \frac{k_2}{k_1}\left(1 + \frac{K_A}{[A]}\right)} \qquad (4\text{–}25)$$

The ordered reaction pathway discussed above (Eq. 4–23) is merely an illustrative example. Other ordered sequences involv-

FIG. 4–6 The dependence of reaction velocity on substrate concentration for reaction involving two substrates. The three situations illustrated are for the random order mechanism [in (1) and (2)] and the completely ordered mechanism in which the formation of EA′ is irreversible (Eq. 4–23) [in (3) and (4)]. In (1) and (2) the curves have a common intersection on the substrate axis and differ in slope, whereas in (3) and (4) the lines are parallel, and hence have no common intersection. An *ordered mechanism* in which the formation of EA′ (Eq. 4–23) is rapidly reversible is shown in (5) and (6).

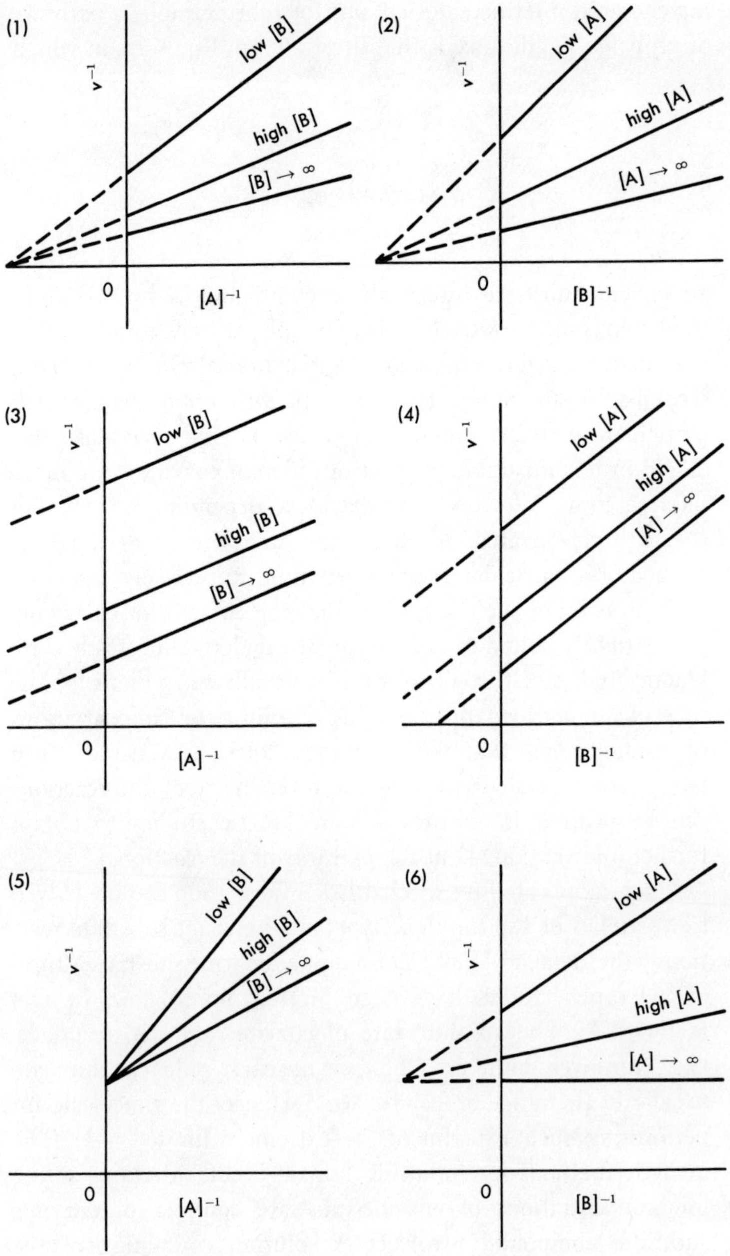
(1)
v^{-1}
low [B]
high [B]
[B] → ∞
0
$[A]^{-1}$
(2)
v^{-1}
low [A]
high [A]
[A] → ∞
0
$[B]^{-1}$
(3)
v^{-1}
low [B]
high [B]
[B] → ∞
0
$[A]^{-1}$
(4)
v^{-1}
low [A]
high [A]
[A] → ∞
0
$[B]^{-1}$
(5)
v^{-1}
low [B]
high [B]
[B] → ∞
0
$[A]^{-1}$
(6)
v^{-1}
low [A]
high [A]
[A] → ∞
0
$[B]^{-1}$

ing chemical intermediates are possible; for example, a pathway of probable significance is that illustrated in Eq. 4–26, in which

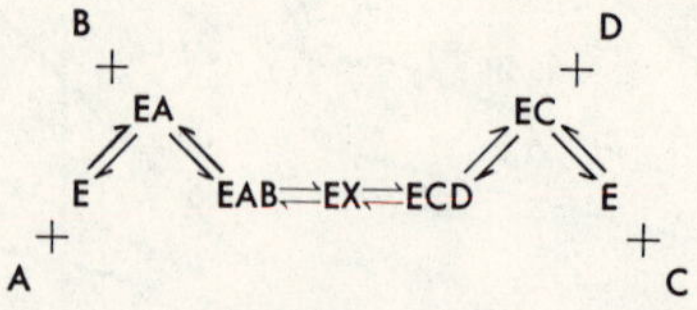

(4–26)

an enzyme-substrate intermediate compound is formed from the triple complex (EAB). This type of pathway is followed in enzyme-catalyzed reactions involving pyridoxal phosphate and a second substrate. Substrates such as pyridoxal phosphate, which participate in many different enzyme-catalyzed reactions (mediated by a comparable number of different enzymes) are often referred to as *cofactors* or *coenzymes* (depending on whether they are regenerated or chemically transformed as a consequence of a particular enzymic reaction, respectively). A common property of such systems is that the cofactor or coenzyme is very tightly bound to the enzyme. In undertaking steady state kinetic studies of these systems, it is usually convenient for the subsequent mathematical analysis to adjust the concentrations of reactants such that the reaction is "quasi-irreversible." In a later section we shall see that the reversibility of the reactions can be utilized in nonsteady state kinetic studies to obtain further information about the pathway of the reaction.

Much more extensive mechanistic information can be derived from studies of the *transient kinetics* of enzyme-substrate reactions (the detailed kinetics of a single enzyme-substrate "turnover"). Such studies have been limited, up to now, by two factors: (1) The absolute rate of enzyme-catalyzed reactions; the "turnover numbers" of enzyme-catalyzed reactions are usually in the range of 10^2–10^4 sec^{-1}. Hence, the time scale for performing such experiments is extremely limited. (2) The analytic methods for following "single event" reactions at the low concentrations of enzyme-substrate complex or enzyme-substrate compound involved. A solution containing of the order of 5 mg of enzyme per milliliter is of the order of $10^{-4}M$. In most instances this represents a large quantity of enzyme.

Kinetic studies of transient reactions require significantly different methods from those utilized in steady state studies. In the past, the overwhelming emphasis in research has been directed towards steady state kinetic studies. We shall consider these two experimental kinetic methods in later sections.

4–6 MULTISITE INTERACTIONS AND THE REGULATION OF ENZYME ACTIVITY

There is a good deal of evidence for the existence of multisited enzyme molecules, as shall be further discussed in Chapter 5. Unlike many solid state inorganic catalysts, multisite enzymes are not made up of an infinite or variable number of sites, but are found to contain a small integral number (and most probably, an even number) of identical or virtually identical sites, and an at least equal number of distinct polypeptide chains. The most exhaustively studied multisite protein is hemoglobin (see Chapter 3). Hemoglobin consists of four polypeptide chains (of two types) and four functionally active heme-iron residues. This is in contrast to the single sited oxygen-carrying protein, myoglobin, which contains a single polypeptide chain and a single heme residue. The polypeptide subunits in some multisite enzymes have each been shown to be composed of identical amino acid sequences. Each site may be composed of one, or more than one, polypeptide chain. In the enzyme *glyceraldehyde-3-phosphate dehydrogenase,* for example, there are four identical polypeptide chains and four sites for the binding of substrate and coenzyme in the active native enzyme molecule. Thus far, all dehydrogenases so investigated have been shown to contain more than one site (and more than one polypeptide chain) per active enzyme molecule.

In our discussions thus far we have assumed that all catalytic sites are independent of one another. However, a good deal of kinetic study bearing on the involvement of multisite interactions in the mechanism of regulation of enzyme activity has been accumulated. A characteristic of these regulatory multisited enzymes is that, under appropriate conditions, the concentration dependence of the reaction rate on either substrate, inhibitor, or activator is not that predicted by the general expression of Eq. 4–15. In such cases, the reaction velocity varies more sharply with concentration of substrate, inhibitor, or activator than is allowed according to the model governing Eq. 4–15. Typical behavior of such systems is illustrated in Fig. 4–7. Indeed, the same type of high-order concentration dependence can be found in the binding of oxygen to hemoglobin (Fig. 4–8), but not to myoglobin. Since the structure of hemoglobin is so much better understood than that of any of these enzymes, much of the reasoning as to the mechanism of regulation of multisite enzyme activity has been based on analogy with studies on hemoglobin. Perhaps the most striking facts concerning oxygen binding to hemoglobin are that both the rates and the equilibria involved in the binding process are dependent on the fraction of binding sites which are already oxygenated, that is, the rate of oxygen binding increases and the dissociation constant of the heme-oxygen complex decreases as the number of binding sites containing oxygen increases. Strikingly, the detailed structure of hemoglobin (Chapter 3) shows that the four heme residues are too far apart in the native molecule to exert

FIG. 4–7 Typical dependencies of reaction velocity on substrate and effector concentrations in allosteric enzyme systems. (a) An attempted plot of the dependence of reaction velocity on substrate concentration according to Eq. 4–5, for the deamination of L-threonine catalyzed by L-threonine deaminase. Upper curve in the absence of, and lower curve in the presence of, the positive effector L-norleucine. (From J.-P. Changeux, *Brookhaven Symposium.* See references, Chapter 8.) (b) The dependence of reaction velocity on the concentration of a negative effector of threonine deaminase activity, L-isoleucine. Curves 1, 2, 3 in the absence of, and curve 4 in the presence of, the positive effector L-norleucine. (From J. Monod, J. Wyman, and J.-P. Changeux. See references, this chapter.) (c) The dependence of deoxycytidylate deaminase on the concentration of the positive effector, deoxycytidylate triphosphate (CTP), at various concentrations of the negative effector, deoxythymidylate triphosphate (TTP) and a fixed concentration of substrate (CMP). (Data of Scarano, as reported by J. Monod, J. Wyman, and J.-P. Changeux. See references, this chapter.)

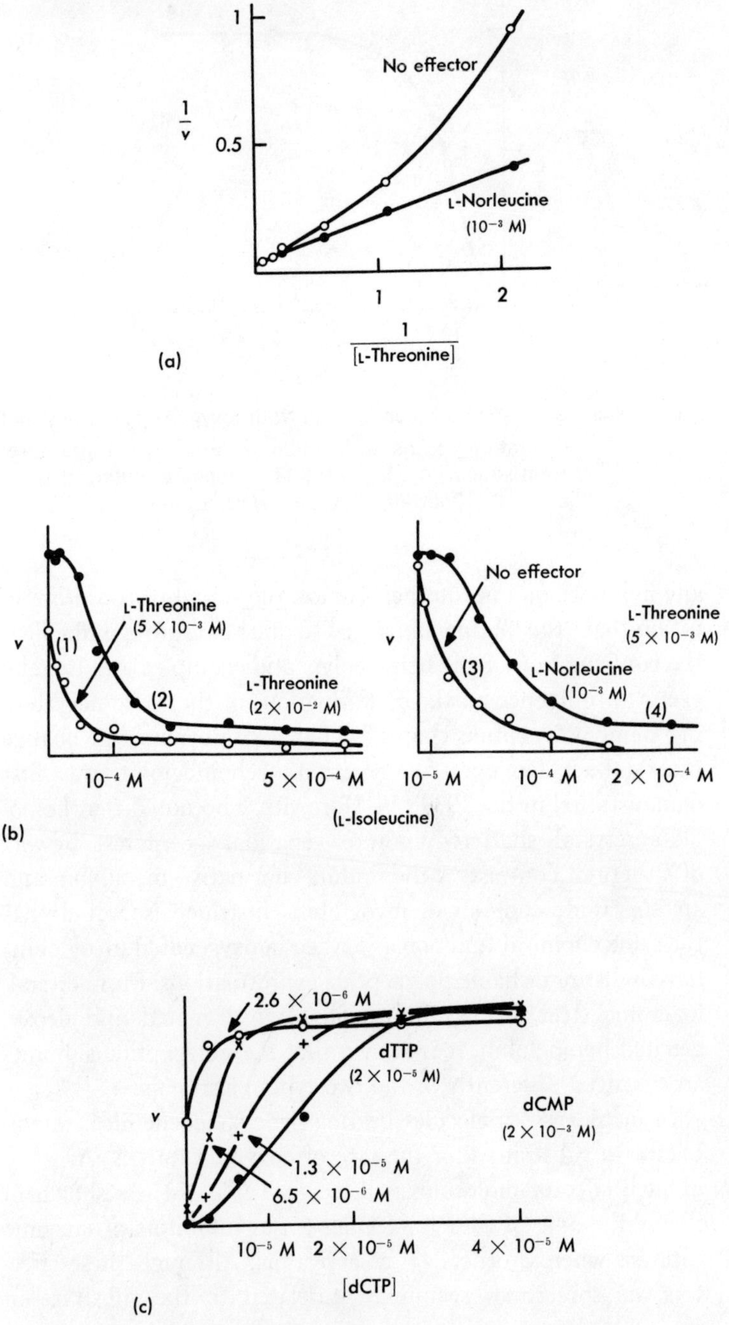
1
0.5
1/v
No effector
L-Norleucine
(10⁻³ M)
1
2
1/[L-Threonine]
(a)
v
L-Threonine
(5 × 10⁻³ M)
(1)
(2)
L-Threonine
(2 × 10⁻² M)
10⁻⁴ M
5 × 10⁻⁴ M
No effector
L-Threonine
(5 × 10⁻³ M)
(3)
L-Norleucine
(10⁻³ M)
(4)
10⁻⁵ M
10⁻⁴ M
2 × 10⁻⁴ M
(L-Isoleucine)
(b)
2.6 × 10⁻⁶ M
dTTP
(2 × 10⁻⁵ M)
dCMP
(2 × 10⁻³ M)
1.3 × 10⁻⁵ M
6.5 × 10⁻⁶ M
10⁻⁵ M
2 × 10⁻⁵ M
4 × 10⁻⁵ M
[dCTP]
(c)

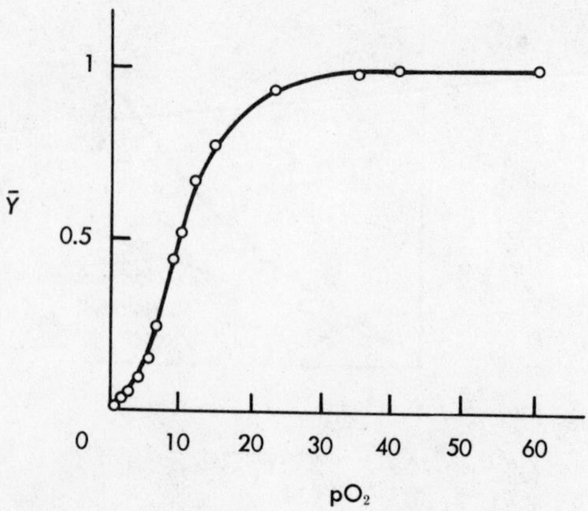

FIG. 4–8 Saturation of hemoglobin with oxygen. The fractional saturation, $\bar{Y}$, as a function of the oxygen pressure. Hemoglobin (4.6%) in 0.6 *M* phosphate buffer, at pH 7 at 19°C. (Courtesy of Dr. R. W. J. Lyster.)

any influence on one another. Hence, the straightforward implication that "the binding of oxygen to one heme group influences the configuration of the heme-polypeptide complex and thereby exerts an influence on the conformations of the three neighboring heme-polypeptide chains." That a conformational change must take place upon oxygenation of hemoglobin was first demonstrated in the 1930's by Horowitz, who noted that hemoglobin crystals shattered upon oxygenation. Of interest, by way of structural contrast, is the finding that native myoglobin and an isocyanide complex of myoglobin (in which isocyanide has the same chemical function as oxygen in oxygenated myoglobin) have indistinguishable polypeptide conformations. From crystallographic structure determinations of oxygenated and deoxygenated hemoglobin, it is known that the polypeptide subunits are arranged differently in the two types of crystals.

In many cases molecules bearing little stereochemical resemblance to substrate alter the rate of enzyme catalysis. A group of such *effector* molecules is listed in Table 4–1. As shown in the table, some of these molecules act as inhibitors of enzymic catalysis whereas others act as activators. Although these effectors are sometimes seemingly unrelated to the substrate in

Table 4-1
Allosteric Systems[a]

Enzyme	Substrates	Effectors	
		Inhibitors	Activators
L-Threonine deaminase	L-**Threonine**	L-**Isoleucine**	L-**Valine**
Aspartate transcarbamylase	L-**Aspartate**, Carbamyl phosphate	**Cytidine triphosphate**	ATP
Deoxycytidylate aminohydrolase	**Deoxycytidylic acid**	**Deoxythymidylic acid**	**Deoxycytidine triphosphate**
Isocitrate dehydrogenase	**D-Isocitrate, nicotinamide adenine dinucleatide**		**5′-adenylic acid, Mg^{2+}**
Hemoglobin	**O_2**		
Phosphorylase	**Glucose-1-phosphate, phosphate, glycogen**	**ATP**	**5′-Adenylic acid**
Fructose-1,6-diphosphate	Fructose-1,6-diphosphate	**5′Adenylic acid**	

[a]Substrates and effectors which exhibit cooperative interactions are indicated in boldface types.

chemical structure, they are usually participants in a common metabolic pathway. It is often the case that a product of reaction in a later step of a biosynthetic pathway acts as an effector of an earlier enzyme activity. In some cases, the substrate itself may be an effector of enzyme catalysis, that is, binding of substrate to a single enzyme site may aid in transforming the aggregate enzyme sites to a catalytically active form. In other cases, substrate has no effect on the catalytic activity of other enzyme sites.

A particular idealized representation of multisite interaction is illustrated in Fig. 4–9. This representation is based on arguments presented by Monod, Wyman, and Changeux. Explicit in their reasoning is the assumption that a multisite enzyme can exist in at least two states but that the conformations of all subunits within a single aggregate are identical (as illustrated in the figure). The role of effector is to drive the equilibrium toward one or the other of these *ordered* aggregate states by complex formation at an allosteric binding site. Effectors which drive the equilibrium in the direction of a catalytically active state and a state of high affinity for the substrate are *activators;* those effectors which drive the equilibrium in the direction of the catalytically inactive state are *inhibitors.* We may differentiate between two types of inhibitors—those which interact with the polypeptide chain at the allosteric site, and those which block the binding of substrate via competitive inhibition at the substrate binding site. If the substrate is itself an effector, then molecules which can function as competitive inhibitors may, in addition, activate the enzyme under conditions of low concentration of substrate (concentrations such that few sites contain bound substrate), since such competitive inhibitors will likewise drive the equilibrium in the direction of the catalytically active state.

If there is a strong binding preference, either for substrate or for inhibitor, in one of the two pre-existent states, then binding to a *single* site can *effect* the catalytic state of the entire aggregate of sites in the complex molecule. In this way, the reaction velocity can exhibit a greater than first-power dependence on the concentration of substrate or inhibitor, contrary to the

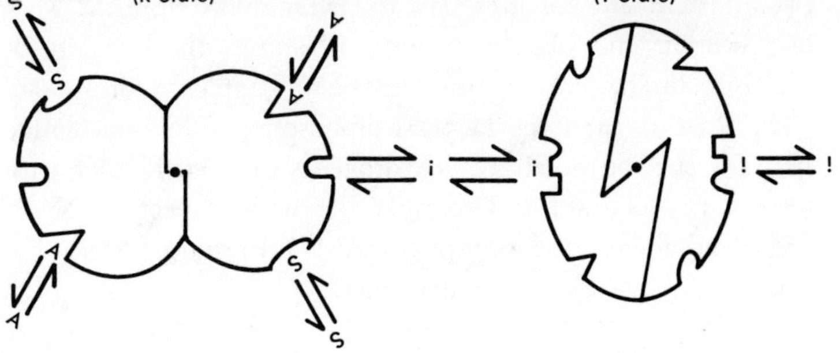

FIG. 4–9 A model for symmetrical multisite interactions in allosteric enzyme systems. The model is described by the following statements: (1) Allosteric proteins are oligomers in which protomers are associated in such a way that they all occupy equivalent positions. This implies that the molecule possesses at least one axis of symmetry. (2) To each ligand able to form a *stereospecific* complex with the protein corresponds one, and only one, site on each protomer. (3) The conformation of each protomer is constrained by its association with the other protomers. (4) Two (at least two) states are reversibly accessible to allosteric oligomers. These states differ by the distribution and (or) energy of interprotomer bonds and therefore also by the conformational constraints imposed upon the protomers. (5) As a result, the affinity of one (or several) of the stereospecific sites toward the corresponding ligand is altered when a transition occurs from one to the other state. (6) When the protein goes from one to the other state, its molecular symmetry (including the symmetry of the conformational constraints imposed upon each protomer) is conserved. (From J.-P. Changeux, *Brookhaven Symposium*. See references, Chapter 8.)

predictions of Eq. 4–15, but in correspondence with a good deal of experimental data, as for example, that illustrated in Fig. 4–7.

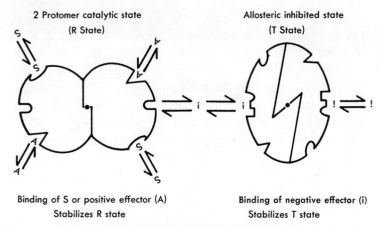

An interesting example of this type of cooperative multisite kinetics is in the reaction of isocitrate with nicotinamide adenine dinucleotide (NAD). This reaction has been shown to involve catalysis by an enzyme containing four substrate sites and two allosteric effector sites. The following quote from the work of Atkinson and his collaborators illustrates the diversity of multisite interactions in this enzyme-substrate-effector system. "The reaction catalyzed by yeast diphosphopyridine nucleotide (NAD) specific isocitrate dehydrogenase is fourth-order with respect to isocitrate and second-order with respect to NAD, Mg^{++}, and adenosine-5′-phosphate (AMP). In dilute systems the reaction follows eleventh-order kinetics."

4–7 STEADY STATE KINETIC METHODS

Invariably, in steady state kinetic procedures, either a reactant or a product concentration is followed as a function of time. Usually a property, physical or chemical, related to reactant and (or) product concentration is the actual experimental variable. A host of analytical methods can be employed for such measurements dependent upon the substrates and products involved. Some methods have proven to be particularly useful. A few of the most common procedures are discussed here.

SPECTROPHOTOMETRIC METHODS

Measurable spectrophotometric differences between reactants and products are common in biological reactions. These spectral changes are usually most pronounced in the ultraviolet and visible spectral regions (the regions involving electronic transitions). Moreover, measurably large changes in the absorption spectra often occur at relatively low concentrations of substrate. An experimentally convenient happenstance in the measurement of many biological reactions is the intense and characteristic ultraviolet absorption spectra exhibited by many of the *coenzymes*. These substances are, in actuality, reactants in the system; the resultant coenzyme product spectrum frequently differs from that of the coenzyme reactant. Hence, the rate of change of the absorption spectrum of the coenzyme is a good indicator of the overall rate of reaction. A common example of spectrophotometric kinetic measurement is illustrated in Fig. 4–10.

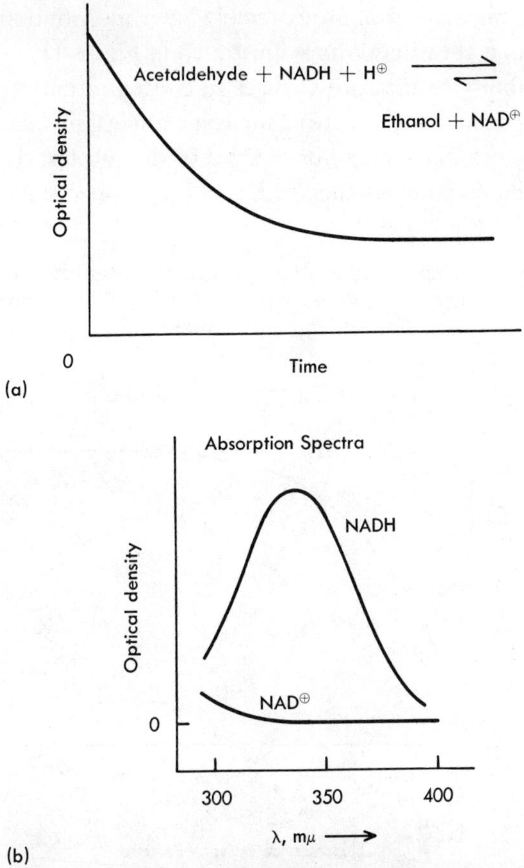

FIG. 4–10 Change in the optical density at 340 mμ accompanying the alcohol dehydrogenase catalyzed reduction of an excess of acetaldehyde in the presence of the reduced coenzyme, NADH, under steady state conditions. The finite absorption at infinite time is due to the reversibility of the reaction.

FLUORESCENCE METHODS

Closely related to the spectrophotometric method is the study of spectral fluorescence, in which the emission of light from photo-induced excited molecules is measured rather than light absorption. Many coenzymes are strongly fluorescent.

Spectral emission from an extremely low concentration of such molecules is measurable as is illustrated in Fig. 4–11.

The ability to measure changes in concentration at very low initial concentrations of reactant is an important factor (aside from the possible scarcity or expense of the substrate). In order to measure parameters such as K_M or K_I, it is necessary to study

FIG. 4–11 The oxidation of alcohol in the presence of the coenzyme NAD^+, as measured by the fluorescence emission from the emergent reduced coenzyme at 436 mμ.

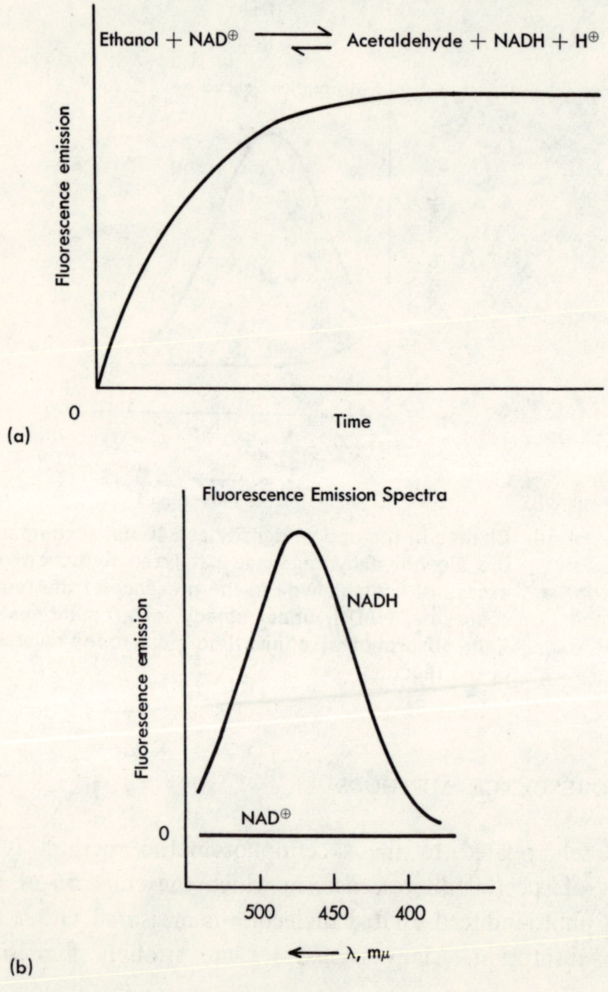

reactions in the concentration range $[S] \sim K_M$ or $[I] \sim K_I$. If the affinity of an enzyme for a substrate or an inhibitor is very great, K_M and (or) K_I will be very small; the reactions will necessarily have to be studied at low concentrations in order to evaluate the parameters. High affinities of enzymes for specific substrates are the rule rather than the exception. It is primarily for this reason that analytical methods are limited.

"pH-STAT KINETICS"

Many enzymic reactions lead to the consumption or liberation of protons. A frequently apropos generalization is that *"synthetic" reactions lead to the consumption of protons and that catabolic (or degradative) reactions lead to the production of protons.* A common procedure for following chemical change is to measure the amount of acid or base required to maintain constant pH in an otherwise poorly buffered reaction mixture. A device for maintaining constant pH (a "pH-stat") is shown schematically in Fig. 4–12. Since most enzymes are optimally active at very low hydrogen ion concentrations (within the range pH 5–9), small changes in hydrogen ion concentration, and hence in substrate concentration, are detectable by this method.

INDICATOR METHODS

If, during the course of enzymic reaction, reactant and (or) product can react rapidly and reversibly with a foreign molecule having an intense absorption spectrum, the [S] and (or) [P] dependent change in spectrum of this foreign molecule can be used as an indicator of chemical reaction. The necessary requirements for such an "indicator" are that its rate of reaction with reactant and (or) product be much greater than the rate of the over-all reaction $S \rightarrow P$, that it have no measurable effect on the rate of catalysis, and that its concentration is sufficiently low compared to substrate so that $[S] - [S \cdot \text{Indicator}] \simeq [S]_{total}$ if it is an indicator of substrate concentration. Both acid-base and oxidation-reduction indicators often fulfill these requirements. The indicator method can be employed to measure product

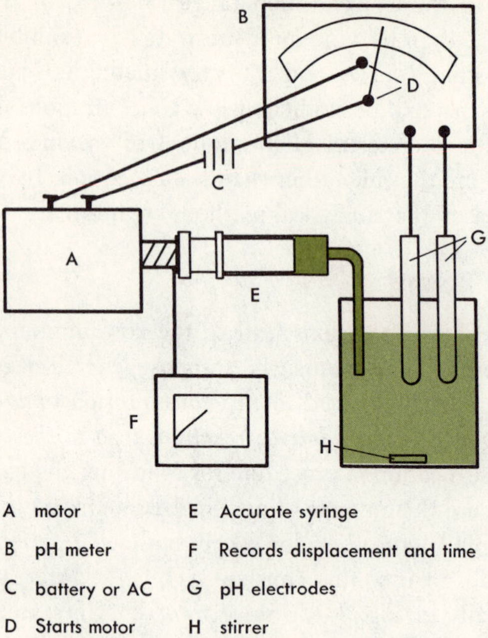

FIG. 4–12 A pH-Stat. The unbuffered reaction produces or consumes hydrogen ions. The resultant change in the pH signals the addition of acid or base to the system to compensate for the change caused by reaction. The amount of titrant added, at nearly constant pH, is recorded as a function of time.

formation even if the indicator reacts irreversibly with the product, provided that the catalytic reaction rate is slow in comparison to the indicator-product reaction rate.

Recently, methods have been devised for following changes in the enzyme site during catalysis, utilizing indicators which are competitive inhibitors. Such methods usually involve the measurement of very small concentrations of enzyme by observing indicator-enzyme site reversible complexes, via changes in either absorption or emission (fluorescence) spectra. From the observed intensity of absorption or emission, the concentration of EI complex and E can be calculated, if K_I is known from experiments in the absence of substrate, as illustrated in Fig. 4–13. With substrate present, the decrease in EI can be extrapo-

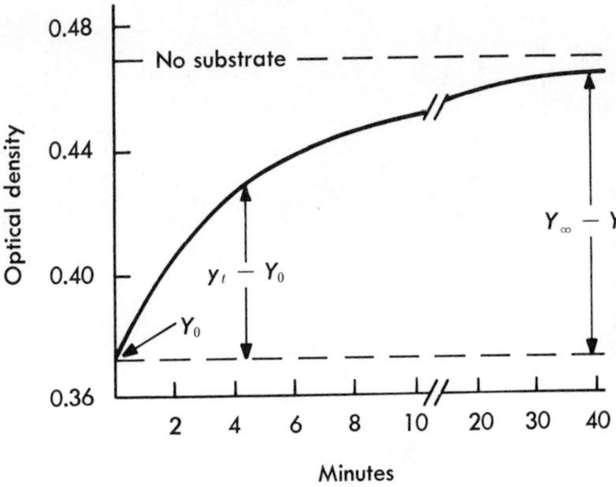

FIG. 4–13 Changes in the extent of enzyme-competitive inhibitor (proflavin) complex formation during the reaction of trypsin with benzoyl-L-arginine amide at pH 7.3, 25°C. The light absorption at 480 mμ is due to the formation of enzyme-proflavin complex. As the reaction proceeds, enzyme-substrate complex concentration decreases giving rise to a larger fraction of bound competitive inhibitor and hence to a higher optical density. The parameter K_M can be determined from the optical density at zero time (Y_0), if the extinction coefficient and the dissociation constant of the enzyme-inhibitor complex are known. K_M and V_{max} can also be determined from a plot of 1/v versus 1/[S] in the usual manner.

lated to zero time, as indicated, and K_M can be determined. Maximal velocity (V_{max}) can then be determined from the time-dependent change in intensity as illustrated in Fig. 4–13, utilizing an integrated rate equation (Section 4–8). The requisite types of indicators can often be synthesized by attachment of an intense chromophore to an organic structure known to function as a specific competitive inhibitor of a particular enzyme.

OPTICAL ROTATION METHODS

Nearly all biological reactants and products are optically active, that is, they contain asymmetric tetrahedral carbon atoms. The magnitude of the optical rotation of such molecules is dependent on the electronic nature of the substituents attached to the asymmetric carbon atoms. In principle, any biological

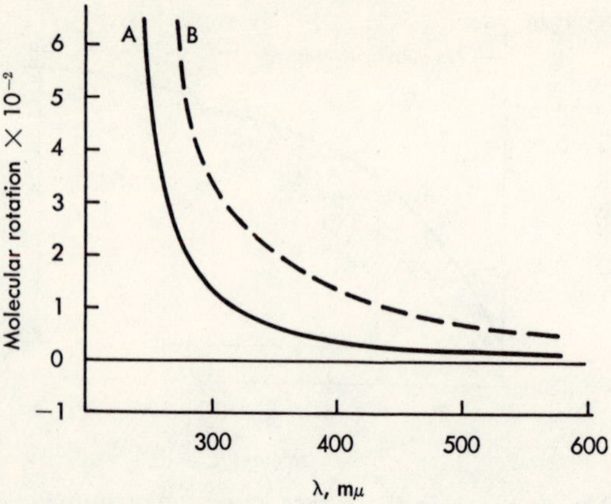

FIG. 4–14 Typical dependence of optical rotation on wavelength of plane-polarized light for nonchromophoric substrates. Curve A: L-Alanine; one center of asymmetry. Curve B: L-Isoleucine; two centers.

reaction involving an asymmetric substrate will lead to a change in optical rotation. Many of the original, and now classic, experiments in enzyme kinetics were performed utilizing this physical property. Often, however, the total amount of chemical transformation required to give a measurable rotation change is quite large and hence optical rotation methods are usually limited to reactions in which both substrate and enzyme are readily available in large quantity. The magnitude of the optical rotation of a substance varies with the wavelength of the polarized monochromatic light beam. The *optical rotatory dispersion* of the substrate is almost invariably of such a nature that at a lower wavelength the optical rotation is enhanced (Fig. 4–14). Greater sensitivity can hence be achieved by measurements in the ultraviolet. A mercury lamp and an appropriate filter or monochromator provides suitably intense monochromatic light for such purposes.

METHODS OF DIRECT CHEMICAL ANALYSIS

Unlike the above methods, in which continuous direct observation of the extent of reaction is possible, any method of direct *chemical* analysis of reactants or products involves the discontinuous sampling of the reaction mixture during the course of kinetic measurement. Aliquots of reaction must be removed and the catalyzed reaction must be rapidly stopped by inactivation of the enzyme, under conditions which have no effect on the concentrations of S and P. Clearly, continuous observation techniques are the methods of choice, if any are practicable.

4–8 ANALYTICAL TREATMENT OF STEADY STATE KINETIC DATA

In enzyme-substrate systems two methods of analysis of the experimental data are common. The distinction depends on whether the *differential* form of the rate equation (for example, Eq. 4–5) or an *integrated* form has been used. In a reaction which is reversible, or if the product is inhibitory, it is desirable to use the differential form of the rate expression and to adapt the experimental method so as to measure the *initial* rate of reaction, v_0 (at which time $[S] = S_0$). Often, however, it is more convenient (due to experimental limitations of precision) to measure substantial changes in substrate (or product) concentration and to utilize an integrated form of the rate equation. If the product of the reaction is inhibitory, this time-dependent inhibition must be made explicit in the differential rate expression prior to integration. Hence, the analytical treatment of reaction rate data in the integrated form is considerably more involved. Nevertheless, this method is often utilized. An interesting example of the treatment of experimental data for the same enzyme-substrate system by both the differential and the integrated method is illustrated in Fig. 4–15. It is notable that the same values of the kinetic parameters (K_M and k_p) are obtained by the two methods. The steady state assumption was expressed mathematically by the condition,

$$\frac{d[\mathrm{ES}]}{dt} = 0$$

in the derivation of rate equations such as 4–5, 4–15, and 4–24. The condition that the concentration of the intermediate (ES)

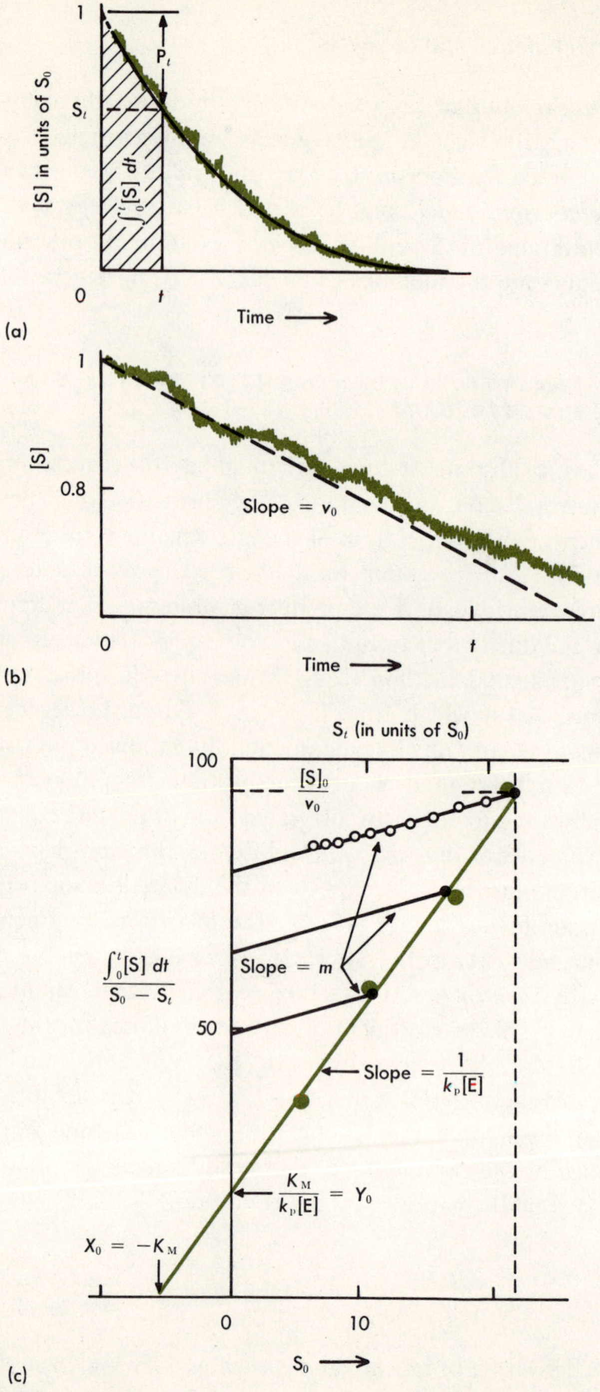

1
P
S
[S] in units of S0
∫[S] dt
0
t
Time
(a)
1
[S]
0.8
Slope = v0
0
t
Time
(b)
S (in units of S0)
100
[S]0
v0
Slope = m
50
Slope = 1/kp[E]
KM/kp[E] = Y0
X0 = −KM
0
10
S0
(c)

remain constant during the course of reaction is clearly invalid: In order to obtain the requisite data for the analytical determination of K_M from the integrated rate equation, the concentration ES must fall appreciably during the measured time course of a single experiment. A less stringent, and more mean-

FIG. 4–15 Alternate methods for the determination of K_M. (a) The curve illustrates the time course for complete reaction at a particular initial concentation of substrate, S_0. (b) The (expanded time scale) curve illustrates the initial reaction velocity. The solid line in (a) represents the best fit of the parameters, via a digital computer program, of the integrated form of the Michaelis-Menten equation, namely,

$$k_P E_0 t = (K_M + S_0) \ln \left\{ \frac{S_0}{S_0 - P_t} \right\} + P_t \left(1 - \frac{K_M}{K_P} \right)$$

(where K_P is the competitive inhibition constant for the product, K_I, of Eq. 4–12) to the experimental data obtained at various S_0. (Data obtained by continuous monitoring of the (substrate-concentration-dependent) optical density at 270 mμ during the α-chymotrypsin catalyzed hydrolysis of methyl hippurate $K_M = 4.0 \times 10^{-3}$ *M*, $K_P = 5.0 \times 10^{-2}$ *M*.)

From a series of experiments as in (b), a secondary plot of S_0/v_0 versus S_0 can be constructed as shown by the colored line in (c), and the parameters K_M and k_P can thereby be determined. Alternatively, the integral, $\int_0^t [S]_t\, dt$, can be evaluated after various times and at various S_0, and a set of parallel lines (each corresponding to a particular S_0) can be generated. The value of the ordinate at the intersection of each line with its S_0 (black circles) is S_0/v_0. From this construction, the parameter K_P is defined by the expression

$$K_P = \frac{Y_0}{2m - (Y_0/K_M)}$$

The open circles in (c) are for the α-chymotrypsin catalyzed hydrolysis of acetyl-L-tyrosinhydroxamide. A similar analysis for the hydrolysis of methyl hippurate yields $K_M = 4.1 \times 10^{-3}$ *M*, $K_P = 5.0 \times 10^{-2}$ *M*. [Data of K. A. Booman and C. Niemann, *J. Amer. Chem. Soc.* **77,** 5733 (1955).]

ingful, condition pertinent to the applicability of the steady state rate equations is the condition of Eq. 4–27.

$$\frac{d[S]}{dt} \gg \frac{d[ES]}{dt} \ll \frac{d[P]}{dt} \qquad (4\text{–}27)$$

When utilizing the differential method, a common procedure is to make a plot of $[S]^{-1}$ versus v^{-1}. Analysis of this plot in terms of the linearity of the slope and the value of v_0^{-1} at $S_0^{-1} = 0$, affords a preliminary indication of the nature of the reaction pathway. A finite value of v_0 at $S_0^{-1} = 0$ is indicative of saturation by substrate; a linear slope indicates compatibility of the system with Michaelis-Menten type models.

When inhibition of the catalytic reaction is being studied, plots of S_0^{-1} versus v_0^{-1} at various concentrations of [I] are again useful preliminaries. The values of v_0^{-1} at $S_0^{-1} = 0$ are then plotted as a function of concentration of inhibitor. From the combined analysis of these two plots, competitive and non-competitive pathways can be distinguished.

The same method utilized in inhibition studies can be applied as a preliminary in the study of two substrate reactions. In this instance, it is desirable to study the initial rate of reaction at a fixed concentration of one substrate (S_1) and to vary the concentration of the other (S_2). The linearity or nonlinearity of the S_2^{-1} versus v^{-1} plots, the values of K_M^1 and K_M^2 at varying concentrations of the other substrate, and plots of v_0^{-1} at $S_1^{-1} \to 0$ versus concentration of S_2, are useful in establishing the existence or absence of an ordered pathway, and the nature of the ordered path if it is existent. Figure 4–16 illustrates double reciprocal plots for various substrates and coenzymes involved in the reversible reaction catalyzed by the enzyme glyceraldehyde-3-PO_4 dehydrogenase:

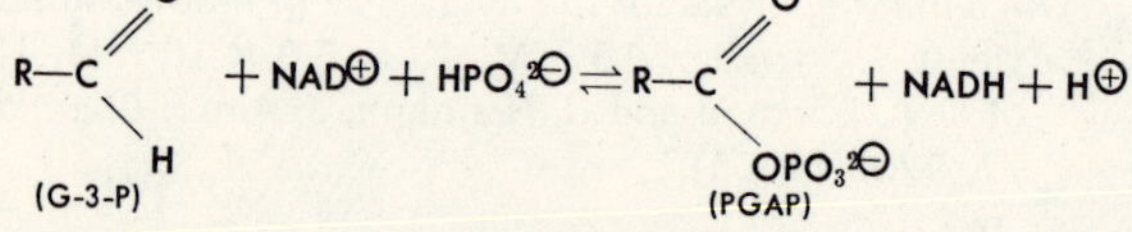

$$R\text{—}C(=O)\text{—}H + NAD^{\oplus} + H_2O \xrightarrow{\text{arsenate}} R\text{—}CO_2H + NADH + H^{\oplus}$$

(where the substrate is glyceraldehyde-3-PO_4: G-3-P)

DETERMINATION OF k_P AND E_0

In order to determine absolute values of the specific rate constant (k_P), a reaction must be found which is stoichiometric with the active site and has a measurable effect on catalytic activity. Irreversible inhibitors of enzyme activity provide the best means of meeting these criteria. A knowledge of both the homogeneity and the molecular weight of the enzyme protein is insufficient for the determination of the concentration E_0. The molecular weight may represent an aggregate of a number of identical enzyme molecules or an aggregate of active and inactive enzyme sites. Only a specific stoichiometric reaction of enzyme site with inhibitor can establish the total mass of protein per unit of catalytic site. The completion of the reaction of enzyme with inhibitor can be ascertained by preincubating the enzyme with known amounts of irreversible inhibitor and measuring the rate of a specific enzyme-substrate reaction at various times after incubation. In this way, the number of grams of enzyme equivalent to one mole of inhibitor can be determined.

It may or may not be the case that the inhibition of enzyme activity is the result of a stoichiometric (1:1) reaction of enzyme and inhibitor. If an approximate molecular weight of enzyme is known from other data, the difference between a 1:1 versus a 2:1 (or 3:1, etc.) inhibitor-enzyme reaction can readily be inferred. Sometimes the actual amino acid residues involved in the reaction with the inhibitor can be detected (see Chapter 5). Thus, for example, in the reaction of diisopropyl fluorophosphate with serine proteases, a unique peptide fragment has been isolated which contains the monophosphorylated product.

Some examples of irreversible inhibition are shown in Fig. 4–4.

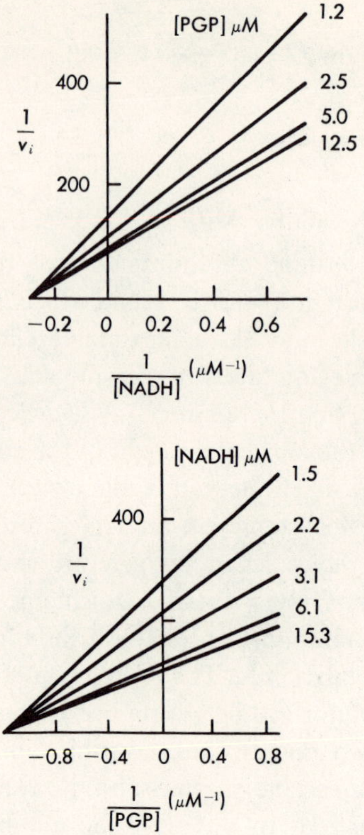
[PGP] μM
1.2
2.5
5.0
12.5
400
200
1/v_i
−0.2
0
0.2
0.4
0.6
1/[NADH] (μM⁻¹)
[NADH] μM
1.5
2.2
3.1
6.1
15.3
400
1/v_i
−0.8
−0.4
0
0.4
0.8
1/[PGP] (μM⁻¹)

(a)

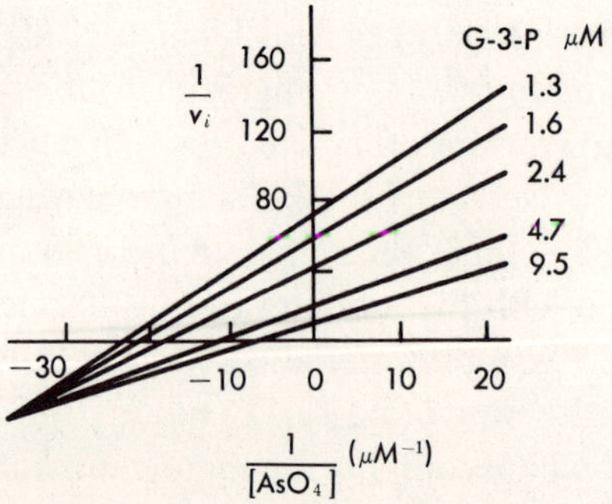
G-3-P μM
1.3
1.6
2.4
4.7
9.5
160
120
80
1/v_i
−30
−10
0
10
20
1/[AsO4] (μM⁻¹)

(b)

4–9 TRANSIENT KINETICS

Since the turnover numbers for virtually all enzyme-specific substrate reactions are rapid ($k_P \gtrsim 10^2$ sec^{-1}), correspondingly rapid rate measurements must be utilized to probe further into the details of the intermediate steps in catalysis. Until recently, the usual procedures for following such reactions involved rapid "flow" techniques. These are commonly of two types, referred to as "continuous flow" and "stopped flow" methods.

"CONTINUOUS FLOW" METHODS

The continuous flow method is quite simple in principle. It is illustrated schematically in Fig. 4–17. Two solutions are rapidly mixed and the mixture is pumped at a predetermined speed through a tube containing an observation chamber. If the flow rates of the two solutions are known, then the precise time after mixing can be determined at the point of observation from the known dimensions of the flow tube. By changing the rate of flow, one can vary the length of time that has elapsed between mixing and entrance of the solution into the observation chamber. Thus, given a method of analysis of reactants or products, the extent of reaction can be determined as a function of time, by variation of the flow rate. Reaction times as short as one millisecond have thus been measured. Continuous flow tech-

FIG. 4–16 (a) Kinetics of the reduction of 3-phosphoglyceroyl phosphate (PGAP) by NADH at pH 7.4 in 0.1 *M* imidazole at 26°C, at an enzyme concentration of 4.6×10^{-11} *M*. The Michaelis constant of each substrate is independent of the concentration of the other. (b) Arsenate-assisted oxidation of glyceraldehyde-3-phosphate (G3P) by NAD. Reciprocal plots versus arsenate concentration in the concentration region in which enzyme is activated by arsenate. The Michaelis constant of arsenate is seen to depend upon glyceraldehyde-3-P (G-3-P) concentration. [From C. S. Furfine and S. F. Velick, *J. Biol. Chem.* **240,** 844 (1965).]

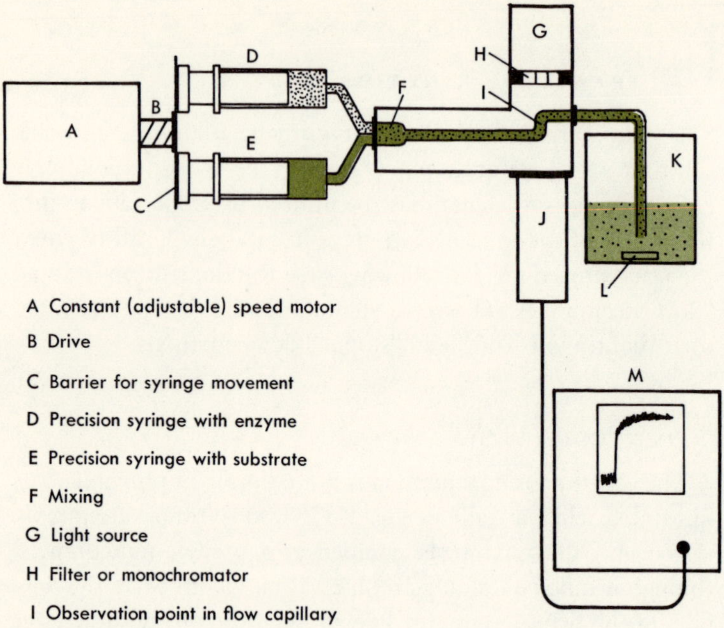

A Constant (adjustable) speed motor

B Drive

C Barrier for syringe movement

D Precision syringe with enzyme

E Precision syringe with substrate

F Mixing

G Light source

H Filter or monochromator

I Observation point in flow capillary

J Photo tube

K Collection container; or stirred solution for rapidly quenching the reaction between D and E

L Stirrer

M Oscilloscope

FIG. 4–17 A continuous flow apparatus for following rapid reactions.

niques suffer the disadvantage of requiring relatively large volumes of solution in order to achieve a uniform rate of flow through the system. Furthermore, in each experiment, only one "time-concentration" point is usually measured. Hence, the method has been limited to enzyme-substrate systems in which both components are available in large quantities. This disadvantage may be obviated appreciably by increasing the number of observation chambers along the flow tube in order to measure a property of the reaction mixture at various reaction times in a single flow experiment. Temperature change and spectral change due to the chemical reaction are convenient properties for such measurements.

"STOPPED FLOW" METHODS

The stopped flow method is illustrated schematically in Fig. 4–18. In this procedure, the continuous time course of the reaction is recorded by means of a rapid detecting device, most commonly, an optical detector coupled to an oscilloscope. The two solutions are forced together through a mixing chamber and into an observation cell, as is illustrated in the figure. In this method, a precisely controlled flow rate is unimportant; the critical requirement is that the two solutions must be rapidly mixed so that the time elapsed between mixing and filling the observation cell is small compared to the time of chemical reaction. Once the flow has stopped, a trigger mechanism records the change in some property of the reactant, product, or the enzyme site itself, as is illustrated in Fig. 4–19. The entire

FIG. 4–18 A stopped-flow apparatus for following rapid reactions.

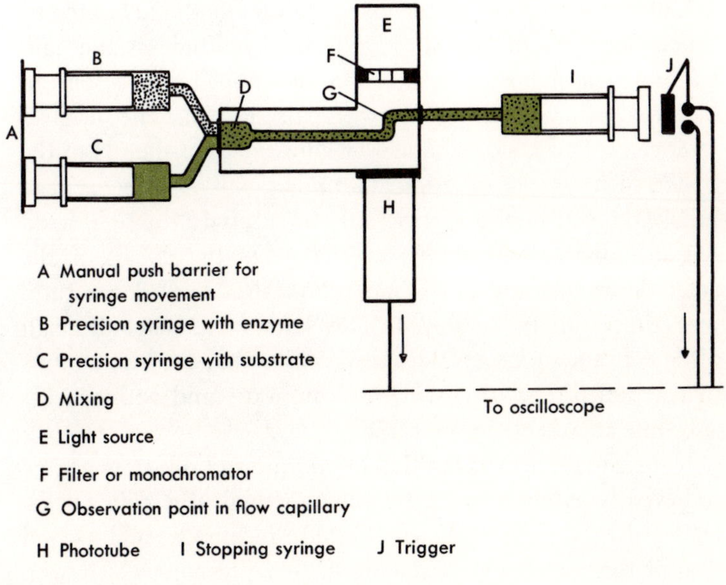

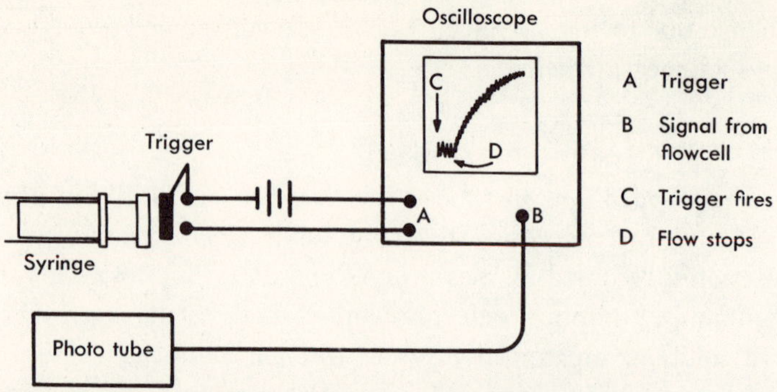

FIG. 4–19 A device for the recording of the transient time course of reaction after flow has stopped (see Fig. 4–18).

time course of reaction can be followed once the flow has stopped and the trigger has fired. The time required to mix, fill the observation cell, stop flow, and fire the trigger can be as short as a few milliseconds. The stopped flow technique has the advantage of being conservative in the consumption of enzyme and substrate, and in not requiring such precise control of the flow rate.

Unfortunately, both of the above flow methods are limited by time resolutions of the same order of magnitude as the half-lives for typical enzyme-substrate complexes. This situation can be improved by a number of expedients: (1) Enzyme turnover rates can often be slowed down by adjustment of the pH of the buffer, such that the experimental time resolution is sufficient to follow the course of a specific enzyme-substrate reaction, without introducing any significant change in the details of the catalytic mechanism. (2) "Pseudosubstrates" with slower turnover times can be employed. This latter expedient, although often convenient, necessitates establishing that the mechanisms of reaction of enzyme with pseudosubstrate and with specific substrate are essentially the same.

If the time course of reaction of enzyme and substrate can be followed by either of these two flow methods, the data can be analyzed by conventional reaction rate techniques and a good deal of the complexity of the algebraic expressions which arise in the steady state treatment can be obviated. Thus, molecular mechanism can be more directly inferred from transient data. Some experimental results are illustrated in Fig. 4–20. Detailed mechanistic information, however, often requires considerably greater time resolution than that allowed by either of these methods.

More recently, time resolutions of the order of 10 microseconds have been achieved in enzyme-substrate reactions, by means of the "relaxation methods" developed by M. Eigen and his collaborators. These new methods introduce the potential for obtaining a still more detailed description of the catalytic reaction mechanism.

4–10 RELAXATION METHODS AND REVERSIBLE SYSTEMS

For convenience in grasping the principles involved, let us consider the reversible system described by Eq. 4–28.

$$\mathrm{E} + \mathrm{S} \rightleftharpoons \mathrm{ES} \tag{4–28}$$

The concentrations of reactants and products are given by the equilibrium expression (Eq. 4–29).

$$\frac{[\mathrm{E}][\mathrm{S}]}{[\mathrm{ES}]} = \frac{\{\mathrm{E}_0 - [\mathrm{ES}]\}\{\mathrm{S}_0 - [\mathrm{ES}]\}}{[\mathrm{ES}]} = K_\mathrm{S} = \frac{k_{-\mathrm{S}}}{k_\mathrm{S}} \tag{4–29}$$

The equilibrium constant (K_S) and hence the equilibrium ratio of products to reactant is a function of particular conditions of state, for example, temperature. The variation of the equilibrium constant with the absolute temperature is determined by the *enthalpy* of the reaction (ΔH), as is given from equilibrium thermodynamics (Eq. 4–30).

$$\frac{d \ln K_\mathrm{S}}{dT} = \frac{\Delta H}{RT^2} \tag{4–30}$$

Let us now consider what occurs if a condition of state, in this case temperature, is suddenly changed. The solvent molecules

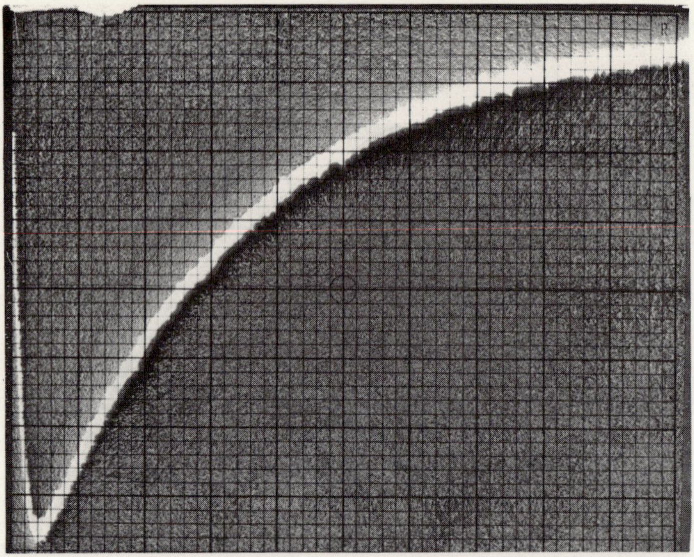

(a)

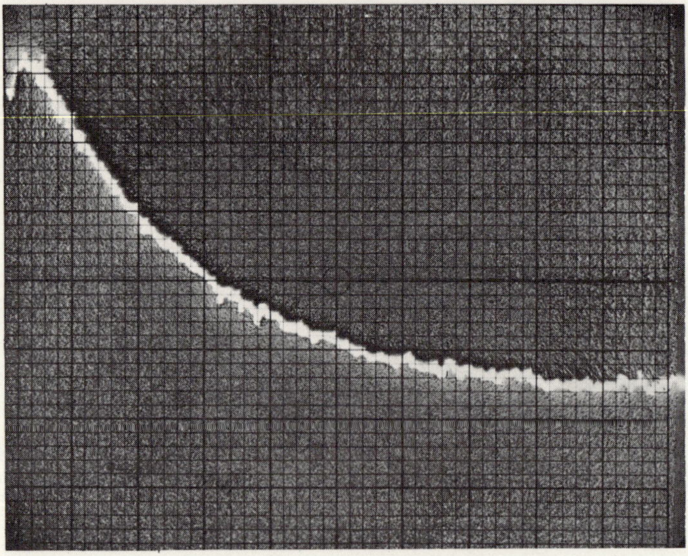

(b)

will suffer a change in kinetic energy such that after a short interval of time, equilibrium is once again established with the surroundings (now at a new temperature). Likewise, there will be a change in the equilibrium constant and hence in the ratio of concentration of product to reactant. In order to achieve this obligatory redistribution of products and reactants, however, the two chemical processes (with specific rate constants, k_S and k_{-S}) will necessarily have to adjust so as to define the new position of equilibrium. If these rates of reaction are slow, in comparison to the rate of thermal re-equilibration of solvent, and if the temperature of the system can be rapidly changed experimentally (relative to the rate of the chemical re-equilibration), a time lag in the redistribution of reactants and products can, in principle, be measured.

SINGLE-STEP REVERSIBLE SYSTEMS (FIG. 4–21)

The net rate of formation of ES is given by the algebraic sum of two rate equations (Eq. 4–31).

$$\frac{d[\mathrm{ES}]}{dt} = k_{\mathrm{S}}[\mathrm{E}][\mathrm{S}] - k_{-\mathrm{S}}[\mathrm{ES}] \tag{4–31}$$

Let us assume that the solvent can be rapidly heated from an initial temperature (T_1) to an elevated temperature (T_2) such that the rate of thermal equilibration of solvent greatly exceeds the rate of either of the two reactions. Let the equilibrium concentrations of ES at the two temperatures be defined by Eq. 4–32.

FIG. 4–20 An oscilloscope recording of a transient reaction obtained by the stopped-flow procedure. Transient changes in the absorption of a substrate chromophore (*furyl-acrylamido-*) in the reaction of N-[β-(2-furyl)-acryloyl]-L-tryptophan methyl ester with a molar excess of α-chymotrypsin; pH 5.30, 330 mμ. (a) At a scan speed of 500 msec per major division. (b) The same reaction at a scan speed of 20 msec per major division. The extent of time between mixing and observation is of the order of a few milliseconds. The time resolution *thereafter* is of the order of 0.1 msec.

$$\mathrm{X}_{\mathrm{eq}} = [\mathrm{ES}]_{\mathrm{eq}}^{T_2} - [\mathrm{ES}]_{\mathrm{eq}}^{T_1} \tag{4-32}$$

Under the conditions of our experiment,

$$[\mathrm{ES}]_{t=0} = [\mathrm{ES}]_{\mathrm{eq}}^{T_1} \equiv [\mathrm{ES}]_0$$

where $t = 0$ is the *instant* of heating from T_1 to T_2. We can define a concentration variable (X_t) such that at any time,

$$\mathrm{X}_t = [\mathrm{ES}]_t^{T_2} - [\mathrm{ES}]_0$$

When the chemical system reaches equilibrium at T_2 we have

$$K_{\mathrm{S}}^{T_2} = \frac{\{\mathrm{E}_0 - \mathrm{X}_{\mathrm{eq}}\}\{\mathrm{S}_0 - \mathrm{X}_{\mathrm{eq}}\}}{[\mathrm{ES}]_0 + \mathrm{X}_{\mathrm{eq}}}$$

$$\mathrm{X}_{\mathrm{eq}} = \frac{\mathrm{E}_0\mathrm{S}_0 - K_{\mathrm{S}}^{T_2}[\mathrm{ES}]_0}{K_{\mathrm{S}}^{T_2} + \mathrm{E}_0 + \mathrm{S}_0 - \mathrm{X}_{\mathrm{eq}}} \tag{4-33}$$

If, and only if, the thermal perturbation from T_1 to T_2 is small (i.e., the equilibrium constant, K_{S}, is not very different at T_1

FIG. 4–21 Typical time dependence for a change in a physical property which is linear with concentration following an *instantaneous* change in temperature, in a system involving a single chemical transformation.

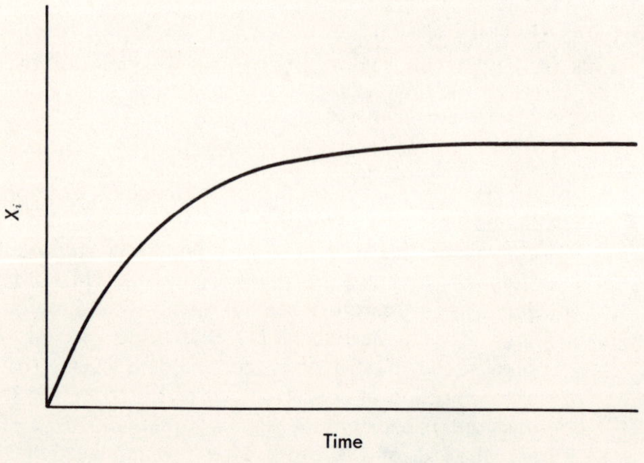

and T_2), X_{eq} will be negligible in the denominator of Eq. 4–33. Under this condition, X_{eq} is given by Eq. 4–34.

$$X_{eq} \cong \frac{E_0S_0 - K_S^{T_2}[ES]_0}{K_S^{T_2} + E_0 + S_0} \tag{4–34}$$

Returning to the rate equation (4–31), we obtain for the conditions specified above,

$$\frac{d[ES]}{dt} = \frac{dX}{dt} = k_S(E_0 - X_t)(S_0 - X_t) - k_{-S}([ES]_0 + X_t)$$

$$\frac{1}{k_S}\left(\frac{dX}{dt}\right) = \{E_0S_0 - K_S^{T_2}[ES]_0\} - X_t(S_0 + E_0 + K_S^{T_2}) \tag{4–35}$$

Substitution of Eq. 4–34 into Eq. 4–35 yields the *linearized* rate equation (Eq. 4–36)

$$\frac{1}{k_S}\frac{dX}{dt} = (X_{eq} - X_t)(K_S^{T_2} + E_0 + S_0)$$

or

$$\frac{dX}{(X_{eq} - X_t)} = \{k_S(S_0 + E_0) + k_{-S}\}\, dt \tag{4–36}$$

The integrated solution to Eq. 4–36 is hence Eq. 4–37.

$$X_t = X_{eq}[1 - \exp\{-k_{-S} - k_S(S_0 + E_0)\}t] \tag{4–37}$$

This equation (Eq. 4–37) can be expressed as

$$X_t = X_{eq}\{1 - \exp(-t/\tau)\} \tag{4–38}$$

or in differential form by Eq. 4–39.

$$\frac{dX}{dt} + \frac{1}{\tau}X_t = \frac{1}{\tau}X_{eq} \tag{4–39}$$

Note that τ has the dimensions of time. It is referred to as the *relaxation time.* Owing to the linear form of the rate equation (Eq. 4–38), the parameter τ can be evaluated from experimental measurements of any property which is directly related to the concentration variable, X_t (e.g., a spectral change where Δ(optical density) = KX_t), as illustrated in Fig. 4–21.

The reader can readily verify that for a unimolecular process

$$A \underset{k_B}{\overset{k_A}{\rightleftharpoons}} B$$

Eq. 4–38 is valid, and that

$$\tau = \frac{1}{(k_A + k_B)}$$

Likewise, for a reversible bimolecular (double displacement) reaction, we obtain for τ, Eq. 4–40.

$$A + B \underset{k_{CD}}{\overset{k_{AB}}{\rightleftharpoons}} C + D$$

$$\tau = \frac{1}{\{k_{AB}(A_0 + B_0) + k_{CD}(C_0 + D_0)\}} \tag{4–40}$$

Two other features of τ are noteworthy:

(1) For any reversible process (e.g., $E + S \rightleftharpoons ES$), τ is dependent on the initial concentrations of the participants in the bimolecular step ($E + S \rightarrow$), but independent of the concentration of the unimolecular participant (ES).

(2) Equations 4–38 and 4–39 are mathematical expressions of a *first-order* reaction. Hence, $1/\tau$ can be evaluated in the same way as a first-order specific rate constant (e.g., from the slope of a plot of ln (X_{eq}/X) versus time). Note, however, that for a reversible system, this apparent specific rate constant (k')

$$k' = \frac{1}{\tau} = k_S(S_0 + E_0) + k_{-S}$$

is the sum of two specific rate constants, and hence, that the rate of approach to equilibrium is determined, in the limit, by the faster rate process and not by the slower. For irreversible pathways, it is often the fashion to relate, mechanistically, the experimentally derived specific rate constant to a single rate limiting slow process in a complex pathway. Such reasoning is clearly invalid in a reversible pathway.

MULTIPLE-STEP REVERSIBLE SYSTEMS (FIG. 4–22)

Enzyme-substrate systems invariably involve a sequence of reversible reaction steps in the catalytic pathway, for example, the pathway represented by Eq. 4–41.

$$\underset{\text{I}}{\mathrm{E + S}} \overset{k_1}{\rightleftharpoons} \underset{\text{II}}{\mathrm{ES}} \overset{k_2}{\rightleftharpoons} \underset{\text{III}}{\mathrm{EP}} \overset{k_3}{\rightleftharpoons} \underset{\text{IV}}{\mathrm{E + P}} \tag{4–41}$$

In a reaction sequence involving a stoichiometrically significant concentration of an intermediate (in the present instance, ES and EP), a *measurable* concentration variable cannot be de-

FIG. 4–22 Typical time dependence for a change in a physical property which is linear with concentration following an *instantaneous* change in temperature, in a system involving three chemical transformations in linear sequence.

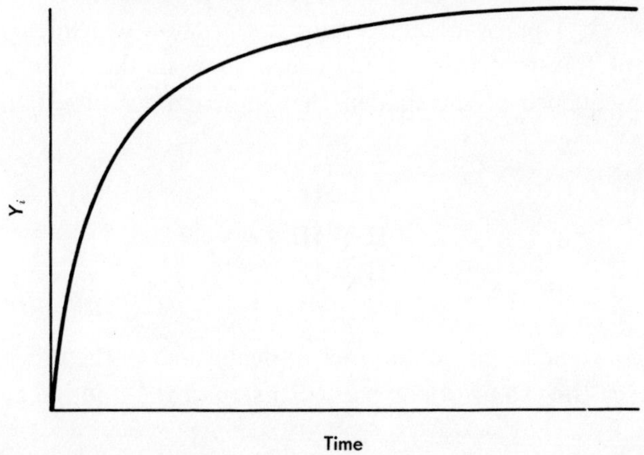

fined by the straightforward procedure utilized previously (Eq. 4–32), since participants in one step (e.g., $E + P \rightleftharpoons EP$) affect the concentrations of participants in all other steps. Multiple-step reaction sequences can, and have been analyzed by application of the "temperature-jump" technique discussed earlier for single-step reversible systems. The theoretical procedure for the analysis of the more complex system differs, in essence, only in the definition of the reaction variables, all or most of which are dependent on the extent of reaction in *every* step. Thus the reaction variables (which we shall define in the following paragraphs) are complex concentration variables relating to a sequence of chemical transformations. Likewise, as we shall see, a sequence of reactions leads to a series of different relaxation times (τ_i), each of which may relate to a collection of rate constants relevant to a variety of steps in the reaction sequence. By suitable mathematical analysis, all of the individual rate constants can be derived from the data, provided that the individual reaction rates lie within the experimental limits of time resolution.

If the equilibrium concentrations of all species are only slightly perturbed by a small change in the condition of state of the system (i.e., each individual equilibrium constant is only slightly altered by the temperature change), the approximation previously employed in Eq. 4–34 ($E_0, S_0 > X_{eq}$) will again be valid. We can define three (and only three) independent concentration variables x_1, x_2, and x_3 for the pathway illustrated by Eq. 4–41. These variables pertain to the (time-dependent) extent of transformation in individual steps in the pathway as a consequence of the sudden change in temperature, namely,

$$\begin{array}{ll} \text{I} \rightarrow \text{II} & x_1 \\ \text{II} \rightarrow \text{III} & x_2 \\ \text{III} \rightarrow \text{IV} & x_3 \end{array}$$

For convenience of notation let us define as Δx_k, the difference between the extent of one particular step of the reaction at any time, x_k, and the extent of this particular step at equilibrium at T_2 $(x_k)_{eq}$.

$$\Delta x_k = (x_k)_{eq} - x_k$$

For the process I $\rightleftharpoons$ II (Eq. 4–41) and utilizing the assumption of Eq. 4–34, we note both the following (Eq. 4–42)

$$\begin{aligned} E_0S_0 - K_1[ES]_0 = {} & K_1\{(x_1)_{eq} - (x_2)_{eq}\} \\ & + E_0(x_1)_{eq} + S_0\{(x_1)_{eq} - (x_3)_{eq}\} \end{aligned}$$

and

$$\left(\frac{1}{k_1}\right)\frac{dx_1}{dt} = E_0S_0 - K_1[ES]_0 - K_1(x_1 - x_2) - E_0x_1 - S_0(x_1 - x_3) \tag{4–42}$$

Combining these two equations, we obtain the change in x_1 with time as a function of all three of the reaction variables (Eq. 4–43).

$$\frac{dx_1}{dt} = k_1\{S_0(\Delta x_1 - \Delta x_3) + E_0\Delta x_1\} - k_{-1}(\Delta x_1 - \Delta x_2) \tag{4–43}$$

By similar arguments, the temporal variations in x_2 and x_3 can be derived (Eq. 4–44).

$$\frac{dx_2}{dt} = k_2(\Delta x_2 - \Delta x_1) + k_{-2}(\Delta x_2 - \Delta x_3)$$

$$\frac{dx_3}{dt} = k_3(\Delta x_3 - \Delta x_2) + k_{-3}\{E_0\Delta x_3 + P_0(\Delta x_3 - \Delta x_1)\} \tag{4–44}$$

Note that *each* rate equation contains a linear combination of the three arbitrarily defined concentration variables x_1, x_2, and x_3,

$$\frac{d(x_i)}{dt} = a_{i1}\Delta x_1 + a_{i2}\Delta x_2 + a_{i3}\Delta x_3$$

or, in general,

$$\frac{d(x_i)}{dt} + \sum a_{ik}x_k = \sum a_{ik}(x_k)_{eq} \tag{4–45}$$

These sets of *linear* equations can always be transformed into an equal number of *linear* equations with a new set of (in this case, three) concentration variables, Y_i, such that each Y_i has a time dependence expressible in the form of Eq. 4–46.[1]

$$\frac{d(Y_i)}{dt} + b_{ii}Y_i = b_{ii}(Y_i)_{eq} \tag{4–46}$$

Each Y_i is a linear combination of terms in x_1, x_2, and x_3. Note that Eq. 4–46 is identical in form to Eq. 4–39, with $b_{ii} = 1/\tau_i$. These new concentration variables (Y_i) relate to time-dependent changes in the entire system rather than to particular reversible processes. Likewise, the inverse of the relaxation times, $1/\tau_i$, relate to combinations of specific rate constants involving all steps in the reaction. If the experimental conditions of concentration are such that $(c_i)_0 > x_k$, always obtains, the observable concentration changes (or changes of physical properties) can be correlated with changes in Y_i according to Eq. 4–46. If they are sufficiently different in time, three distinct relaxation times (τ_1, τ_2, and τ_3) should be observed for a system with three reversible steps (as for example, in Eq. 4–41), as indicated in Fig. 4–22. If some, or all, of the relaxation times are of comparable magnitude, the observable changes in Y_{total} will be a linear combination of the individual rate equations, namely,

$$Y_{total} = A \exp\left(-\frac{t}{\tau_1}\right) + B \exp\left(-\frac{t}{\tau_2}\right) + C \exp\left(-\frac{t}{\tau_3}\right)$$

and the change in Y_{total} will not follow a single exponential rate law. Once the (minimum) number of distinct relaxation times (τ_i) has been established, a corresponding (minimum) number of reversible steps can be defined. From the resultant minimal reaction pathway (for example, Eq. 4–41), the relationship between τ_i and the individual specific rate constants is rigorously specified by the necessary relationship between the x_k and Y_i

[1] The linear transformation of equations such as 4–43 and 4–44 into Eq. 4–46 is described both in standard texts on matrix algebra and in the excellent article on relaxation methods by Eigen and de Maeyer (see references).

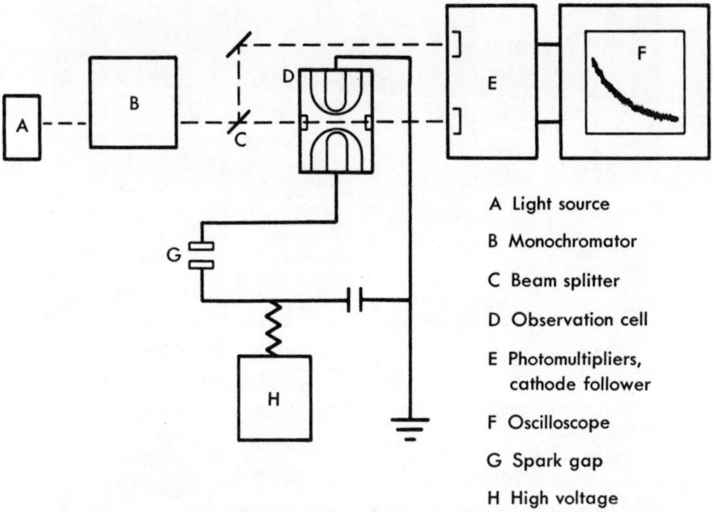

FIG. 4–23 A schematic diagram of a *temperature jump* apparatus.

FIG. 4–24a The time course of optical density change following a *temperature jump* for a system involving a single relaxation time. Relaxation effect in Ni(II)-pyrophosphate system; $\Sigma Ni^{2+} = 9.62 \times 10^{-4}$ *M*, $\tau = 900$ μsec; abscissa scale is 500 μsec per major division, ordinate scale is in arbitrary units of decreasing absorbancy. [From G. G. Hammes and M. L. Morrell, *J. Amer. Chem. Soc.* **86,** 1497 (1964).]

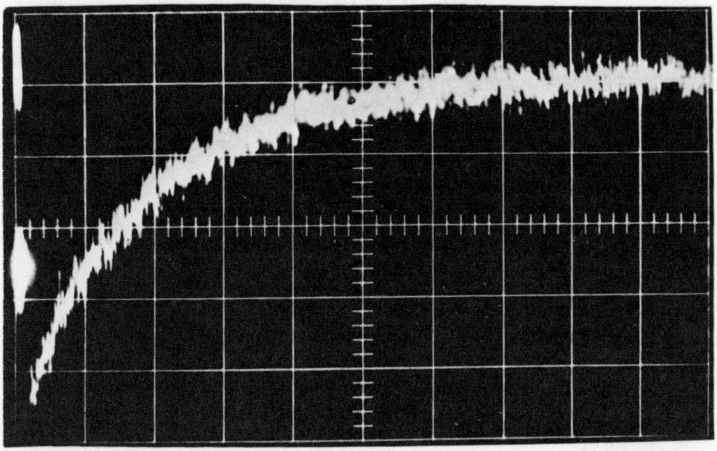

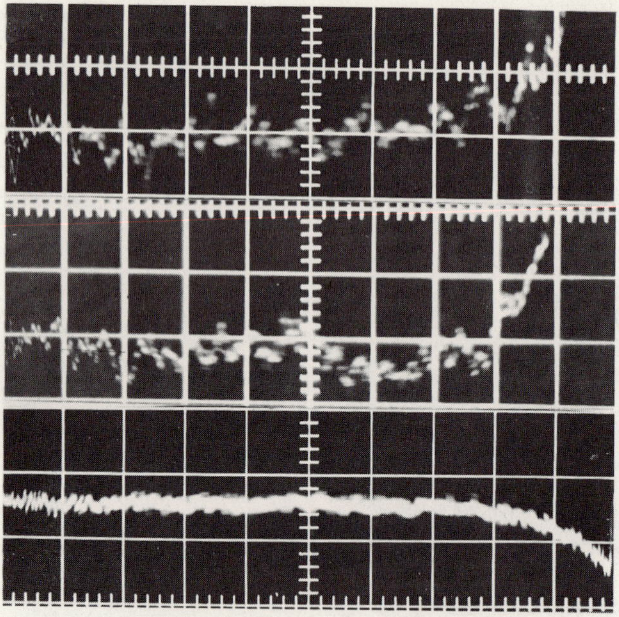

FIG. 4–24b Oscilloscope traces of temperature jump effects observed with ribonuclease and cytidine-3′-phosphate at pH 6.7. All solutions contained 0.1 *M* KNO_3 and 2×10^{-5} *M* chlorophenol red indicator. Top picture, enzyme isomerization: 1.41×10^{-4} *M* ribonuclease $[E_0]$; abscissa scale: 1 msec per large division; $\tau_1 = 0.792$ msec. Middle picture, enzyme-substrate interaction: $[E_0] = 1.35 \times 10^{-4} M$, $[S_0]$(cytidine-3′-phosphate) $= 2.87 \times 10^{-4}$ *M*; abscissa scale: 100 μsec per large division; $\tau_2 = 90$ μsec. The enzyme-substrate interaction causes the initial rapid decrease in absorbance; the very small absorbance increase at longer times is the beginning of the relaxation process associated with enzyme isomerization shown on the top frame. Bottom picture, intramolecular process: $[E_0] = 1.22 \times 10^{-4} M$, $[S_0] = 1.10 \times 10^{-3}$ *M*; abscissa scale: 100 μsec per large division; $\tau_3 = 113$ μsec. The vertical scale in all cases is in arbitrary units of decreasing absorbancy. [From G. G. Hammes and R. E. Cathou, *J. Amer. Chem. Soc.* **86**, 3240 (1964).]

(Eqs. 4–45 and 4–46). It is again to be noted that the rates of bimolecular processes, and hence the various τ_i to which they relate, will be dependent on initial (equilibrium) concentrations whereas the rates of unimolecular processes (such as ES $\rightleftharpoons$ EP in Eq. 4–41) will be concentration invariant.

TEMPERATURE-JUMP MEASUREMENTS

A schematic diagram of a temperature-jump apparatus is illustrated in Fig. 4–23. The response time (heating time) depends on electrical discharge through the reaction cell and hence requires an electrolytic solution. It is therefore particularly suitable for the measurement of enzymic reactions, which invariably require a significant electrolyte concentration ($\sim 0.1\ M$) for optimal activity.

Actual experimental data involving changes in the transmission of monochromatic visible light due to chemical re-equilibration are shown both for a single-step process and for a multi-step process in Fig. 4–24. Note that three relaxation times are observable in Fig. 4–24b, the reaction of the enzyme *ribonuclease* with substrate.

REFERENCES

Gutfreund, H., *An Introduction to the Study of Enzymes,* Wiley, New York, 1965. A detailed text on the fundamental principles underlying enzyme kinetics, the actual methodology commonly utilized, and the mechanistic inferences thus derived. This book is recommended as an extensive supplement for students with special interests in enzymology; it contains an excellent bibliography of the enzyme kinetic literature.

Dixon, M., and Webb, E. C., *Enzymes,* 2nd ed., Longman, Green, London, 1964. This large volume contains a detailed treatise on enzyme-substrate-modifier reactions in the steady state.

Neilands, J. B., and Stumpf, P. K., *Outlines of Enzyme Chemistry,* 2nd ed., Wiley, New York, 1958. A well-written elementary text on all aspects of enzyme chemistry including intermediary metabolism. Unfortunately, it is considerably out of date on descriptive details, particularly in regard to enzyme (protein) structure and protein synthesis.

Perutz, M. F., Bolton, W., Diamond, R., Muirhead, H., and Watson, H. C., "Structure of Hemoglobin," *Nature* **203,** 687

(1964). A detailed analysis in terms of molecular structure of the effect of oxygen binding on the three-dimensional structure of hemoglobin. An early observation, that hemoglobin crystals shatter upon exposure to oxygen [Haurowitz, F., *Hoppe-Seylers Z.* **254,** 266 (1938)], provided a strong motivation for the development of the X-ray crystallographic methods of protein structure analysis.

Gerhart, J. C., and Pardee, A. B., "The Effect of the Feedback Inhibitor, CTP, on Subunit Interactions in Aspartate Transcarbamylase," *Cold Spring Harbor Symp. Quant. Biol.* **28,** 491 (1963). The key paper introducing the phenomenon of allosteric regulation of enzymic activity.

Monod, J., Wyman, J., and Changeux, J.-P., "On the Nature of Allosteric Transitions," *J. Mol. Biol.* **12,** 88 (1965). A long article on allosteric enzymes, containing an excellent review of the behavior of multisite, multipeptide chain, functional proteins. The authors present a general hypothesis, relating the enzymic activity of multisite proteins to the macromolecular conformation, which is to a large extent based on the crystal structures of hemoglobin and of oxygenated hemoglobin.

Eigen, M., and de Maeyer, L., "Relaxation Methods," in *Investigations of Rates and Mechanisms of Reactions* (Friess, S. L., Lewis, E. S. and Weissberger, A., eds.), Vol. VIII, Part II, p. 896, Interscience, New York, 1963. The definitive treatise (over 150 pages) on the theory and methods of measurement of very rapid reactions. Requires a background in physical chemistry, differential equations, and the elements of matrix algebra for a thorough understanding; however, the effort is most worthwhile for students with an interest in the mechanisms of rapid catalyzed reactions (particularly, enzyme-catalyzed reactions). The appearance of this fundamental article in a large, multi-authored and expensive volume is seriously regretted.

Eigen, M., and Hammes, G. G., "Elementary Steps in Enzyme Kinetics," *Advan. in Enzymol.* **25,** 1 (1963). A very much simplified presentation of the theory underlying relaxation kinetic measurements, as well as some fundamental inferences concerning enzymic mechanisms are contained in this excellent article.

Roughton, F. J. W., and Chance, B., "Rapid Reactions," in *Investigations of Rates and Mechanisms of Reactions*, Friess *et al.*, eds., Vol. VIII, Part II, p. 703, Interscience, New York, 1963. A comprehensive article on flow methods for the measurement of rapid reactions, with particular emphasis on enzyme-substrate reactions.

FIVE ~ THE ANATOMY OF ENZYMES

HAVING CONSIDERED SOME OF THE FORMALISM OF ENZYME kinetics, the next steps in our discussion will be to consider the experimental kinetic data for some specific enzyme-substrate systems, and to discuss reaction mechanisms. This sequence of discussion is usual in studies of reaction mechanisms involving catalysis by small molecules. A distinction between catalyzed organic reactions in homogeneous solution and enzyme-catalyzed reactions is that, in the former, the chemical structures of all the participants in the reaction are often either known or restricted to a limited number of alternatives. The present level of sophistication in organic chemistry permits us not only to write the structures of all (stable) participants in the chemical reaction, but also to make accurate assignments as to the often rigid atomic configurations within these molecules. In enzyme catalysis we are faced with a structurally more complex molecule. For this reason, we must first consider some aspects of the chemical structure and conformation of enzyme proteins.

The first chemical structure of a protein, that of the hormone insulin, was solved by Sanger and his collaborators. This determination was a major step towards an understanding of biological function in terms of detailed amino acid sequence. The demonstration that the complete covalent structure of a protein could be elucidated lent tremendous impetus to the determination of amino acid sequences in other proteins. The fact that the insulin structure could be solved at all was an indication that all insulin molecules from a single species were probably identical, or nearly identical, in amino acid sequence, as will become evident from the method of structure elucidation described below. Having determined the sequence in beef insulin, Sanger and his collaborators could demonstrate that insulin molecules from different mammalian species were only slightly different in amino acid sequence. For example, beef, sheep, and hog insulins differ from one another by only a single amino acid (as illustrated in Fig. 5–1).[1]

5–1 THE PRIMARY SEQUENCE OF AMINO ACIDS IN PROTEINS

Many of the essential techniques in current use can still be illustrated by the original determination of sequence for insulin. The small size of the insulin molecule (molecular weight approximately 6000) was an important factor in selecting it as a model protein for sequence studies. Moreover, Sanger had pre-

[1] Since the frequency of spontaneous mutations in the genetic code is very much greater than the reciprocal of the evolutionary time involved in the differentiation of mammalian species, it is evident that *selection* rather than *mutation* must be the dominant biological factor determining the existent protein sequences. This in turn is an indication of the importance, at the biological level, of the very specific sequence of amino acids in any particular protein. The possibility that not all of the amino acid residues in a polypeptide sequence play a crucial role in enzymic catalysis does not necessarily indicate that such specific sequences are not essential. Aggregation-disaggregation, solubility in cellular fluids, adsorption by cell particulates, internal conformational stability, and viscosity and diffusibility are all pertinent intracellular properties of enzymes which may be affected by regions of the protein which do not, a priori, influence the catalytic activity in an *in vitro* system. Structural properties which do not affect catalytic activity will be of interest to us in only a secondary fashion in this discussion; we should not, however, draw the conclusion on the basis of *in vitro* studies alone, that much of the amino acid sequence in proteins is of little biological significance.

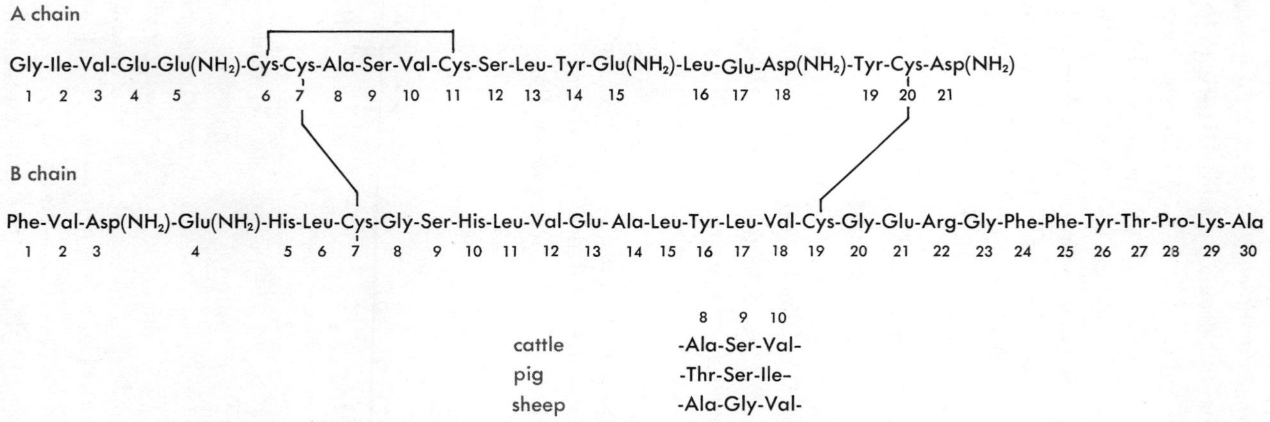

FIG. 5–1 The primary sequence of amino acids in beef, sheep, and hog insulins.

viously shown that the molecule could be split approximately in half by cleaving the disulfide bonds. Hence the problem of determining the sequence could be solved separately for each of the two, relatively small, polypeptide chains (30 and 21 residues each).

The first step in the procedure was to degrade each of the chains by cleaving them with a specific proteolytic enzyme (an enzyme which hydrolyzes peptide bonds). The enzyme *trypsin* was found to be more selective than any other enzyme available, cleaving peptide bonds at the carboxyl terminus of arginine and lysine residues exclusively (Fig. 5–2). From the known amino acid composition of each of the two large polypeptides, the theoretical number of "tryptic" peptides can be calculated a priori, assuming that insulin molecules have a unique sequence. This theoretical number of peptides could actually be separated from tryptic digests of each of the large polypeptide chains by chromatographic methods. Treatment of these resultant tryptic peptides with dinitrofluorobenzene, a chromophoric reagent which attacks free amino groups (but not peptide nitrogen), resulted in the labeling of each of the amino terminal residues as the corresponding yellow dinitrophenyl derivative (Eq. 5–1). The same reagent can be used on

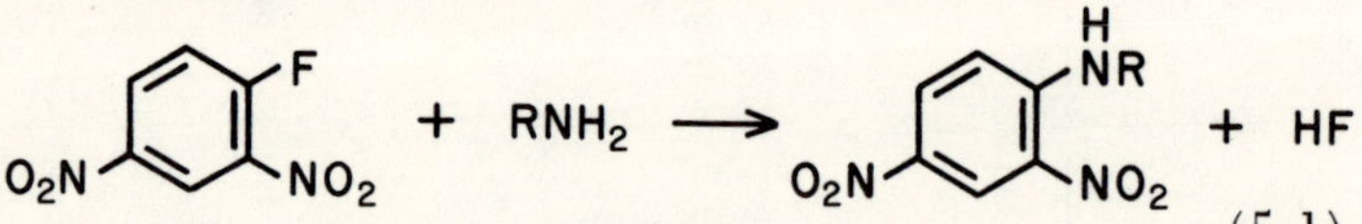

(5–1)

each of the undigested polypeptides to determine the amino terminal ends of the two polypeptide chains. A complete acid hydrolysis of each peptide to yield its constituent amino acids was carried out, and the amino acid *composition* of each peptide was determined. This resulted in the accumulation of information of the type illustrated in step 2 of Fig. 5–3. Two types of new information are necessary to carry elucidation of the structure beyond this point: the determination of amino acid sequence within each tryptic peptide, and the determination of the (linear) order of tryptic peptides in the polypeptide

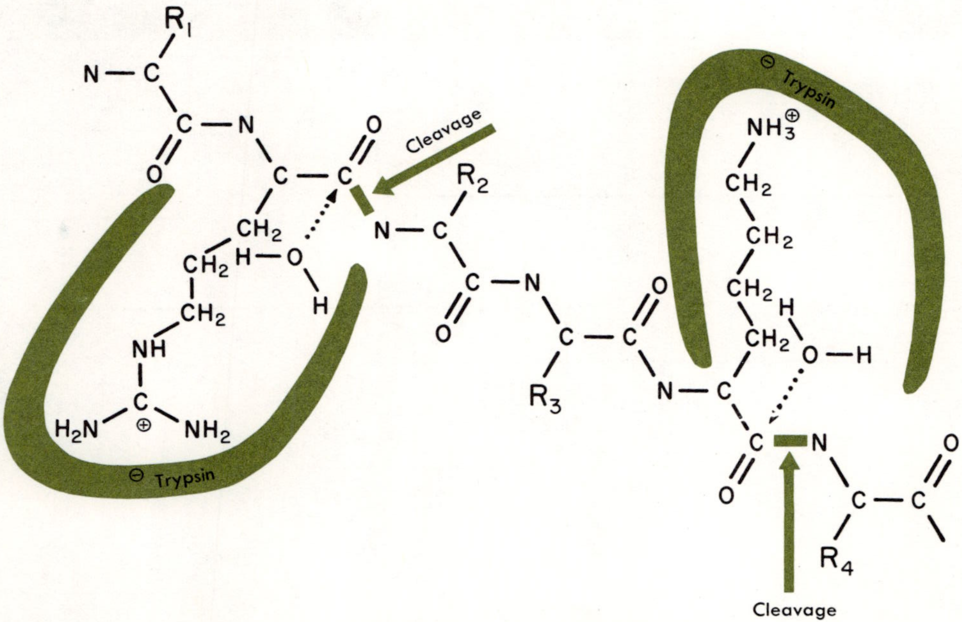

FIG. 5–2 Substrate specificity in trypsin-catalyzed hydrolysis.

FIG. 5–3 An experimental procedure for determining the sequence of amino acid residues in the polypeptide B chain of insulin.

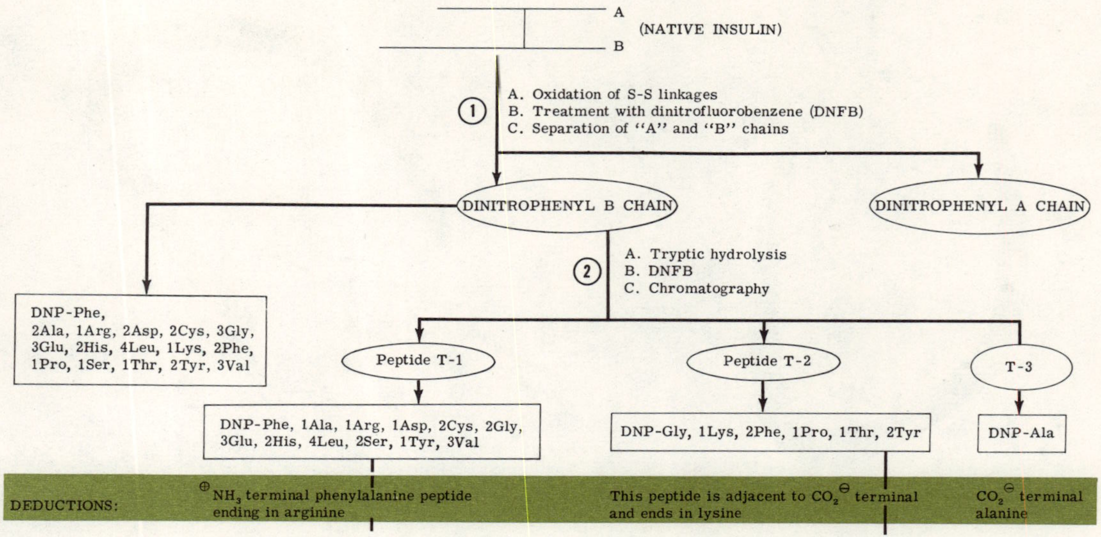

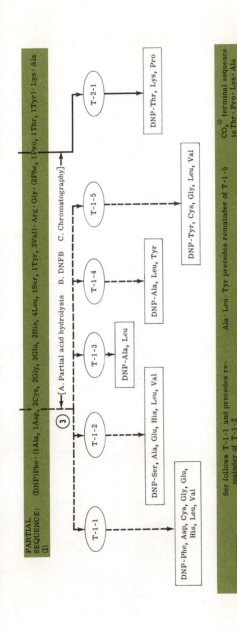

PARTIAL SEQUENCE: (I) (DNP)Phe · (1Ala, 1Asp, 2Cys, 2Gly, 3Glu, 2His, 4Leu, 1Ser, 1Tyr, 3Val) · Arg · Gly · (2Phe, 1Pro, 1Thr, 1Tyr) · Lys · Ala

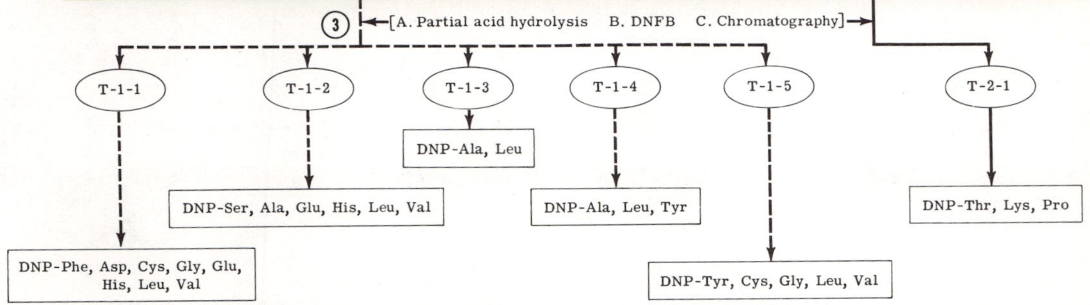

DEDUCTIONS:

Ser follows T-1-1 and precedes remainder of T-1-2
One Glu of T-1 unaccounted for

Ala · Leu · Tyr precedes remainder of T-1-5

$CO_2^{\ominus}$ terminal sequence is Thr · Pro · Lys · Ala

PARTIAL SEQUENCE: (II)

Phe · (Asp, Cys, Gly, Glu, His, Leu, Val) · [Glu, Ser · (Glu, His, Leu, Val) · Ala · Leu · Tyr · (Cys, Gly, Leu, Val)] · Arg · Gly · (2Phe, Tyr) · Thr · Pro · Lys · Ala

(continued)

FIG. 5–3 (cont.)

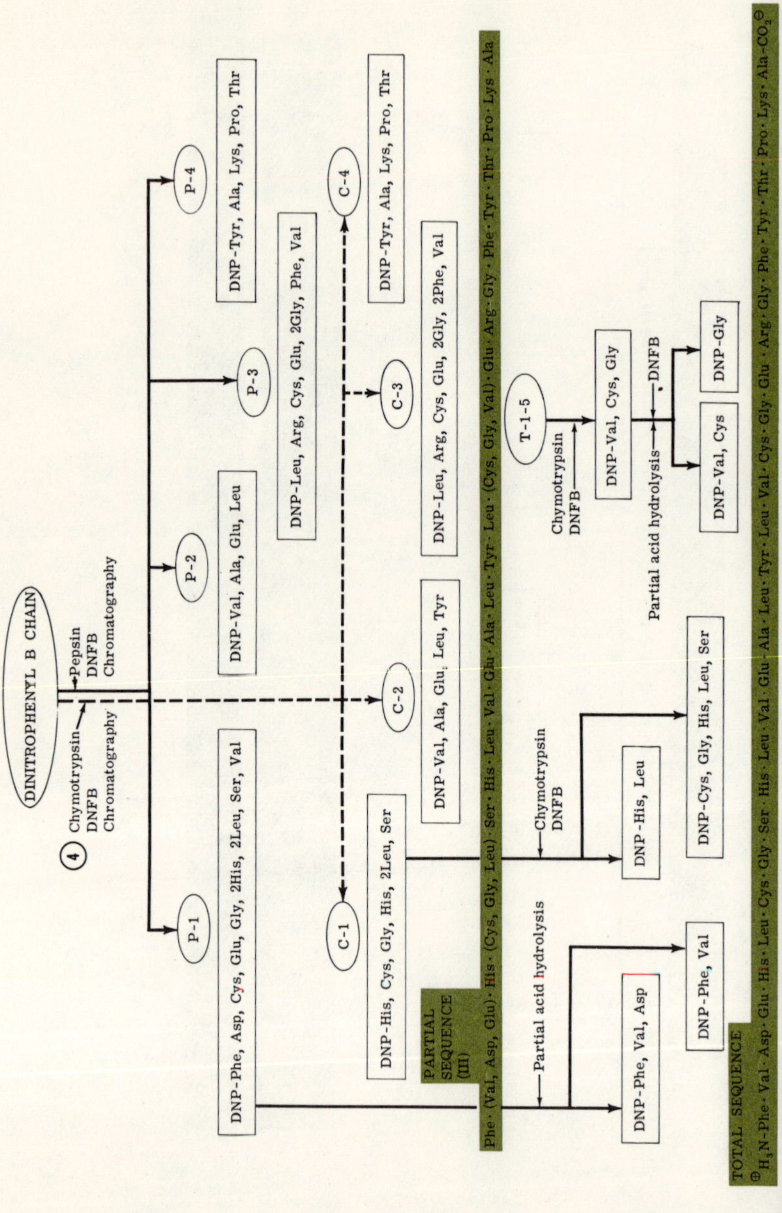

DINITROPHENYL B CHAIN
4
Chymotrypsin
DNFB
Chromatography
Pepsin
DNFB
Chromatography
P-1
DNP-Phe, Asp, Cys, Glu, Gly, 2His, 2Leu, Ser, Val
P-2
DNP-Val, Ala, Glu, Leu
P-3
DNP-Leu, Arg, Cys, Glu, 2Gly, Phe, Val
P-4
DNP-Tyr, Ala, Lys, Pro, Thr
C-1
DNP-His, Cys, Gly, His, 2Leu, Ser
C-2
DNP-Val, Ala, Glu, Leu, Tyr
C-3
DNP-Leu, Arg, Cys, Glu, 2Gly, 2Phe, Val
C-4
DNP-Tyr, Ala, Lys, Pro, Thr
PARTIAL SEQUENCE (III)
Phe · (Val, Asp, Glu) · His · (Cys, Gly, Leu) · Ser · His · Leu · Val · Glu · Ala · Leu · Tyr · Leu · (Cys, Gly, Val) · Glu · Arg · Gly · Phe · Tyr · Thr · Pro · Lys · Ala
Partial acid hydrolysis
DNP-Phe, Val, Asp
DNP-Phe, Val
Chymotrypsin
DNFB
DNP-His, Leu
DNP-Cys, Gly, His, Leu, Ser
T-1-5
Chymotrypsin
DNFB
DNP-Val, Cys, Gly
Partial acid hydrolysis
DNFB
DNP-Val, Cys
DNP-Gly
TOTAL SEQUENCE
$^{\oplus}H_3N$-Phe · Val · Asp · Glu · His · Leu · Cys · Gly · Ser · His · Leu · Val · Glu · Ala · Leu · Tyr · Leu · Val · Cys · Gly · Glu · Arg · Gly · Phe · Tyr · Thr · Pro · Lys · Ala-$CO_2^{\ominus}$

sequence. Sanger attacked the former problem by the method of *partial acid hydrolysis*. The individual tryptic peptides were incompletely hydrolyzed to yield a variety of products, as is illustrated in step 3 of Fig. 5–3. Isolation of a number of these partial hydrolysis products, and determination of their amino acid composition, established the sequence of amino acids within one particular tryptic peptide.[2] Partial acid hydrolysis is seldom used today to establish amino acid sequence within peptides. The newer techniques involve the principle of "sequential" rather than "random" degradation of the peptide chain. Sequential degradation can be accomplished by two methods, namely, by continuous degradation of the peptide chain with an enzyme which cleaves *only* at the amino or carboxyl terminus of the peptide, or by stepwise chemical degradation. The enzyme carboxypeptidase always cleaves peptides at the carboxyl terminal amino acid residue; leucine aminopeptidase always cleaves at the amino terminal residue. Thus, at low concentrations of either of these peptidases, the time-dependent liberation of free amino acids from the peptide chain can be followed, and from this, the sequence can be established.

Enzymes such as those mentioned above are of limited use because some amino acid residues are resistant to attack and because interpretation is complicated when different residues are removed at different rates, as is usually the case. A reaction with phenylisothiocyanate which can be carried out sequentially from the amino terminus of the peptide is illustrated in Fig. 5–4. This is called the Edman degradation, and is currently the most useful tool in determining the sequence of peptides.

The linear order of the tryptic peptides can only be established by resorting to a procedure for cleaving the polypeptide chain at residues other than those cleaved in tryptic hydrolysis, as illustrated in step 4 of Fig. 5–3. The procedure used by

[2] It is interesting to note that this method of determining sequence by partial "random" degradation could, in principle, have been applied to the entire polypeptide chain. However, separating the large number of peptides that would have resulted would probably prove very difficult experimentally.

Sanger involved the degradation of the insulin polypeptides with two other enzymes, chymotrypsin and pepsin. Neither of these enzymes is as specific as is trypsin. Both are sufficiently specific, however, so that a limited number of bonds are cleaved to yield primarily large peptides. With the amino acid sequence of the individual tryptic peptides at hand, it becomes possible to *recognize*, in individual isolated "chymotryptic peptides," the "overlap" of two or more tryptic peptides and hence the linear order of these tryptic peptides can be established. When a sufficient number of large peptides have been isolated to identify all of the "overlaps," the entire linear sequence within the polypeptide chain is established.

The actual number of "overlapping" peptides which must be accumulated in order to determine the sequence of the tryptic peptides unambiguously may seem surprisingly small, when compared to the "random" probability of finding the requisite overlapping peptides. The reason for this apparently small number is the unique and peculiar sequence of amino acid residues in any particular polypeptide chain. In general, specific polypeptide sequences will have some distinctive sequential features which facilitate recognition of the overlaps, although, a

FIG. 5–4 The Edman degradation.

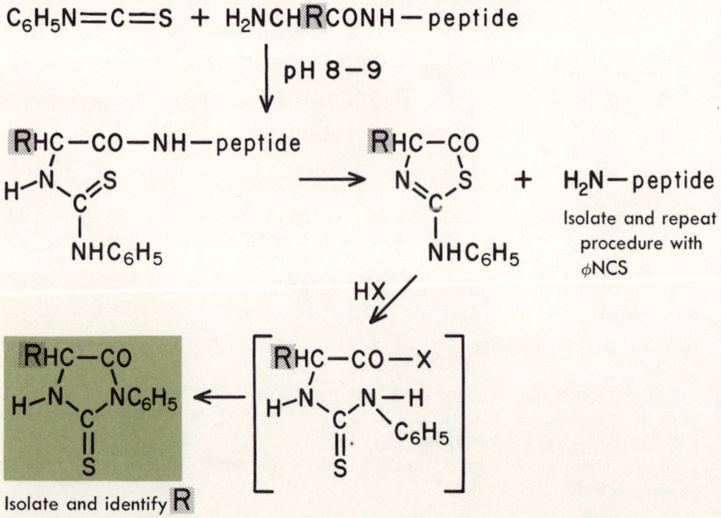

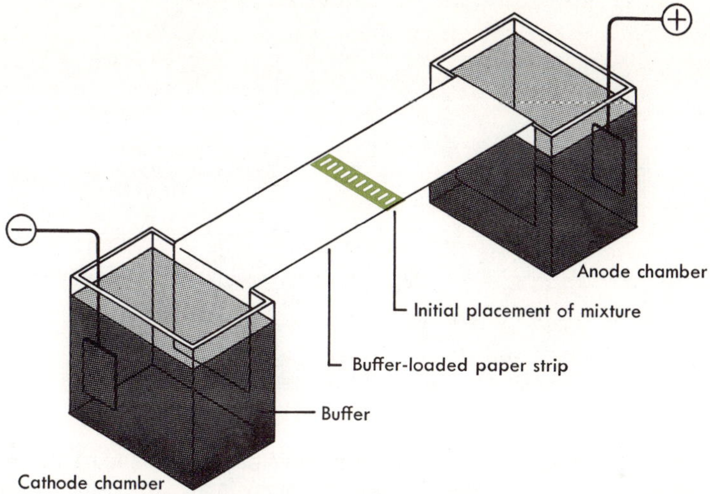

(a)

FIG. 5–5 A variety of chromatographic techniques, useful in the analysis of amino acids and peptides. (a) Schematic version of a high-voltage paper electrophoresis apparatus; (b) the resultant ninhydrin-developed "one-dimensional" chromatogram.

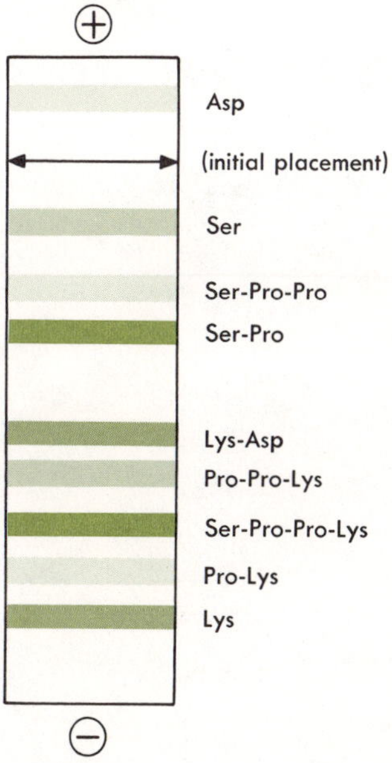

(b)

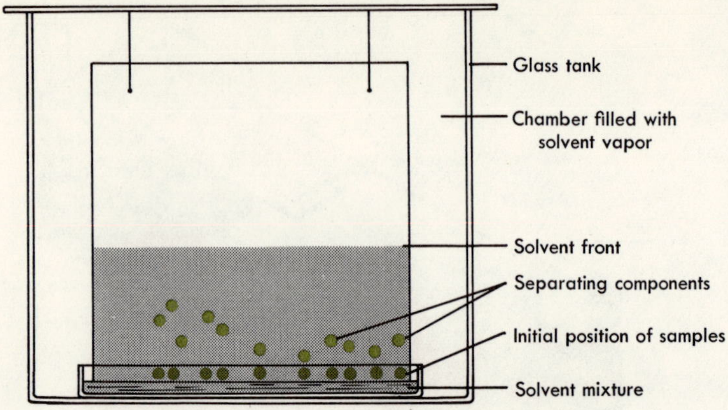

(c)

FIG. 5–5 (cont.) (c) Schematic version of an "ascending" solvent paper chromatography apparatus; (d) the resultant developed "two-dimensional" chromatogram following chromatography in two different solvent systems, as indicated.

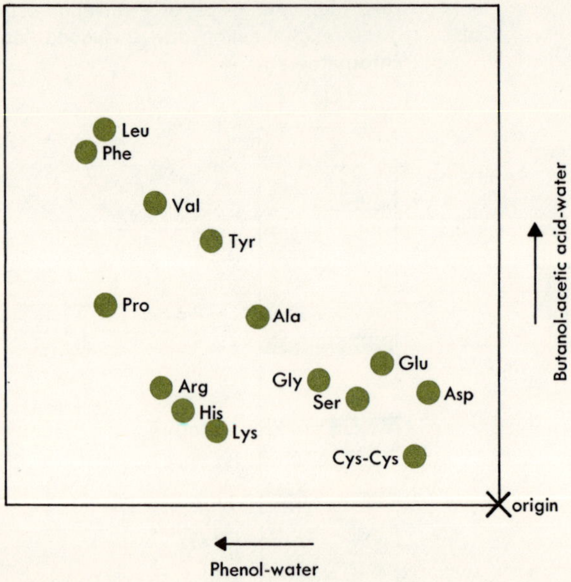

(d)

priori, we cannot predict what these distinctive features will be. Recognition of the overlapping regions via the composition of overlapping peptides will, on the average, be a much easier task than might be anticipated on the basis of probability theory.

The crucial techniques in establishing sequence are those involving the separation of the various degraded peptides, and the separation of the individual amino acids following total degradation of these peptides. Chromatographic techniques either on paper or on columns have been the principal methods for achieving these separations. A number of these techniques are illustrated in Fig. 5–5.

FIG. 5–5 (cont.) (e) An amino acid analyzer. The amino acids are separated by solvent (buffer) extraction from an ion-exchange resin column. The column effluent is reacted with ninhydrin and the location of colored amino acid products in the effluent is compared with the behavior of standard amino acids, as are shown in the chromatogram (f).

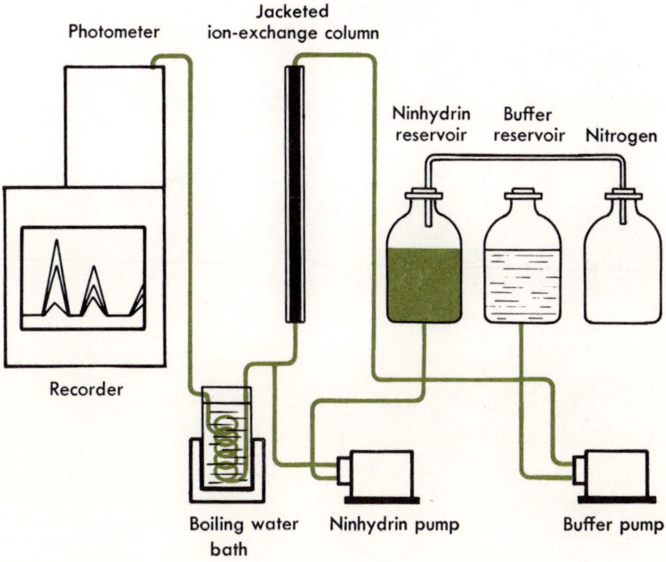

(e)

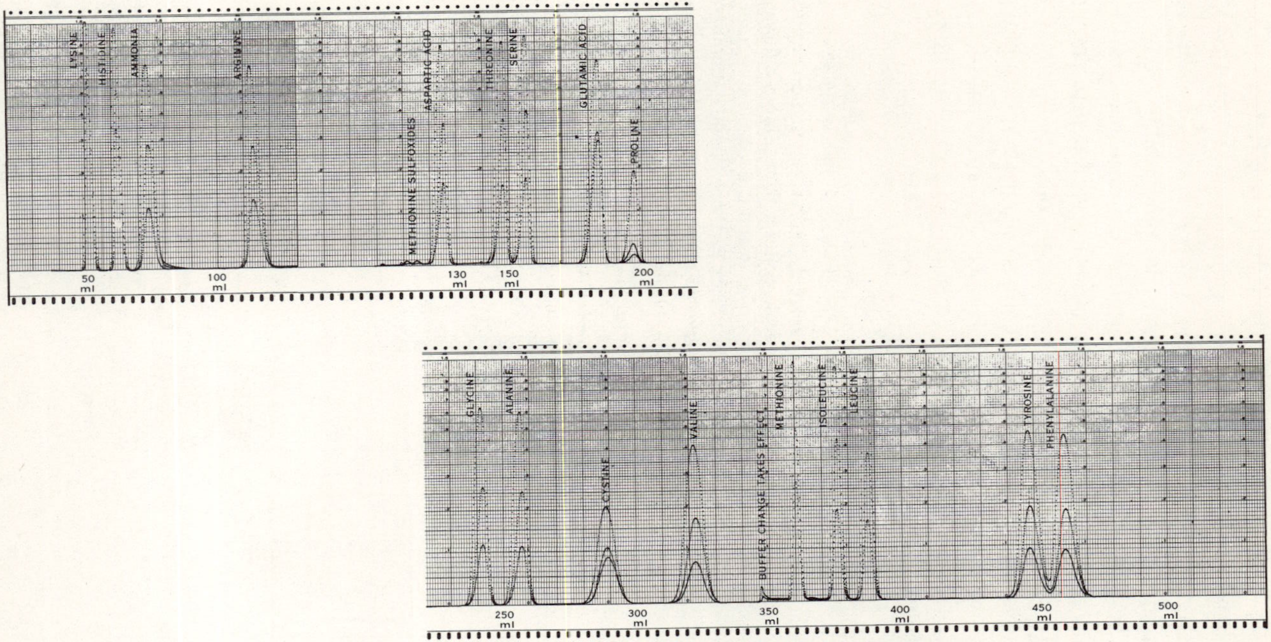

FIG. 5–5 (cont.) (f) Chromatogram from an amino acid analyzer.

5–2 DISULFIDE BRIDGES

As has already been noted (Fig. 5–1), the insulin molecule is held together by S—S (disulfide) bonds as well as by peptide bonds. Disulfide bonds are the only type of covalent cross-links known to occur in enzymes.[3] In insulin we note (Fig. 5–1) that these disulfide bonds are of two types: (1) disulfide linkages within a single polypeptide chain, and (2) disulfide linkages connecting two different polypeptide chains. There are proteins, besides insulin, which contain both of these types of disulfide linkages; other proteins contain only one of these two types (either the interchain or the intrachain linkage), and some proteins contain no disulfide bonds at all. The locations of disulfide linkages in a variety of proteins of known sequence is illustrated in Fig. 5–6. Disulfide linkages appear to be associated with maintaining an active conformation of the enzyme. Cleavage of these bonds, either by oxidation or by reduction, invariably results in the loss of enzyme activity.

The pairing of disulfides and distinction of intra- and interchain linkages can be accomplished by comparing the proteolytic degradation products of the native protein with those of the oxidized, disulfide-cleaved protein. Different peptide fragments are formed upon degradation, as is shown in Fig. 5–3 and in Fig. 5–7.

5–3 THE SPECIFIC AGGREGATION OF POLYPEPTIDES INTO OLIGOMERIC STRUCTURES

As has already been noted in Chapter 4 (Section 4–5), many active enzymes are composed of a number of polypeptide subunits. The forces of interaction holding these subunits together may involve covalent disulfide linkages, hydrogen bond-

[3] Peptides are known in which other covalent linkages exist, for example, the γ-peptide linkage in glutathionine (γ-glutamylcysteinylglycine) and the serine hydroxyl and lysyl-ϵ-amino linkages to carbonyl groups in various naturally occurring antibiotics. Such peptides are apparently synthesized via enzymic pathways other than those usually involved in protein biosynthesis. With larger functional proteins no such linkages have as yet been found.

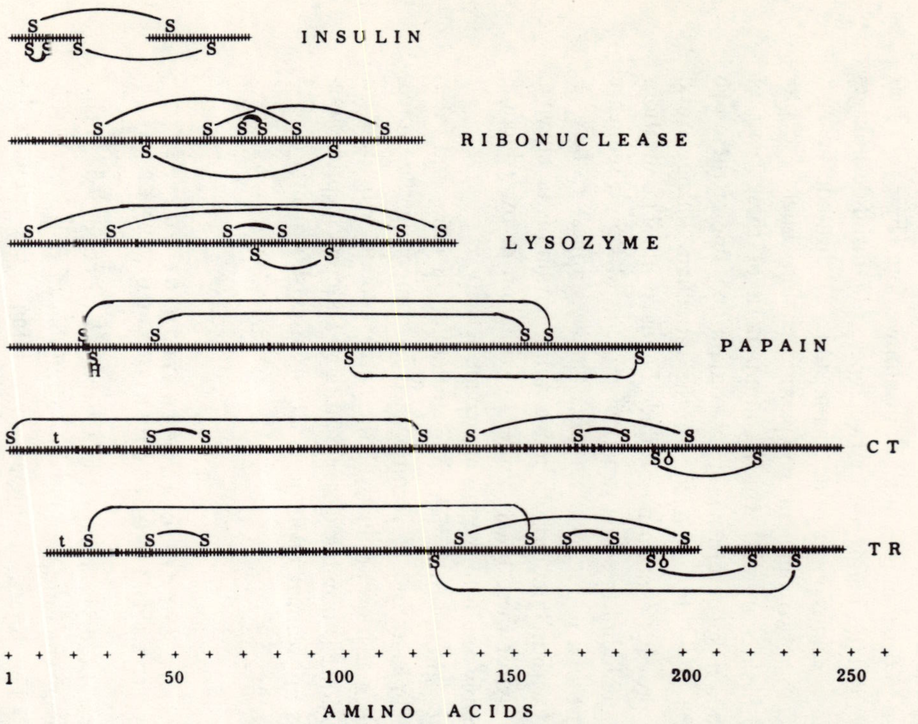

FIG. 5–6 A variety of disulfide linkages in proteins. The short vertical lines indicate the approximate number of amino acid residues in the polypeptide. "CT" and "TR" are chymotrypsinogen and trypsinogen. (Courtesy of Dr. J. Brown.)

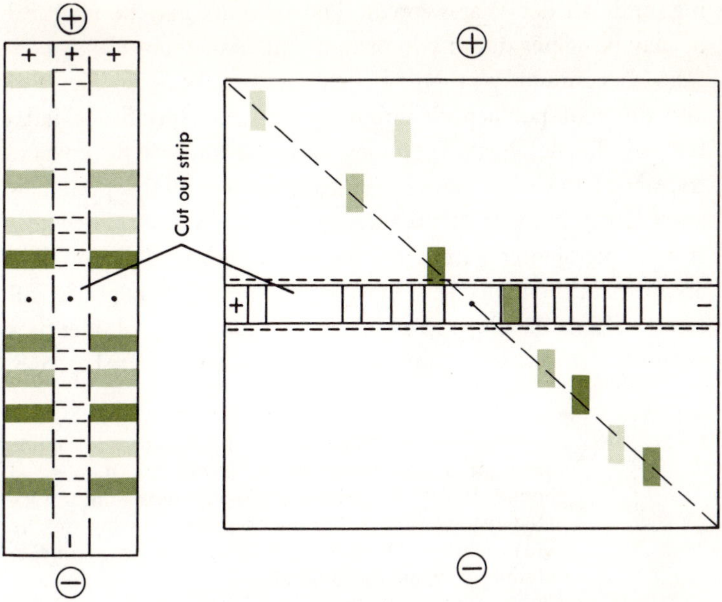

FIG. 5–7 The "diagonal" method for locating disulfide linkages in peptides. The polypeptide is degraded to shorter peptides with a proteolytic enzyme and chromatographed electrophoretically as in Fig. 5–5a. The chromatogram is then treated with performic acid vapors which results in the cleavage by oxidation, of disulfide linkages (yielding negatively charged sulfonates) in the peptides. A vertical strip of the treated chromatogram is then sewed into the middle of a new sheet of chromatographic paper in the *horizontal* direction (as shown in the right hand illustration) and the electrophoretic chromatography is repeated. Peptides which remain unchanged by performic acid oxidation (and therefore contain no disulfide linkages) will have the same mobility as previously and therefore should form a perfect *diagonal* pattern on the second chromatogram. Any peptides which contained disulfide linkages will cleave and on the average gain one unit of negative charge upon oxidation, and will therefore have altered electrophoretic mobility. These peptides will lie *off* the diagonal and can so be identified. (Courtesy of Dr. J. Brown.)

ing, and van der Waals forces. The subunits may be identical or may be aggregates of two or more different types of polypeptides. A common method of distinguishing between identical and different polypeptide subunits is by the specific degradation of the native polypeptides with the proteolytic enzyme trypsin. If the total amino acid composition of the protein is known, the number of peptides generated by tryptic degradation is predictable (from the known number of lysines and arginines in the molecule). Fewer different tryptic peptides will result if a large oligomeric protein is composed of identical subunits than if the subunits differ, as is illustrated in Fig. 5–8.

FIG. 5–8 The estimation of the number of different polypeptide chains in a multichain protein from the number of peptide fragments following tryptic digestion. Anticipated results from a tetrameric protein having (left to right) *identical*, *two pairs* (as in hemoglobin), and *four different* polypeptide subunits.

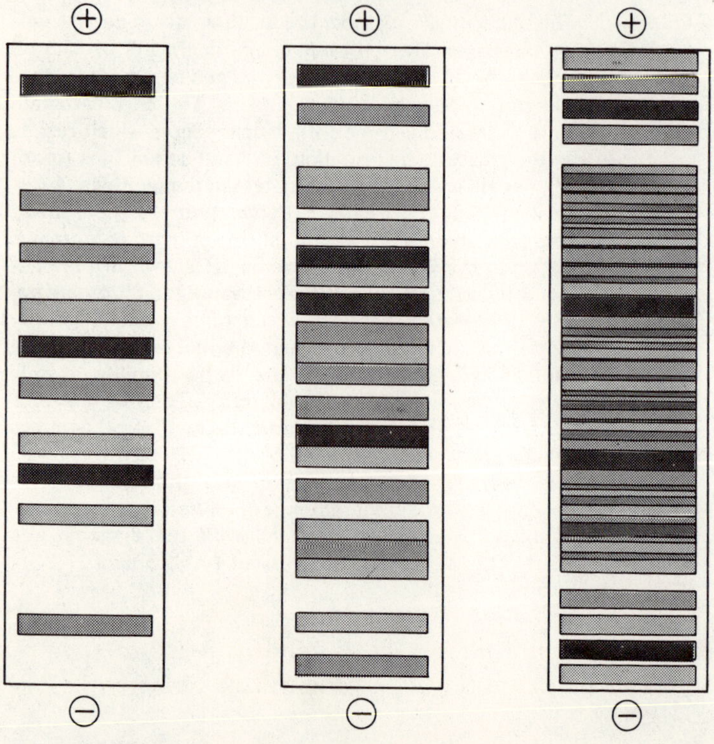

Qualitative determination of the number of different N-terminal amino acid residues, for example, by the Sanger method (Fig. 5–3), gives an indication of the number of different kinds of polypeptide subunits. Quantitative determination of the *number* of end groups allows one to calculate the number of subunits per complex even if the subunits are identical.

5–4 ACTIVE AND INACTIVE PROTEIN CONFORMATIONS

The factors governing the conformation of an active protein have been classified into four types. Hereafter in discussions, we shall refer to these four types of factors by the following commonly used nomenclature:

1. *Primary structure:* all the *covalent* bonds (peptide bonds and disulfides).
2. *Secondary structure:* the coiling and folding of the peptide chain, for example, the α-helix.
3. *Tertiary structure:* the *superfolding* of the helix, or other organized structure, into a higher state of organization.
4. *Quaternary structure:* the organization of subunits in a multiunit protein.

Primary structure is investigated by methods that are straightforward, albeit experimentally elaborate. Secondary, tertiary, and quaternary structure are not so readily elucidated. To date, the only method for determination of complete structure is X-ray crystallography. Even when the coordinates of virtually every atom of the molecule are known, as they are in myoglobin and lysozyme, it is by no means straightforward to assess which specific forces of interaction govern these coordinates. Nevertheless, it is possible (by chemical techniques) to establish which amino acid residues are in close proximity to which others in the region of the active site, by examining the effects of modification of such residues on either the "binding" or activity of substrates and (or) competitive inhibitors.

A fundamental distinction must be made in enzymic catalysis between the effect of environment on catalytic groups within the enzyme site, and the effect of environment on the structural integrity of the active conformation. Often, a change in environment leading to altered activity of the enzyme can be plausibly explained either on the basis of gross changes in the conformation of the protein, or on the basis of changes in the chemical structure of catalytic centers. Chemical modification of specific residues of the protein can lead to inactivation due to destruction of a catalytic center, or because the site in the chemically modified protein becomes inaccessible to the substrate, as illustrated in Fig. 5–9.

In many instances it is possible to modify discrete regions of the protein without any significant modification of the gross protein structure. An interesting example is found in hemoglobin, where oxygenation of the heme groups results in an altered arrangement among the polypeptide subunits but has little effect on the overall three-dimensional structure of each of the individual subunits.

In contrast to this small change of protein structure over a limited region is a process generally referred to as *denaturation*.

5–5 PROTEIN DENATURATION

The specific active conformation of an enzyme protein is the result of a large number of interactions (as has been discussed in detail in Chapters 2 and 3). When the environment is changed so that some of these interactions are weakened (or disappear completely), the cooperative nature of many of these interactions causes the facile break-up of a large portion of the structure. It is probable that many active enzyme conformations are only marginally stable. That is, given a small decrease in total free energy of interaction due to the loss of a few specific interactions, the active conformation is no longer favored and further breakdown of the structure can ensue. The transition from an *ordered* conformation to an essentially disordered *random* conformation, is usually referred to as *denaturation*. There is considerable doubt that the so-called random denatured state is always entirely disordered. Hence an exact

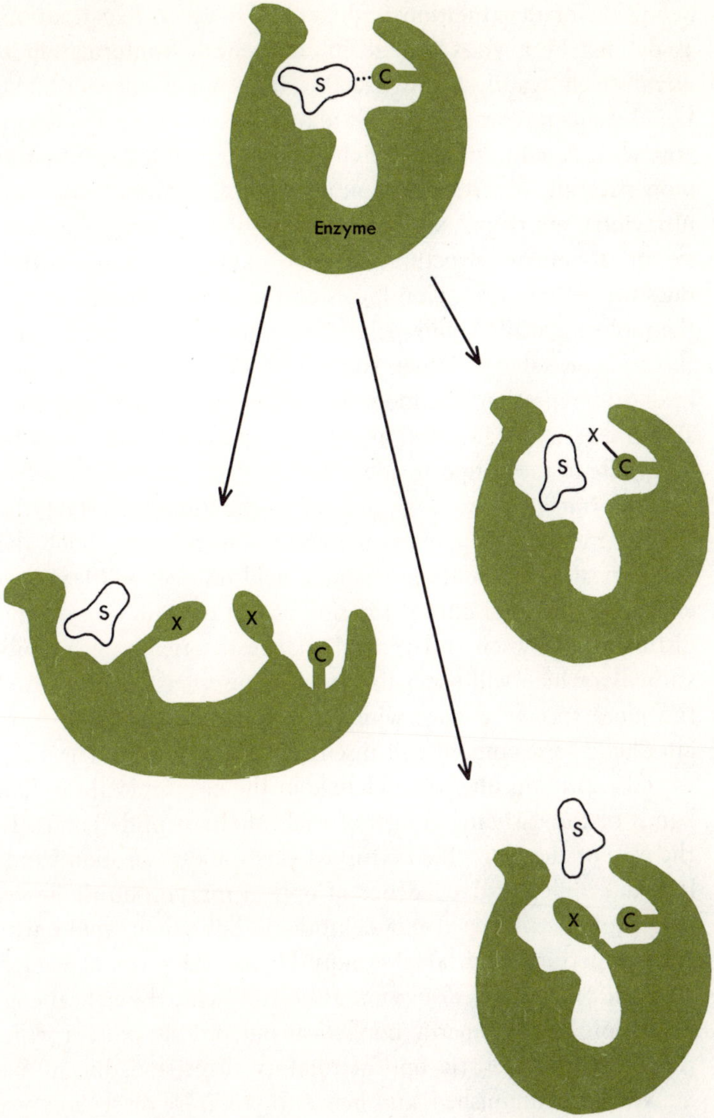

FIG. 5–9 Various modes of inactivation of an enzyme by specific chemical modification. The modifying reagent (X), a catalytically active enzyme constituent (C), and the substrate (S) are shown.

definition of denaturation is difficult. However, denaturation, as defined by a gross change in polypeptide conformation is qualitatively readily distinguishable from the small conformational changes referred to in the above section (5–4). Concomitant with denaturation, large changes in a number of physical properties are observable: typically optical rotation, viscosity, ultraviolet spectrum. Native enzyme proteins have, to a large extent, a compact structure in which nonpolar side chain residues interact via dispersion forces and tend to be buried within the molecule while highly polar or charged residues extend into the aqueous solvent. When the polypeptide chain is "randomized" in orientation, the molecule will be much more extended and much less rigid and moreover, aqueous solvent will be accessible to the nonpolar side chains.

Upon transition from the ordered to the disordered state, the change in orientation of neighboring substituents about the asymetrical α-carbon atom of amino acid residues will lead to a change in the total optical rotation of the protein. One should distinguish between this general change in orientation of substituents, which will occur in any order-disorder transition, and the more specific change which occurs in the transition from an α-helix to a random coil discussed in Chapter 3. The helix to coil transition involves a change in the electronic absorption bands of the carbonyl groups of each of the peptide bonds. In the case of the helix, the perturbed electronic absorption bands lead to a different dependence of optical rotation on the wavelength of the polarized monochromatic light than would arise from unperturbed carbonyl groups. Hence, there is a characteristic *optical rotatory dispersion* associated with the helical conformation. In a primarily nonhelical but orderly protein structure, the characteristic optical rotatory dispersion due to the helix will be diminished, and hence the rotatory dispersion profile will resemble that of the random coil. The total optical rotation, however, will reflect whatever orderly arrangement exists, and hence the transition from order to disorder will be manifest in a change in optical rotation rather than in a change in optical rotatory dispersion. Indeed, such is the case in two enzymes, ribonuclease and chymotrypsin (see Chapter 8), in which there is very little helical structure.

The amino acids tyrosine and tryptophan contain strong chromophores in the ultraviolet ($\sim$280 mμ). In this wavelength region, the spectrum of the protein is essentially due to the contributions from these two chromophores. The characteristic spectra of such chromophores are dependent on the polarity of the solvent environment. When these nonpolar residues are buried within the protein molecule, they are inaccessible to solvent and hence the spectra reflect the polarity of the internal protein environment. In native enzyme structures, these spectra appear to be remarkably constant (indicative of a similar internal environment in all the proteins). When the polypeptide chain is randomized, the chromophores interact with the aqueous solvent and the spectra resemble those observed in aqueous solution with corresponding small molecule chromophores. This leads to a rather large and easily measurable spectral change in the region 270–290 mμ. A number of typical denaturations as measured by optical rotation and by spectroscopy are illustrated in Fig. 5–10.

Protein denaturation can be brought about by a great variety of changes in environment. Typical changes involve the following:

(1) The transfer of protein to a strongly hydrogen-bonding solvent, such as concentrated guanidine hydrochloride or urea.

(2) The transfer of protein to a less polar environment such as mixed aqueous-organic solvents (including concentrated aqueous urea). Urea may act as a denaturant via two mechanisms, that is, by the rupture of hydrogen bonds and by the breaking up of hydrophobic nonpolar interactions.

(3) Change in temperature: When the temperature is raised sufficiently to break some of the hydrogen or hydrophobic bonds, abrupt structural transitions take place (see for example Figs. 3–9 and 5–10).

(4) The addition of *detergents:* By introducing an organic molecule which contains both a hydrophobic, nonpolar residue and a highly polar residue (a *detergent*), the constituent hydrophobic side chains of the protein can be brought to the solvent interface due to the strong interaction with the hydrophobic residue of the detergent molecule and the strong interaction of the polar residue with aqueous solvent. This detergent effect

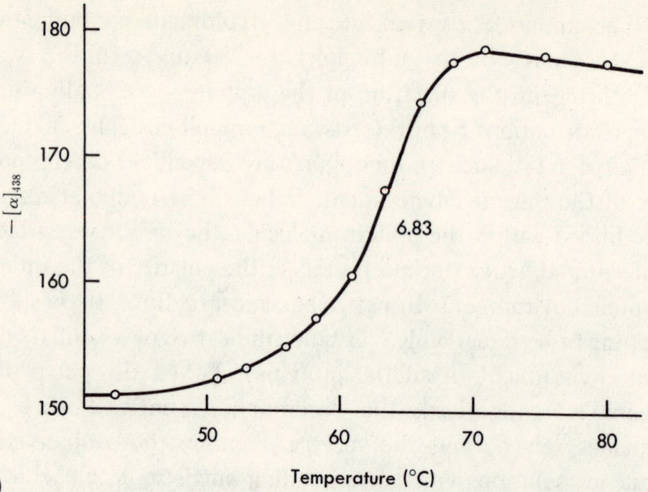

(a)

FIG. 5–10 Changes in physical properties accompanying protein denaturation. (a) The optical rotation of ribonuclease solutions (pH 6.83) as a function of temperature. (b) The optical density of ribonuclease at 287 mμ, as a function of temperature. [Data of J. Hermans, Jr., and H. A. Scheraga, *J. Am. Chem. Soc.*, **83**, 3283 (1961).]

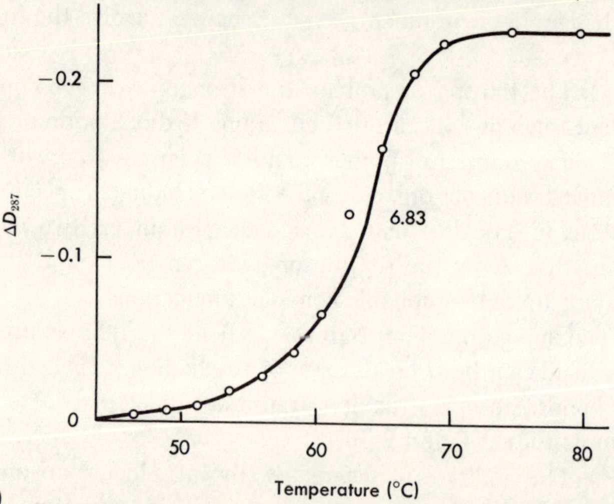

(b)

opens the structure and tends to randomize the polypeptide chain.

(5) Change of the ionic environment: In the compact, orderly structure, most of the charged residues will lie on the surface of the protein in contact with solvent. If there is a preponderance of one type of charged group (e.g., NH_3^+ or CO_2^-), the stability of the protein molecule may depend on the ionic strength of the aqueous medium. At very low salt concentrations, repulsions between like charges may destabilize the compact structure and tend to favor a more extended random structure. Specific ions may also have profound effects due to the formation of complexes with particular side chains.

(6) Change of pH: The excess of total positive or negative charge of the protein is determined by the pH of the medium.

FIG. 5–10 (cont.) (c) The optical rotatory dispersion of myoglobin in the native conformation (lower curve) and in 3 *M* guanidine hydrochloride where it is completely denatured (upper curve). (Courtesy of Dr. B. Littman.)

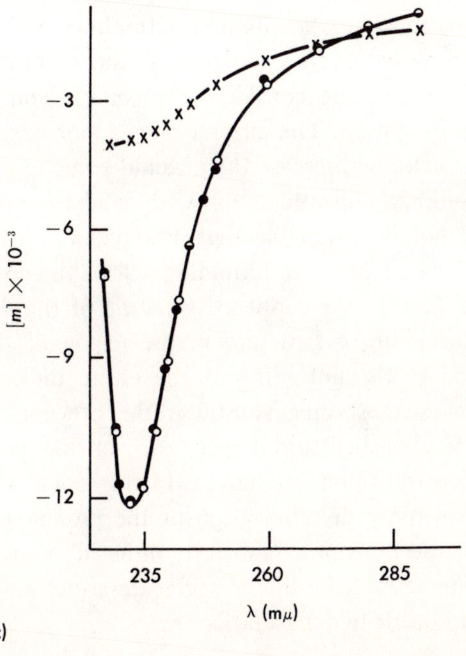

Aspartic acid, glutamic acid, lysine, histidine, and the amino and carboxyl groups at the ends of polypeptide chains have pK_a's in aqueous solution within several pH units of neutrality. At pH values where there is a large excess of positive or negative charge, coulombic repulsions may destabilize the compact structure. Indeed, most proteins are stable only over a rather narrow range of pH (usually but not always in the range of neutrality). Pepsin is an example of a protein that is only stable at pH values below 5.

It is interesting to note that substrates often contribute to the stability or instability of particular enzyme proteins. For example, many enzymes such as ribonuclease, the proteolytic enzymes trypsin and chymotrypsin, the *dehydrogenases* involving nicotinamide adenine dinucleotide, and enzymes that contain pyridoxyl phosphate are stabilized against denaturation in the presence of substrate or coenzyme. In other enzymes, particularly some enzymes involved in metabolic control (i.e., the allosteric enzymes referred to in Section 4–5), there is evidence that the presence of an allosteric substrate molecule or an analogue which binds at the substrate site *may* loosen the protein structure and facilitate denaturation. Stabilization of the ordered native enzyme structure by substrate or coenzyme is plausible when one considers the general compactness of protein conformation. The enzyme site cannot be a completely compact structure; otherwise there would be no opportunity for substrate binding and interaction with catalytic groups. Therefore, it is not unreasonable that the region of the enzyme protein most susceptible to unfolding will be the open structure of the site. Due to the cooperative nature of the denaturation process, unfolding of structure in the region of the site will rapidly lead to the unfolding of the entire molecule. In the presence of stereospecific substrate, the previously open site structure is abolished and hence, the starting point for the denaturation transition is removed. In the case of the allosteric enzymes, substrate destabilization of the protein has been explained on the basis of a "substrate induced" loosening of the subunit interaction, the looser aggregate structure being more readily susceptible to denaturation.

There has been much speculation as to whether the active enzyme conformation is the thermodynamically most stable conformation which an enzyme protein can assume in its physiological solvent environment. Much evidence has been gathered, particularly from studies on reversible denaturation, to substantiate the argument that the active conformation is thermodynamically most stable. Much of the experimental evidence is derived from studies on the well-characterized enzyme ribonuclease (which will be discussed in detail in Chapter 8). The strongest arguments favoring this view are as follows: (1) In many instances, denaturation (particularly by urea) is readily reversible. (2) Reduction of the four disulfide linkages of ribonuclease (RS—SR → 2RSH) results in unfolding of the polypeptide chain. Reoxidation (2RSH → RS—SR), in solvents where the native structure is stable, results in reformation of the proper four disulfide linkages to yield the native active enzyme.

It is often the case, however, that denatured proteins are not readily renatured (although it has not always been ruled out that some covalent changes might accompany denaturation). Moreover, unlike ribonuclease, there are enzyme proteins which do not contain disulfide linkages to aid in the stabilization of a rigidly ordered structure. In many instances, the *rate* of renaturation of denatured protein is very slow under all experimental conditions investigated. The rate of synthesis of new proteins from the constituent amino acids, on the other hand, is often a very rapid process *in vivo*. It thus appears that, whether or not the native conformation is the most stable conformation, it arises somehow in the process of protein synthesis. One might envisage, for example, that the polypeptide chain folds sequentially as it is synthesized, and that the final conformation is virtually complete at the termination of the synthesis. When a completed polypeptide chain is unfolded by denaturation, many alternate refolded conformations may be possible; if this is so, formation of the proper active structure may be either a slow (random) process, or may not occur at all, if the catalytically active conformation is not the thermodynamically most stable conformation.

5–6 THE EFFECT OF MODIFICATION OF SPECIFIC AMINO ACID RESIDUES ON THE ACTIVITY OF ENZYMES

We may now proceed to examine the influence of specific chemical modifications on the activity of enzymes, and thereby consider the role of specific side chains in various catalytic mechanisms. Recently, a variety of chemical reagents have been synthesized for the chemical modification of the side chains of specific amino acid residues. Many of these reagents have been designed on the basis of a priori information on the potential significance of particular residues to the catalytic process, and these specific reagents will be considered later in this chapter (and also in Chapter 8).

One specific modification, long known to have a profound effect on enzyme activity, merits special attention. This is the reversible protonation of weakly basic residues. By careful control of the pH, the effect of protonation of individual, or small classes, of basic residues on enzyme activity can be examined. Such effects have been studied with nearly every enzyme whose kinetic behavior has been investigated.

THE EFFECT OF pH ON ENZYME ACTIVITY

Enzyme activity is invariably pH dependent. The effect of pH on enzymic activity can often be represented by one of the two types of curves illustrated in Fig. 5–11, when allowance is made for any protonic dissociation constants pertinent to the substrate alone. These two types of pH-activity profiles are commonly referred to as *bell-shaped* and *sigmoid-shaped* curves. Note that in both profiles, *the maximal velocity* (i.e., the velocity at substrate saturation) is plotted as a function of pH. This is essential, since very often the binding of substrate to enzyme site is itself pH dependent due either to changes in the detailed conformation of the site (Section 8–1) or to the pH-dependent charge of the enzyme protein. The latter effect will, of course, be more profound with substrates which are themselves charged. At extremes of pH, denaturation of the enzyme often occurs. We must be cautious not to confuse denaturation of the enzyme with the reversible protonation of specific basic residues. When proper account is taken of any reversible or irreversible denaturation at the extremes of pH, the effect of pH on enzyme activity can very often be accounted for on the basis of one of the two catalytic models shown in Eq. 5–2. In the process of fitting the experimental

$$\begin{array}{ccc} HES \overset{K_{EH}}{\rightleftharpoons} ES + H^{\oplus} & H_2ES \overset{K_{EH_2}}{\rightleftharpoons} HES \overset{K_{EH}}{\rightleftharpoons} ES & \\ \downarrow & \downarrow & (5\text{–}2) \\ E + P & HE + P & \\ \text{Sigmoid-shaped profile} & \text{Bell-shaped profile} & \end{array}$$

profile of pH versus maximal activity to the appropriate model, the dissociation constants, K_{EH} and so on, are determined. Typical profiles for various values of these dissociation constants are illustrated in Fig. 5–11. The experimentally determined dissociation constants can then be compared with the known dissociation constants of particular conjugate acids of amino acid residues. All such dissociation constants are listed in

FIG. 5–11 Typical variations in enzyme activity as a function of the pH of the medium. The drop in activity at the extremes of pH (indicated by the dotted line portions of the curves) are due to protein denaturation. The pH region in which denaturation occurs is characteristic of the particular protein structure and may vary greatly from one enzyme to another.

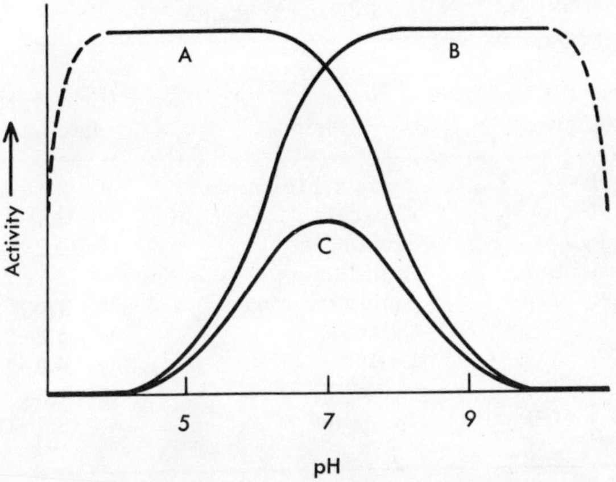

Table 5-1. In correlating particular values of the dissociation constants with particular acidic species, one should bear in mind the following: (1) The presence of other fixed charges in the immediate vicinity of a dissociating species will always perturb the dissociation constant of that species. The perturbation will be such that fixed cations will always increase the dissociation constant and fixed anions will always decrease the dissociation constant. (2) The environmental polarity will affect dissociations of the type

$$HA \rightleftharpoons H^{\oplus} + A^{\ominus}$$

where neutral acidic groups dissociate to charged acidic and basic species, whereas positively charged acids of the type

$$BH^{\oplus} \rightleftharpoons B + H^{\oplus}$$

will be relatively unaffected by the solvent polarity. The second of these two effects may be of no great significance if the aqueous environment is freely accessible to the enzyme site. Moreover, any perturbation of a dissociation constant due to the second of these two phenomena may be investigated by a further study of the effect of pH on enzymic activity under conditions of less than optimal substrate concentration, since the bound substrate should perturb the dissociation constant if

Table 5-1
Dissociation Constants of Weakly Acidic Amino Acid Residues in Peptide Linkages

Acidic group	Residue	pK_a (Unperturbed by adjacent charge)
$-CO_2H$	Carboxyl terminal	2.1–2.4
$-CO_2H$	Asparate	3.7–4.0
$-CO_2H$	Glutamate	4.2–4.5
—Imidazolium$^{\oplus}$	Histidine	6.7–7.1
$-NH_3^{\oplus}$	Amino terminal	7.6–8.0
—SH	Cysteine	8.8–9.1
$-NH_3^{\oplus}$	Lysine	9.3–9.5
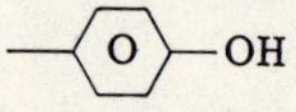	Tyrosine	9.7–10.1

this second phenomenon is of consequence. The first phenomenon (that of a fixed charge at the enzyme site) can be investigated by studying substrates of similar structure but different charge. Once the effect of pH on activity can be restricted to a small number of weakly dissociating species, the residues potentially involved may be investigated further by the more detailed chemical methods to be described in the following sections.

It is interesting to note that for a very wide variety of enzymes, pK_a's in the range 4–9 have been implicated in activity. In this pH range, the only dissociable groups in an enzyme composed entirely of amino acids are the carboxylic acids (aspartic and glutamic acids), the terminal α-carboxyl, the imidazole of histidine, the thiol group of cysteine, and the N-terminal α-amino group (the phenolic hydroxyl of tyrosine and the ϵ-amino group of lysine lie just above this range). Perhaps the most common residue implicated in enzymic activity, on the basis of pH studies, is the imidazole residue of histidine, whose conjugate acid has a pK_a of about 7.

The detailed role of acids and bases in enzyme catalysis is discussed in the following chapter.

In order to reveal particular amino acid side chains involved in catalytic activity, many investigations have employed selective chemical modification of specific side chains of an enzyme protein, and the effect of these modifications on the catalytic activity. When such chemical modification leads to inactivation (or a lowering of activity), it is important to ascertain whether the modification has resulted in the destruction of a catalytic center, in a change in the conformation of the active site either by steric interference with the substrate, or in the modification of the three-dimensional protein structure. Studies of catalytic activity at high concentrations of substrate often aid in distinguishing a change in substrate affinity from a loss of catalytic reactivity. Studies of protein structure, such as those mentioned in the previous sections, are often utilized for examining gross changes in protein conformation.

In attempting to modify specific side chains, three general types of reactions are frequently utilized. We shall now discuss these reactions in detail.

SPECIFIC MODIFICATION REACTIONS, AS DEDUCED FROM STUDIES WITH MODEL PEPTIDES

By examination of synthetic peptides or other analogues, conditions are sometimes found wherein only one or a small number of the twenty types of amino acid side chains undergo chemical modification. The intact native enzyme may then be subjected to similar conditions. The rate of chemical modification of such residues may (and often does) vary, depending upon the location of particular residues within the protein structure. For example, a side chain which is "buried" inside the protein may react very much more slowly than one which extends into the aqueous solution. A rather frequently utilized and plausible hypothesis is that those residues in the native enzyme which react at rates comparable to those observed with model peptides are accessible to the aqueous solvent in the native enzyme structure. Thus, even when a specific residue is repeated in the polypeptide sequence, it is often possible to specifically modify one, or a small number of these residues, by careful control of the extent of reaction. Since the active site must be accessible to the aqueous environment, it is not surprising that residues involved in catalytic activity are often the first of their type to be modified.

Chemical reagents often react with more than one of the twenty different amino acid side chains. Under specified conditions, however, such a reaction may become selective for only one type of residue. A good example is the alkylation reaction with iodoacetamide (Eq. 5–3). This reagent will react with good electron donors such as the sulfhydryl of cysteine, the thioether of methionine, the imidazole nitrogen of histidine, and the ϵ-amino group of lysine. The reaction with methionine is virtually independent of pH, since there is no competition between reagent and solvent protons for the electron pair on the sulfur atom. The same reagent will favor attack by the electron pair of a (basic) nitrogen (or mercaptide anion) only when there is little competition between reagent and the solvent protons, that is, at pH's near or above the pK_a of the conjugate acid.

$$B{:}H^{\oplus} \rightleftharpoons H^{\oplus} + B{:} \xrightarrow{ICH_2CONH_2} B^{\oplus}\text{—}CH_2CONH_2 + I^{\ominus} \quad (5\text{–}3)$$

Barring steric interference, the velocity of the alkylation reaction will depend on the ability of the particular atom to donate electrons, a property that can often be correlated with basicity. For the two nitrogen bases, histidine and lysine, the relative velocities should correlate with the pK_a's of the conjugate acids (see Table 5–1), since the binding of a proton also depends on the electron donating ability of the base. The velocities will also be dependent on the fraction of the base which is unprotonated. In the absence of cysteine, which reacts most rapidly, the reaction can be made relatively specific for methionine, histidine, or lysine by controlling the pH of the solution. The use of this reagent in the specific modification of amino acid residues of ribonuclease and α-chymotrypsin, which have no cysteine, are discussed in Chapter 8.

If a particular residue is involved in catalytic activity, this can be ascertained by the following type of experiment. If the amino acid composition of an enzyme protein is known, the modification reaction can be followed by measuring the change in total number of amino acid residues of a particular type as

FIG. 5–12 Typical dependencies of enzyme activity on the number of amino acid residues of a particular kind which have been modified by chemical reaction.

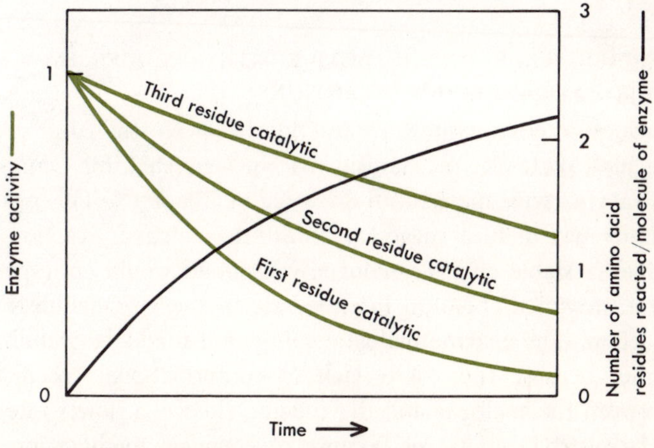

a function of time; the activity of the enzyme can be followed simultaneously. Modification of a particular residue may be correlated with changes in catalytic activity. The above procedure can lead to presumptive evidence for the involvement of such a residue in either the catalytic mechanism or the binding of substrate.

Often, residues important for activity are much more reactive than residues of the same type that are not involved in the catalysis. However, sometimes a residue important for catalysis is not modified until the later stages of the total reaction. In such a case, a more gradual diminution of enzymic activity will be observed (as is illustrated in Fig. 5–12). Once information of the type shown in Fig. 5–12 has been accumulated, the specific residue involved in the catalytic activity often can be revealed by repeating the experiment with a radioactive label (or an analogous but distinguishable chemical reagent) *after* the nonessential residues have been modified. Partial digestion of the enzyme protein to peptide fragments is then carried out, as previously described, and the peptide fragment containing the covalently bound radioactive label can usually be isolated and the amino acid sequence in the region of the active site can be determined.

This type of chemical modification has been utilized extensively in the investigation of the role of methionine, cysteine, histidine, and lysine residues in catalytic activity.

SPECIFIC LABELING OF ACTIVE SITE RESIDUES BY SUBSTRATE-ANALOGUE REACTIONS

Once an enzyme-substrate reaction has been studied in some detail, a particular mechanism may be suggested, for example, the acyl-enzyme mechanism discussed in Chapter 4. This mechanism may in turn suggest a "substrate-analogue" reaction in which a stable enzyme-analogue compound would correspond to a presumed chemical intermediate in the enzyme-substrate reaction. The enzyme-analogue compound might be stabilized either because there is a lack of conformational specificity between the analogue and the enzyme (hence, a slower rate of subsequent reaction) or because of chemical modification of the catalytic apparatus. Good examples of the former are nonspecific acylating reagents (such as acyl imidazoles and nitrophenyl esters) and phosphorylating reagents for enzymes which function via a presumed acyl-enzyme mechanism. These compounds mimic the *reactivity* of substrates (they are, in fact, chemically far more reactive than typical substrates) but lack the configurations specific to any particular enzyme. They form metastable intermediates, or stable products, with active centers of many enzymes. The particular reactive residue to which such reagents become attached can be established by the usual procedures of isolating degraded peptides containing either a chemical or a radioactive label. This type of *active center labeling* has been utilized extensively in enzymic reactions involving acyl or phosphoryl transfer. When the "enzyme-analogue intermediate" has only transient stability, it may sometimes be isolated by denaturation of the protein during the course of reaction. Near pH neutrality all such acyl or phosphoryl derivatives have been found to be stable, once the protein is denatured.

When the presumed enzyme-analogue (or -substrate) complex disappears at rates characteristic of enzyme turnover numbers, ingenious procedures have been devised for "trapping" the intermediate by rapid chemical modification of catalytic centers involved in subsequent steps of the reaction. A good example (which will be discussed in detail in Chapter 8) is the case of enzyme-substrate reactions which are presumed to proceed via a Schiff's base intermediate (Eq. 5–4). Such compounds are

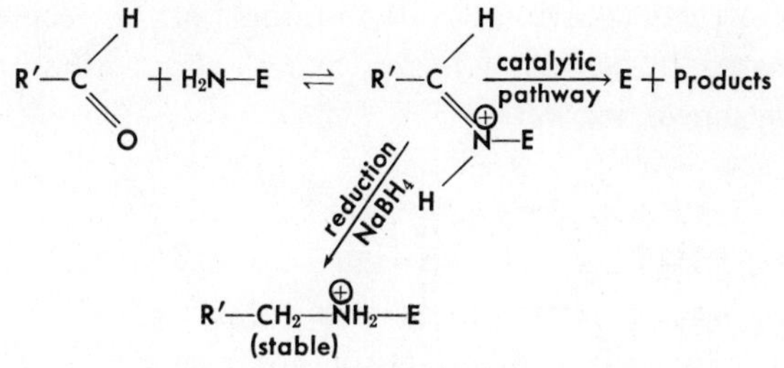

(5–4)

reactive and, in the presence of suitable catalytic species on the enzyme, may be transformed to a variety of products. If an alternative rapid reaction of the intermediate is available, the intermediate may be converted to a stable compound incapable of further enzyme-catalyzed reaction. Schiff's bases are readily stabilized by reaction with strong reducing agents such as sodium borohydride to form the very stable amine derivative (Eq. 5–4).

The particular modified side chain and the sequence of amino acids in its vicinity once again can be established by the usual procedures. This latter method has been used extensively to demonstrate the importance of particular lysine residues in enzyme catalysis.

SPECIFIC LABELING OF ACTIVE SITES BY STEREOSPECIFIC REAGENTS

In contrast to the method mentioned above, which makes use of analogous *chemical* structures in the substrate and the substrate-analogue reagent, this method utilizes the affinity of the enzyme for a particular configuration of the substrate as a means of labeling specific residues which are in proximity to the bound substrate. The general method involves the synthesis of a molecule stereospecific for the active site (as deduced from specificity studies with substrates and (or) competitive inhibitors) and containing, in addition, a chemically reactive center capable of forming a stable covalent derivative with one or more amino acid residues. The high affinity of such a reagent for a specific site greatly increases the probability of reaction with an appropriate residue near the reactive center of the reagent. A particularly noteworthy example of this technique is that devised by Schoellman and Shaw, who synthesized the molecule shown below.

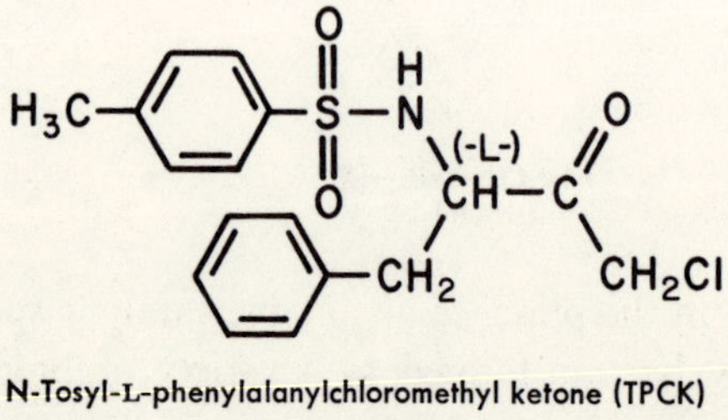

N-Tosyl-L-phenylalanylchloromethyl ketone (TPCK)

This molecule has the configuration of specific substrates of α-chymotrypsin. In addition, it contains an *alkylating center* ($-CH_2Cl$). Once this compound (TPCK) is bound to the active site, alkylation can occur with any group capable of donating an electron pair (i.e., displacing chloride). Such an alkylation reaction has been shown to occur with α-chymotrypsin, and moreover to result in complete loss of catalytic activity. The reaction is stoichiometric (one mole of TPCK per mole of α-chymotrypsin). The site of alkylation has been demonstrated to be at only one of the two imidazole nitrogens, in only one of the two histidine residues of the enzyme. The use of this reagent was particularly fortunate, in this instance, since catalytic activity has been demonstrated to depend upon the particular side chain residue (histidine) which is modified. Were the modified residue not essential to catalysis, the method would still have indicated that the modified residue was close to the active site. This specific reaction is discussed in further detail in regard to the mechanism of action of α-chymotrypsin (Section 8–1). It is of interest to note here that the histidine residue which is modified by the alkylation reaction is separated by approximately 140 amino acid residues from a specifically reactive serine residue (established by the substrate analogue method described above). Other enzymes have been modified using this principle of a stereospecific reagent, in particular by modification with stereospecific alkylating reagents.

The above discussion has presented a general outline of methods currently in use for revealing catalytic residues in enzymes. Some specific results are briefly summarized in Table 5–2. Since a great variety of protein conformations can be obtained from the different amino acid sequences of different proteins, it is perhaps unnecessary to state that other techniques for chemical modification have been developed, based on the individualities of different enzymes. It is not within the scope of this book to treat such methods exhaustively. A number of examples are given in Chapter 8, in which mechanisms of selected enzyme-catalyzed reactions are discussed.

Table 5-2
Specific Chemical Labeling of Catalytic Residues

Enzyme	Labeling agent	Chemical reaction involved	Amino acid residue labeled
Mammalian pancreatic proteases (chymotrypsin, trypsin, thrombin, elastase) Esterases (cholinesterase liver esterases) Bacterial proteases (subtilisin)	$\underbrace{[(CH_3)_2CHO]_2}_{R}-P(=O)-F$	$R_2-P(=O)-F + EOH \longrightarrow R_2-P(=O)-OE + HF$	Serine ($-CH_2OH$)
Chymotrypsin, Subtilisin	p-NO_2 phenyl acetate Acetyl imidazole $R-CH{=}CH-C-C(=O)-$(imidazol-1-yl)	$R-C(=O)X + EOH \longrightarrow R-C(=O)OE + HX \xrightarrow{\text{denature}}$	Serine
Phosphoglucomutase	Glucose-6-PO_4	$ROP(=O)(O^{\ominus})OH + EOH \longrightarrow EOP(=O)(O^{\ominus})OH + ROH \xrightarrow{\text{denature}}$	Serine
Alkaline phosphatase	$H_2PO_4^{\ominus}$	$HOP(=O)(O^{\ominus})OH + EOH \longrightarrow EOP(=O)(O^{\ominus})OH + H_2O \xrightarrow{\text{denature}}$	Serine
Phosphorylase, Glutamate-Aspartate Transaminase	Pyridoxal phosphate	$E-NH_2 + RCHO \rightleftharpoons E-N{=}CHR \xrightarrow[NaBH_4]{} E-N^{\oplus}H_2-CH_2R$	Lysine ($-(CH_2)_4-NH_2$)

Acetoacetate decarboxylase	Acetoacetate	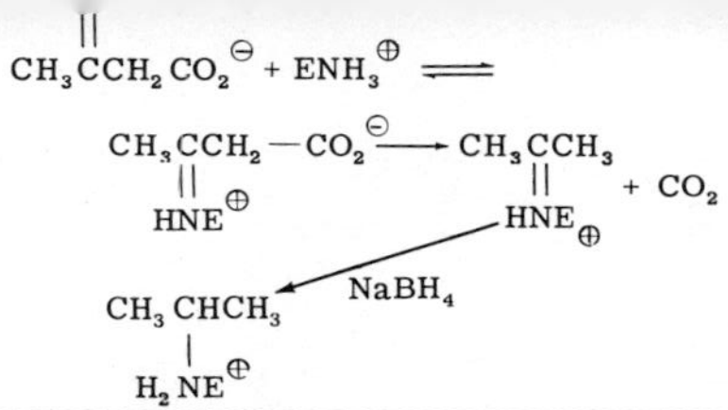	Lysine
Aldolase	Acetaldehyde	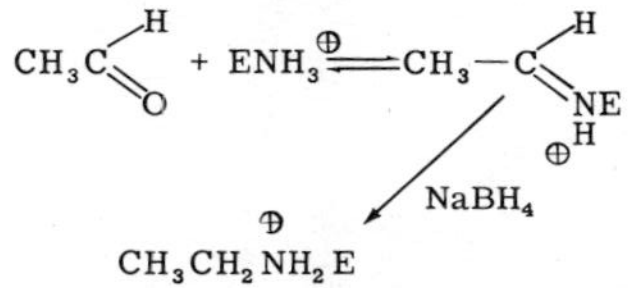	Lysine
Glyceraldehyde-3-phosphate dehydrogenase, lactic dehydrogenase, alcohol dehydrogenase,	$ICH_2CO_2^{\ominus}$	$ICH_2CO_2^{\ominus} + E\ddot{S}H \longrightarrow ES{-}CH_2CO_2^{\ominus}$	Cysteine ($-CH_2SH$)
Glyceraldehyde-3-phosphate dehydrogenase, papain	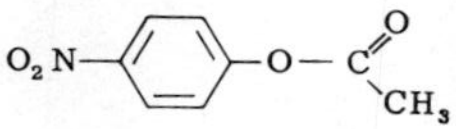	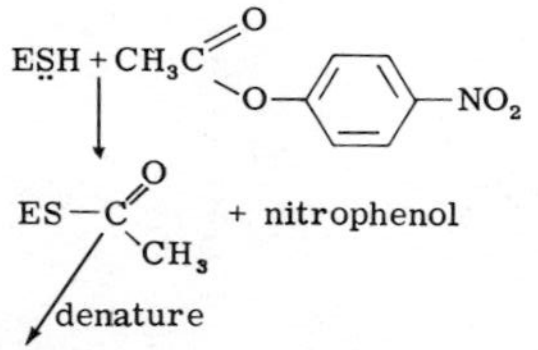	Cysteine
Chymotrypsin, trypsin	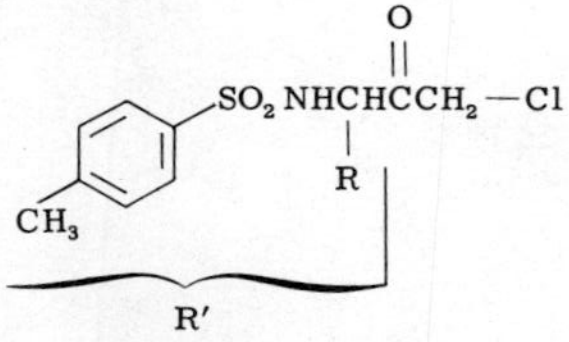	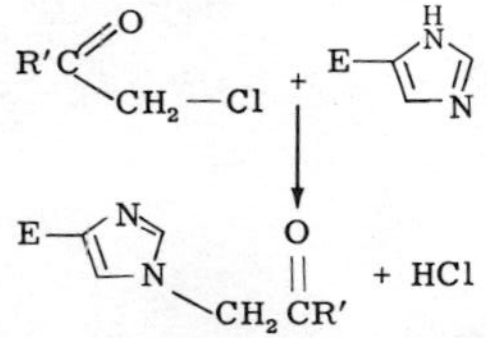	Histidine

Once the specific side chains involved in, or influencing, the catalytic activity of an enzyme have been revealed, one may begin to speculate more fruitfully on the details of the catalytic mechanisms involved.

REFERENCES

Thompson, E. V. P., "The Insulin Molecule," *Scientific American*, May 1955.

Stein, W. H., and Moore, S., "The Chemical Structure of Proteins," *Scientific American*, February 1961.

Stein, W. H. and Moore, S., "Chromatography," *Scientific American*, March 1951.

The first two of these well-illustrated informative articles deal with the methods, original and current, for the determination of the linear sequence of amino acids in proteins, and for the locations of the disulfide bridges. The article on chromatography describes the various experimental devices.

Sanger, F., "The Structure of Insulin," in *Currents in Biochemical Research*, D. E. Green (ed.), Interscience, New York, 1956. A good review article on the primary structure of insulin by the principal investigator.

Kopple, K. D., *Peptides and Amino Acids*, Chapter 4 and Appendix, W. A. Benjamin, New York, 1966. This elementary text contains a good description of the methods utilized in the determination of the primary structure of proteins.

Koshland, D., "The Active Site and Enzyme Action," *Adv. in Enzymol.* **22**, 45 (1960). An excellent review of the methods which had been used up to 1960, for the labeling and identification of active site constituents.

Goodwin, J., Harris, I., and Hartly, B., *Structure and Activity of Enzymes*, Academic Press, London, 1964. A more detailed and comprehensive account through 1963 of the methods utilized and of the results of active site labeling.

Laursen, R. A., and Westheimer, F. H., "The Active Site of Acetoacetate Decarboxylase," *J. Am. Chem. Soc.* 88, 3426 (1966). A masterful demonstration of the art of locating the catalytic center of an enzyme by first deducing the catalyzed mechanism and then trapping an enzyme-substrate intermediate.

SIX — MECHANISTIC PATHWAYS AND MODELS OF ENZYME CATALYSIS

ENZYMES CATALYZE A SUBSTANTIAL FRACTION OF THE KNOWN types of reactions of carbon, hydrogen, oxygen, nitrogen, phosphorus, and sulfur compounds. It would, in a sense, be presumptuous to attempt a discussion, in one short chapter, of the variety of potential chemical mechanisms involved in biosynthetic and catabolic processes. Certain common and striking features of enzyme catalysis are, however, demanding of a restrictive set of causes. With a recognition of these features as a guide, we may attempt a sensible, if not comprehensive, discussion of enzyme mechanisms. Four common features are as follow:

(1) *The rates (i.e., "turnover numbers") of enzyme reactions do not vary greatly despite the very great variety in the types of chemical reactions.* The rates of enzyme-catalyzed reactions can be broken down into two classes: reactions involving only "electron transfer," and reactions involving both "electron and proton (hydrogen) transfer." The latter class of reac-

tions is by far the larger of the two. Under optimal conditions, the "turnover number" of reactions of this class is approximately 10^3 molecules of substrate per molecule of enzyme per second (most such enzymes have turnover numbers within an order of magnitude of this mean value). The former (smaller) class of "electron transfer" reactions are catalyzed by enzymes having very much larger turnover numbers ($\cong 10^8$ sec^{-1}).

(2) *Chemical catalysis by enzyme proteins is mediated by a limited number of different functional groups.* The inert side chains of the amino acids glycine, alanine, phenylalanine, leucine, valine, isoleucine, and proline are not involved in chemical catalysis. Of the remaining amino acids, a number are of very similar type (aspartate and glutamate, asparagine and glutamine, threonine and serine). Hence the *chemical* "elements" for the construction of a catalytic site are highly restricted. Some amino acid side chain residues, most notably those of histidine, serine, cysteine, and lysine are known to be intimately involved in the catalytic process of a variety of enzymes. Other residues may be involved as well. For example, there is a good deal of evidence suggesting catalytic functions for the carboxyl groups of either glutamate or aspartate.

(3) *The rates of enzyme-catalyzed reactions show either pH dependent optima or maxima.* Usually the maxima or optima lie fairly close to pH neutrality (± 2 pH units).

(4) *Native enzyme molecules are very large in comparison to their substrates.* Although enzyme proteins can be partially degraded by a variety of methods, there is no documented evidence for the catalytic activity of any substantially smaller polypeptide fragment at a rate comparable, even within orders of magnitude, to that observed with the intact enzyme.

In order to facilitate discussion of mechanism, it is useful to define some specific types of catalytic chemical processes. By convention, these definitions are given with respect to the role of the catalyst, rather than to the substrate molecule which is undergoing chemical transformation.

6–1 NUCLEOPHILIC AND ELECTROPHILIC CATALYSIS

NUCLEOPHILIC CATALYSIS

An atomic center which has a strong tendency to donate an electron pair is termed a *nucleophile.* Likewise, a catalyzed reaction that proceeds via the donation of an electron pair from the catalyst to the substrate is termed *nucleophilic catalysis,* if this "nucleophilic attack" either partially, or completely governs the rate of the reaction. Some examples of nucleophilic catalysis by molecules containing functional groups analogous to specific amino acid side chains are shown in Fig. 6–1.

In all of the illustrative examples shown in Fig. 6–1, the transfer of an electron pair from the nucleophile is a "rate-influencing" factor. In some of the examples there are additional steps in the reaction sequence. Whenever these additional steps occur at rates very much faster than the nucleophilic reaction step, they will not be discernible from the kinetics of the overall reaction. When nucleophilic attack is rate-influencing, the reaction rate will depend on the electron-donating ability of the nucleophile (i.e., a good electron donor will be a good nucleophilic catalyst). One indication of electron donating ability is the "base-strength" of a nucleophile. This is so because the tendency towards formation of the conjugate acid can usually be correlated with the tendency for a nucleophile to donate an electron pair. It is not surprising, therefore, that there exists a relationship between the pK_a of the conjugate acid of a nucleophilic base and its effectiveness as a catalyst of reaction. It has been shown, particularly by the studies of Eigen and his collaborators, that in the reactions of Eq. 6–1,

$$B + H_3O^{\oplus} \underset{v_r}{\overset{v_f}{\rightleftharpoons}} BH^{\oplus} + H_2O$$

$$A^{\ominus} + H_3O^{\oplus} \underset{v_r}{\overset{v_f}{\rightleftharpoons}} HA + H_2O \qquad (6\text{–}1)$$

the forward rate (v_f) is controlled by the rate of diffusion of the participants, whereas the reverse (dissociation) reaction rate is controlled by chemical factors. For a set of homologous con-

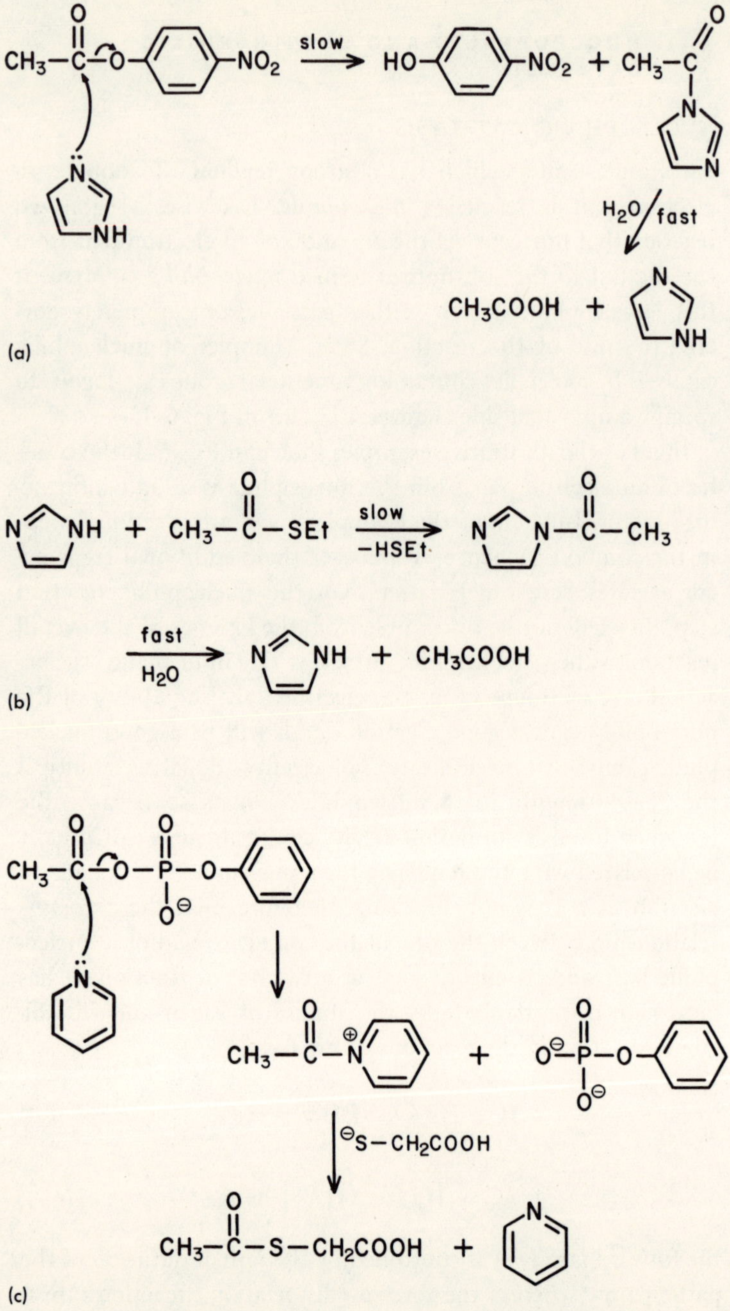

FIG. 6–1 Some examples of nucleophilic catalysis.

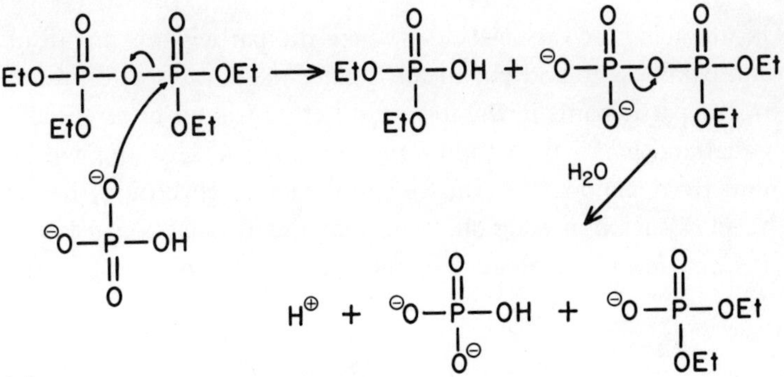

(d)

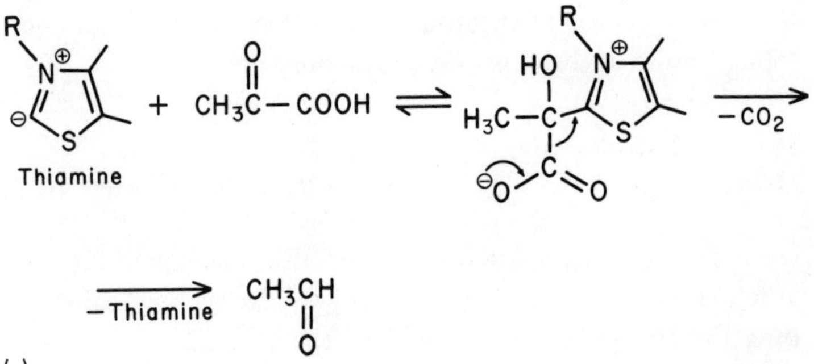

(e)

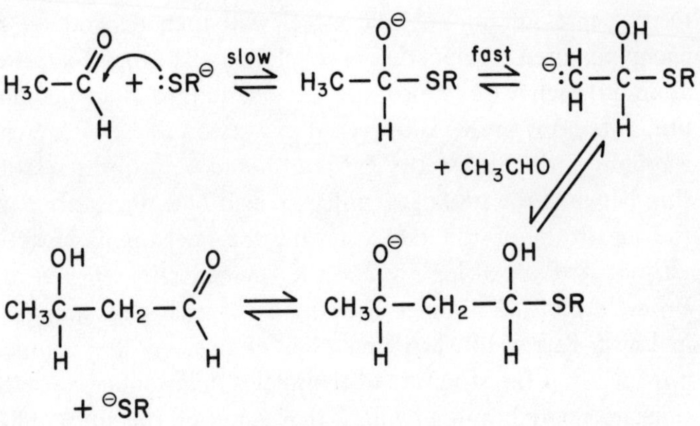

(f)

FIG. 6–1 (cont.)

jugate acid-base dissociations (where the participants are all of comparable size and chemically related), the rates of diffusion of the participants in the forward reaction will all be essentially equal and hence the equilibrium constants, $K_a = k_r/k_f$, will be directly dependent on the chemical factors governing the reverse dissociation reaction. Note that this dissociation reaction is electronically analogous to the reversal of a nucleophilic attack.

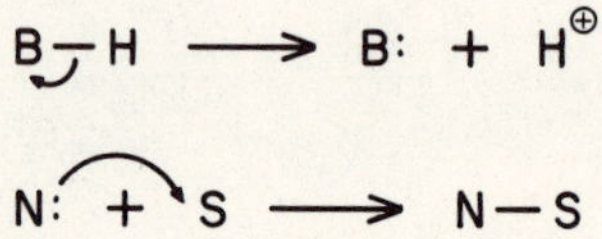

At any specified temperature, the pK_a of an acid is an energetic measure of this dissociation reaction.

$$\Delta G^0 = -2.303\, RT\, \mathrm{p}K_a = RT \ln(k_r/k_f)$$
$$= RT \ln k_r - \text{const} \quad (6\text{–}2)$$

where ΔG° is the *standard free energy* for the reaction of Eq. 6–1. When a set of nucleophiles are all of similar geometrical structure, the energy barrier imposed by the substrate towards the approach of the nucleophile (for the formation of a bond between the nucleophilic atom and the substrate atom) will not differ significantly from nucleophile to nucleophile. The relative rates of nucleophilic attack will then depend on the energy required for donation of an electron pair to the substrate atom in each case. Since the dissociation of a nucleophile-proton bond is simply the reversal of nucleophilic attack (when a proton is the "substrate"), there should be a linear relationship between the pK_a of a conjugate acid of a nucleophile and the logarithm of the rate constant for nucleophilic reaction (Eq. 6–2). Plots of log k versus pK_a are generally referred to as *Brønsted plots*. A number of examples of such plots are shown in Fig. 6–2a,b,c. Linear correlations of this type are common, provided that the structure of the nucleophiles under consideration are related, and provided that none of the nucleophiles contain functional groups which sterically hinder the nucleo-

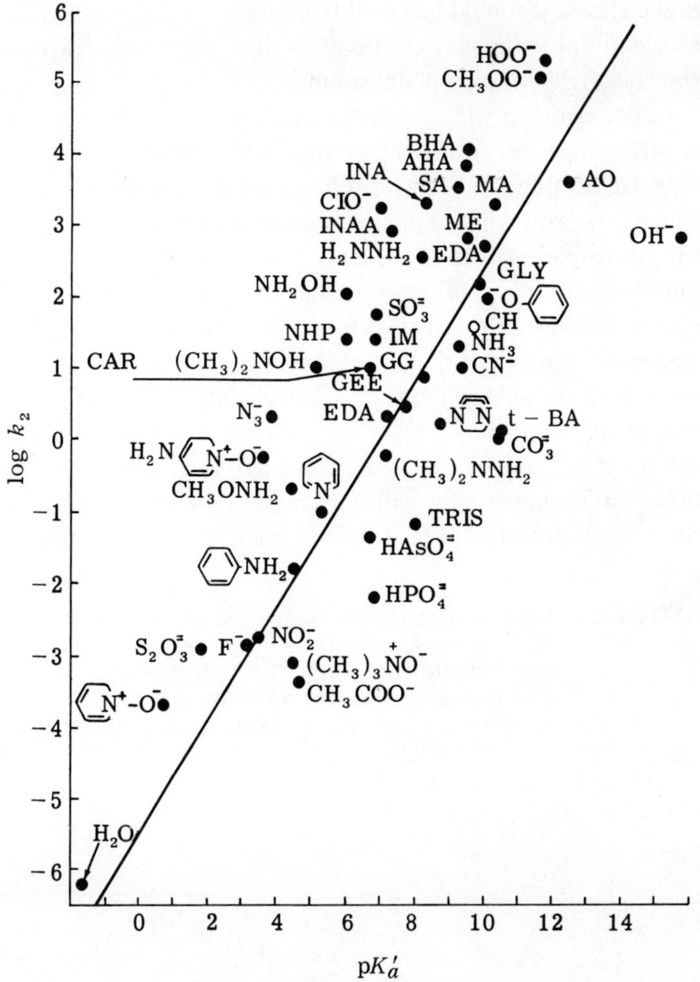

FIG. 6–2a Rates of nucleophilic reactions with *p*-nitrophenyl acetate in aqueous solution at 25° plotted against the basicity of the attacking reagent. Abbreviations: GEE, glycine ethyl ester; GLY, glycine; GG, glycylglycine; IM, imidazole; AHA, acetohydroxamic acid; BHA, *n*-butyrylhydroxamic acid; CH, chloral hydrate anion (at 30°); INA, isonitrosoacetone; SA, salicylaldoxime; MA, sodium mercaptoacetate; ME, mercaptoethanol; INAA, isonitrosoacetylacetone; EDA, ethylenediamine, NHP, N-hydroxyphthalimide; CAR, carnosine; AO, acetoxime; *t*-BA, *t*-butylamine. No statistical corrections have been made. (Figure 6–2a, b. c from T. C. Bruice and S. Benkovic, *Bioorganic Mechanisms*, Benjamin, New York, 1966.)

philic attack. It should be noted that when the nucleophiles are of significantly different electronic or stereochemical structure, the linear relationship no longer holds.

In comparing a wide variety of nucleophiles, the distinction is often made between *basicity* and *nucleophilicity*. A prominent factor influencing this distinction is the interatomic electronic repulsion which must be overcome in the formation of the nucleophile-substrate bond as compared with the nucleophile-proton bond. For example, the bond between sulfur and carbon is longer than that between oxygen and carbon. The nucleophilic attack of thiolate anion (RS^-) on carbon, as compared with the analogous attack of alcoholate anion (RO^-) on carbon, involves both smaller interelectronic repulsions and a greater separation of the positive nuclei. This distinction between sulfur and oxygen will have a less pronounced effect in the formation of bonds to hydrogen, than in the formation of

FIG. 6–2b Brønsted plot for the reaction of oxygen nucleophiles with *p*-nitrophenyl acetate (in water at 25° or 30°) (1) $CCl_3CH(OH)O^-$; (2) $C(CH_2OH)_3(CH_2O^-)$; (3) N-acetylserinamide anion; (4) $F_3CCH_2O^-$; (5) $C_6H_5O^-$; (6) $Cl_3CCH_2O^-$; (7) $F_2CH—CF_2CH_2O^-$; (8) $HC{\equiv}CCH_2O^-$; (9) $ClCH_2CH_2O^-$; (10) CH_3O^-; (11) OH^-; (12) *p*-$ClC_6H_4O^-$; (13) *p*-$CH_3OC_6H_4O^-$.

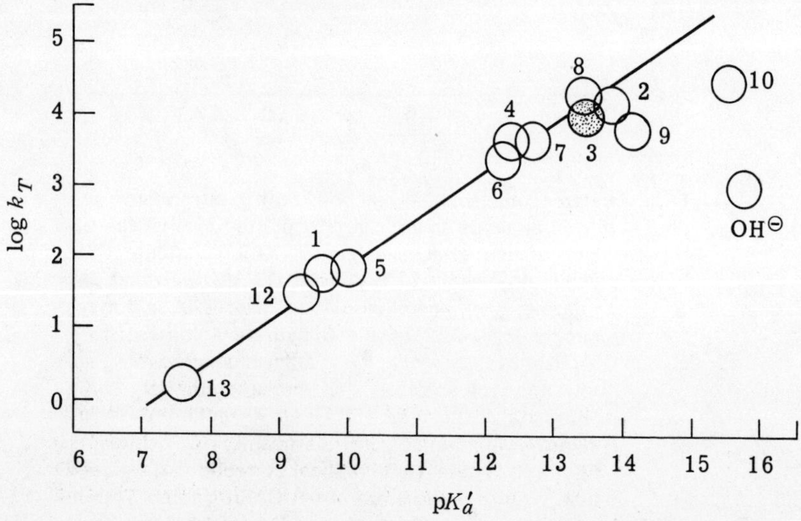

bonds to carbon (where the electronic repulsion barrier is greater). Hence, the nucleophilicity of sulfur compounds is greater than that of oxygen compounds having similar pK_a's (basicities).

In considering the effectiveness of enzyme catalysis, we must take into account not only the base strength of a potential enzymic nucleophile, but the fraction of the total such residues which are unprotonated at the pH of the aqueous environment. This fraction (α) is given by Eq. 6–3.

$$\alpha = \frac{K_a}{K_a + [H^{\oplus}]} \qquad (6\text{–}3)$$

In considering the effectiveness of two nucleophiles, for example the ϵ-amino group of lysine and the imidazole of histidine,

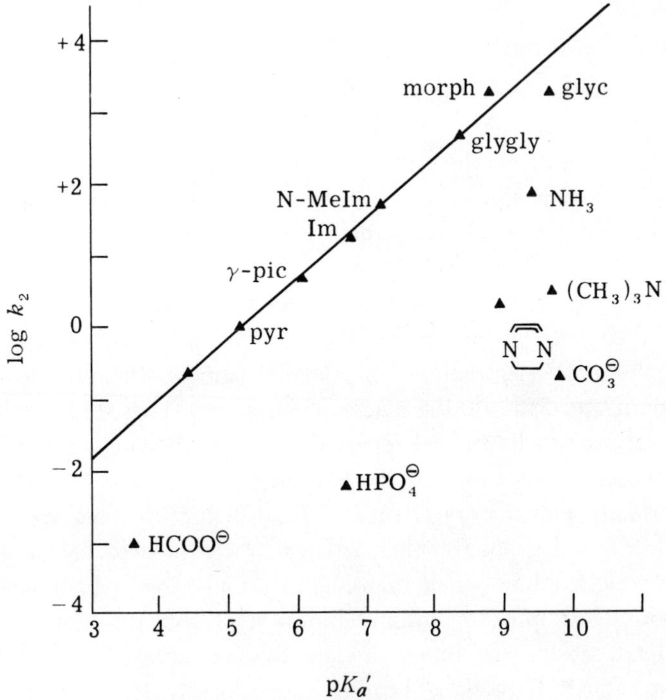

FIG. 6–2c Rates of reactions with acetyl phenyl phosphate monanion at 39° plotted against the basicity of the attacking reagent (abbreviations: pyr, pyridine; γ-pic, γ-picoline; Im, imidazole; N-MeIm, N-methylimidazole; morph, morpholene).

both the relative base strengths of the two residues (lysine is the stronger base since it has the higher pK_a) and the fraction of each of the two types of residues which will be unprotonated must be taken into account. At pH 7, we can predict from Table 5–1 that approximately half the imidazole nitrogens will be unprotonated, whereas less than 1% of the ϵ-amino groups will exist in the basic form. The relative importance of these two factors is given by the slope of the Brønsted plot illustrated in Fig. 6–2. This slope will vary, depending on the nature of the nucleophilic reaction involved. It should be noted that the slope of the Brønsted plot for model reactions related to some of the nucleophilic catalyses of biological interest are considerably greater than unity, indicating a very strong dependence of reaction rate on the basicity of the attacking nucleophile.

ELECTROPHILIC CATALYSIS

In *electrophilic catalysis* the roles of catalyst and substrate are simply the reverse of that defined by nucleophilic catalysis, that is, electrophilic catalysis involves the abstraction of an electron pair from the substrate by the electrophilic catalyst. The distinction between these two complementary forms of catalysis is perhaps trivial, except for the fact that there are at least seven different *nucleophilic* side chains among the amino acid constituents of proteins and only one rather special type of (potential) electrophilic side chain, namely, the protonated conjugate acids of the nucleophilic bases. These latter acid catalysts will be treated separately in the following section on general acid-base catalysis.

Many enzymic mechanisms, nevertheless, involve electrophilic catalysis in the chemical reaction sequence. The necessary electrophile (other than one of the above-noted conjugate acids) is supplied by nonprotein enzymic "cofactors," of which metal cations are among the notable examples. Some metal ion-catalyzed reactions, bearing an analogy to particular enzymic reactions, are illustrated in Fig. 6–3. In each of these examples, the metal ion serves as an electrophilic catalyst.

6–2 GENERAL ACID-BASE CATALYSIS

Most biological reactions involve the transfer of a proton, either from one atom to another within the substrate, or from one substrate to another, during the course of the reaction process. In homogeneous aqueous solution, two types of mecha-

FIG. 6–3 Metal ion catalysis.

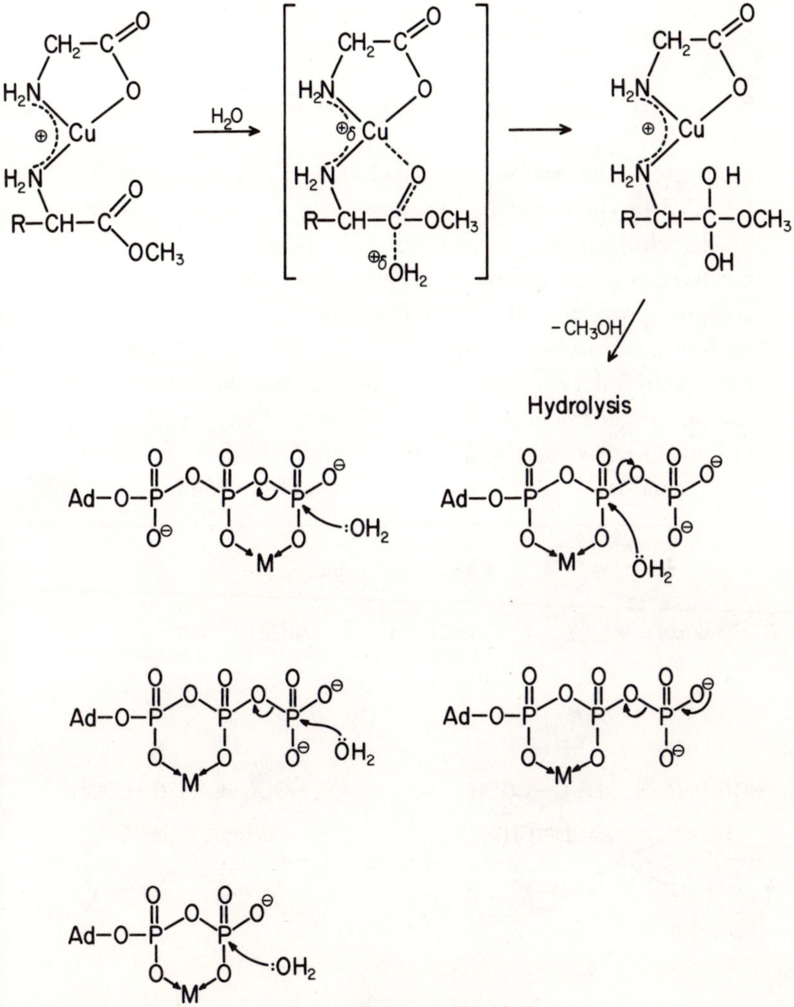

Note: Ad represents adenosyl residue, as in Fig. 2–2.

nisms may be operative as illustrated in Fig. 6–4. If a proton transfer process limits the overall reaction rate, such reactions may be catalyzed by compounds which can transfer protons more effectively. Many nucleophilic or electrophilic reactions may be aided by the transfer of a proton to or from the reactive center of the substrate, as indicated in mechanism I of Eq. 6–4.

$$N: \overset{(+)}{\frown} S - X: \overset{(-)}{\frown} H - A \longrightarrow \overset{\oplus}{N} - S - X - H + :A^{\ominus} \quad (I)$$

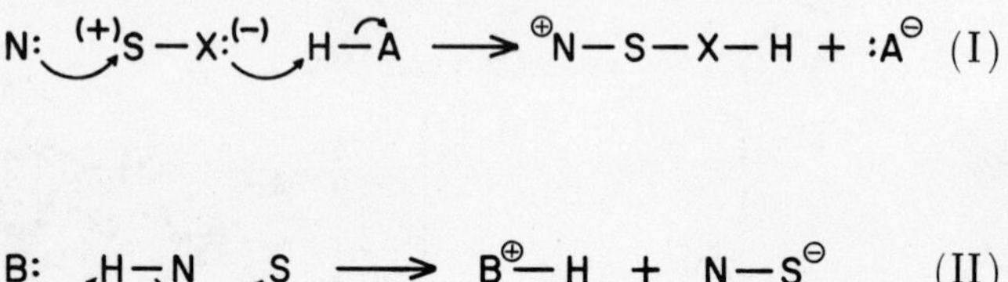

$$(6–4)$$

$$B: \frown H - N \frown S \longrightarrow B^{\oplus} - H + N - S^{\ominus} \quad (II)$$

Conversely, the nucleophilic or electrophilic attack may itself be aided by the transfer of the proton from the nucleophile or to the electrophile (mechanism II). When the rate of the catalyzed reaction is directly dependent on the proton donor or acceptor properties of the catalyst, the reaction is said to be *acid- or base-catalyzed*, respectively. When a variety of acids or bases can function catalytically in a particular reaction, the reaction is *general acid-* or *general base*-catalyzed. In the mechanism illustrated by Eq. 6–4, there may be both a general acid (or general base) and a nucleophilic (or electrophilic) reaction

---

FIG. 6–3 (cont.)

2 Pyridoxal + 2 R–CH–COOH + Al(III) $\underset{+ 2 H_2O}{\overset{- 2 H_2O}{\rightleftharpoons}}$
   NH₂

$\overset{+2 H_2O}{\underset{-2H_2O}{\rightleftharpoons}}$ 2 Pyr–NH₂ + 2 R–C–COOH + Al(III)

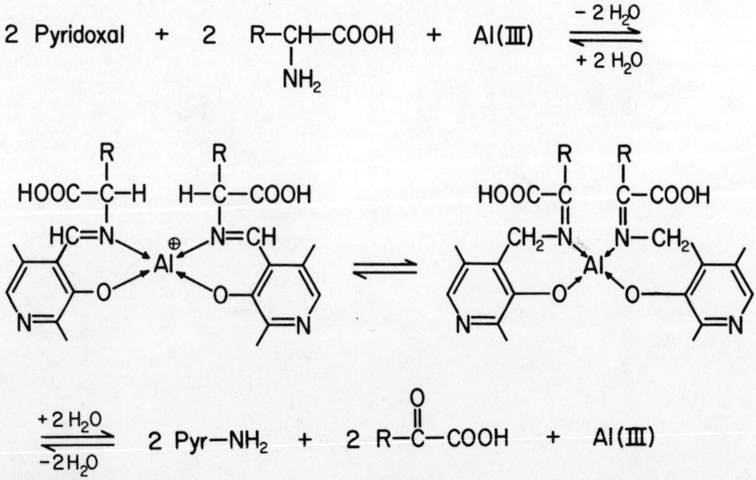

Note: See Section 7–1 for structure of pyridoxal and of pyridoxamine (Pyr-NH₂).

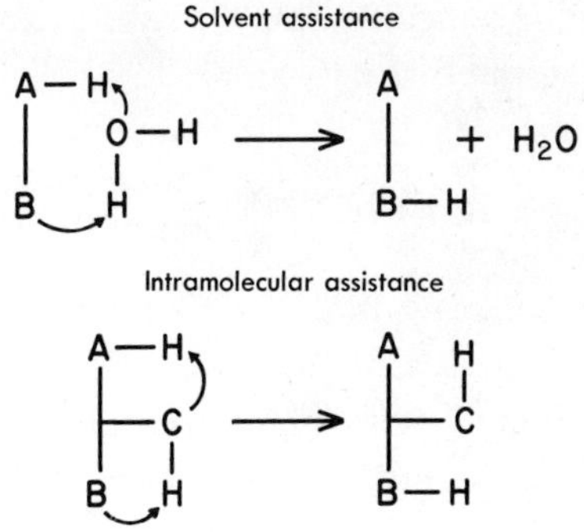

FIG. 6–4 Examples of proton transfer processes in chemical reactions.

step in the mechanism. Under a specified set of conditions, the rate of such a reaction may depend on either of the two catalytic functions. If the difference in rate between the two processes is great, the overall reaction will depend on only the *slower* process. The fundamental distinction between the two rate-controlling mechanisms is between a general *bond-making* or *bond-breaking*, and a *proton transfer process*. Both types of processes are involved in enzyme-catalyzed reactions; one may ask which process is likely to be rate-controlling. In homogeneous aqueous solution, the electrophilic-nucleophilic attack is often, but by no means always, the slower process. In some specific instances (to be discussed below) the accumulated evidence indicates that the two processes are "concerted," that is, that they occur in a single reaction step.

GENERAL ACID-BASE CATALYSIS AND pH DEPENDENCE

All of the conjugate acid constituents of enzyme proteins (as listed in Table 5–1) can function as general acid catalysts, and likewise, all of the conjugate bases can serve as general base catalysts, as is evidenced by an abundance of precedents in catalyzed homogeneous solution reactions. Consider a simple reaction which is general base-catalyzed, for example,

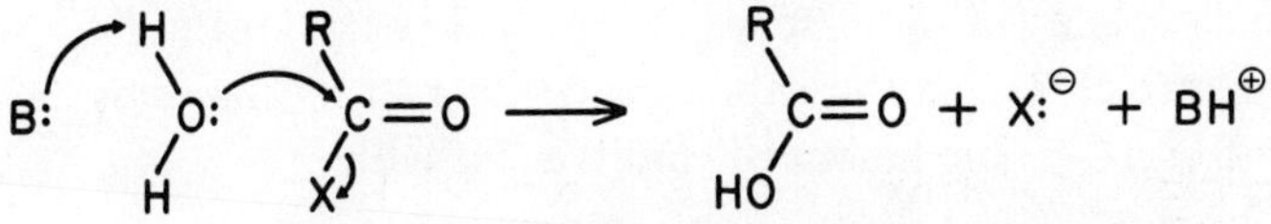

For such a reaction mechanism, the rate equation in aqueous solution will be given by Eq. 6–5:

$$v = k\,[H_2O]\,[S]\,[B:] = k'\,[S]\,[B:] \qquad (6\text{–}5)$$

At any pH, the concentration of the conjugate base is given by Eq. 6–3, namely, $[B:] = B_0\,\{K_a/(K_a + [H^+])\}$ where B_0 is the total concentration of base plus conjugate acid). Hence, the pH-dependent rate is given by Eq. 6–6, as was discussed in Chapter 5.

$$v = k'\,[S]\,B_0 \frac{K_a}{K_a + [H^{\oplus}]} = k''\,[S]\,\frac{B_0}{K_a + [H^{\oplus}]} \qquad (6\text{–}6)$$

Let us now consider a more complex mechanism, involving nucleophilic attack by the conjugate base of an extremely weak acid coupled with general acid catalysis, as for example, Eq. 6–7.

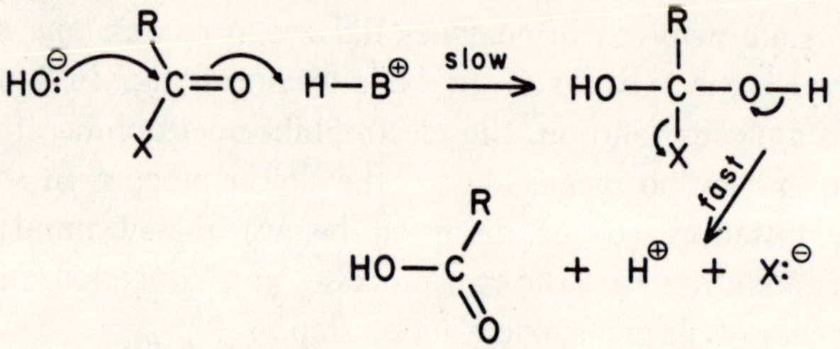

(6–7)

The rate equation in this case is given by Eq. 6–8

$$v = k\,[S]\,[OH^{\ominus}]\,[BH^{\oplus}] = \frac{kK_W\,[S]\,[BH^{\oplus}]}{[H^{\oplus}]} = \frac{kK_W\,[S\,]B_0}{(K_a + [H^{\oplus}])} \qquad (6\text{–}8)$$

where $K_W = [H^+]\,[OH^-]$. If the pK_a of the conjugate acid of the nucleophile (in this case water, $pK_a = 15.8$) is greater than any realizable pH for the experiment, the two mechanisms (Eqs. 6–6 and 6–8) are kinetically indistinguishable.

Likewise, a simple mechanism of general acid catalysis leads to the pH-dependent rate equation given by Eq. 6–9; once

$$v = \frac{k\,[S]B_0\,[H^{\oplus}]}{(K_a + [H^{\oplus}])} \qquad (6\text{–}9)$$

again a mechanism involving *electrophilic* attack coupled with *general base* catalysis is kinetically indistinguishable from the simple general acid catalysis model. Hence any mechanism in which general acid or base catalysis is the sole rate determining factor will lead to pH dependencies as illustrated in Fig. 6–5. If, on the other hand, both the electron transfer and the proton transfer reactions described above are of comparable rate, and if both the electrophile or nucleophile, and the general base or general acid have pK_a's in the experimentally measurable pH range, the dependence of rate on pH may be more complex.

FIG. 6–5 The k_{rate}-pH profiles for participation of single acid and base species. (From T. C. Bruice and S. Benkovic, *Bio-organic Mechanisms*, Benjamin, New York, 1966.)

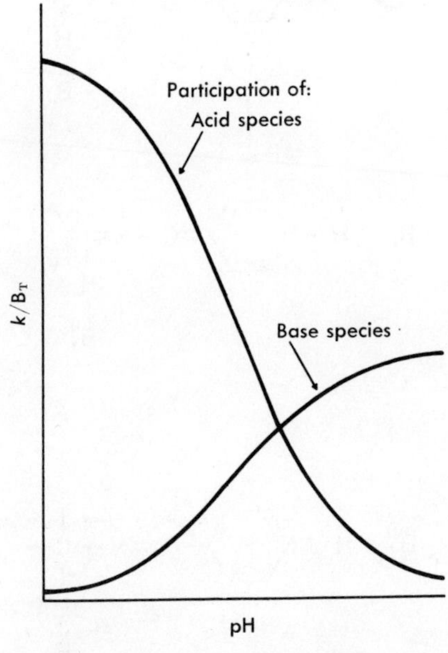

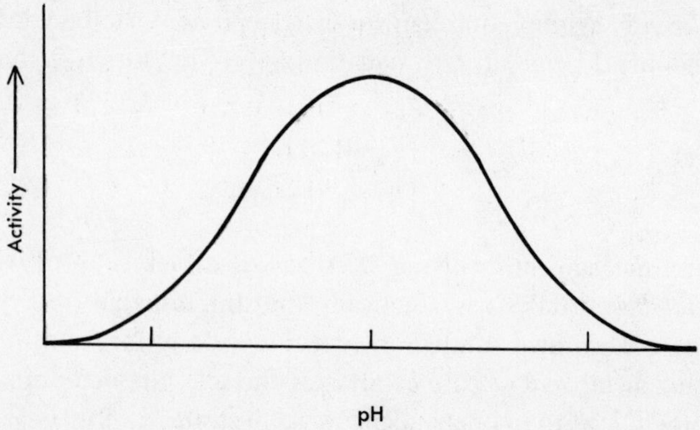

FIG. 6–6 "Bell"-shaped dependence of enzyme activity on pH.

The pH rate profiles, when such conjugate acid-base catalyzed processes are coupled, are illustrated schematically in Fig. 6–6. Note that "bell"-shaped curves can arise from these coupled processes.

General acid *and* base catalysis may occur as a concerted process, for example in a tautomerization reaction (Eq. 6–10).

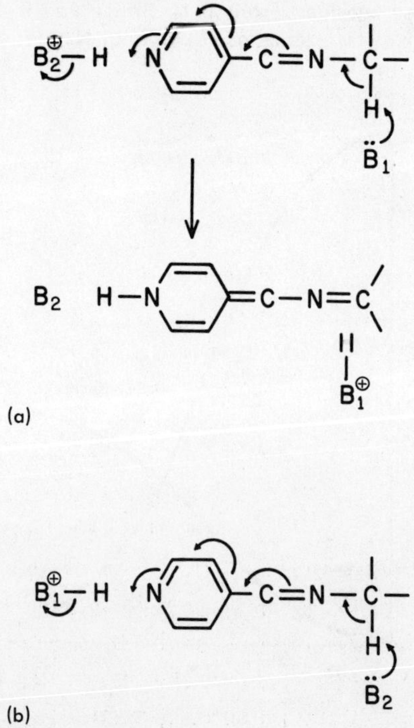

(6–10)

Whenever such "conjugate" acid-base catalysis occurs, a bell-shaped pH-rate curve will be a consequence, the shape of the "bell" being dependent on the difference in pK_a between the general acid (B_1H^+) and the conjugate acid of the general base (B_2). Again it should be noted that interchanging the roles of general acid and general base, as for example in Eq. 6–10(b), results in a rate equation (Eq. 6–11) which is indistinguishable from the former model (Eq. 6–10).

$$v = \frac{k(B_1)_0(B_2)_0K_2\,[H^{\oplus}]}{(K_1 + [H^{\oplus}])(K_2 + [H^{\oplus}])} \qquad (6\text{–}11)$$

6–3 RATES OF PROTON TRANSFER

Recently, by the introduction of ultrarapid reaction rate measurement techniques (Section 4–7), the rate of proton transfer between an acid and a conjugate base in aqueous solution has been measured for a variety of acid-base conjugates.

$$HA + B\colon \rightleftharpoons BH^{\oplus} + A^{\ominus}$$

A brief numerical summary of the rate constants are given in Table 6–1. These quantitative results are of significance to the mechanism of enzyme catalysis. In order for a general acid or general base catalyst to return to its initial state (i.e., to its active form for initiating another round of catalysis) both the forward and the reverse proton transfer reactions must occur; a general base catalyst which becomes protonated during the course of reaction must transfer its proton, either back to substrate, or to a proton acceptor such as the aqueous solvent, in order to regenerate the original catalyst. As can be verified from the table, a general base which can accept a proton at a relatively rapid rate near pH neutrality, will give rise to a conjugate acid with a comparably slow rate of proton transfer, the two rates being related to the base and acid strengths. An optimal situation for enzyme catalysis in aqueous solution will occur when the base strength of the base is equal to the acid strength of its conjugate acid, that is, when the aqueous pK_a is near neutrality. Of the potential general acid-base catalysts in

Table 6-1
Rates of Proton Transfer at Neutrality (25°C, in H_2O at pH 7.0)[a,b]

$$B: + AH \underset{k_r}{\overset{k_f}{\rightleftharpoons}} B{-}H + A:$$

B	AH	v_f, sec^{-1}	v_r, sec^{-1}
$OH^{\ominus}$	$CH_3NH_3^{\oplus}$	4×10^3	2×10^4
$OH^{\ominus}$		2.5×10^3	1.2×10^3
$OH^{\ominus}$	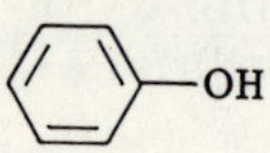	1.4×10^3	2×10^3
$OH^{\ominus}$	$CH_3COCH(H)COCH_3$	4×10^{-3}	$\sim 10^{-2}$
$CH_3CO\overset{\ominus}{C}HCOCH_3$	$H_3O^{\oplus}$	1.2	1.4×10^{-2}
$CH_3CO_2^{\ominus}$	$H_3O^{\oplus}$	4.5×10^3	4×10^3
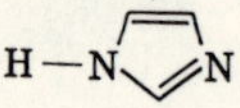	$H_3O^{\oplus}$	1.5×10^3	0.8×10^3
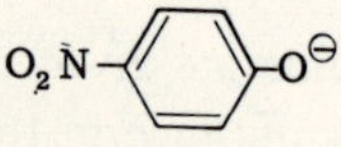	$H_3O^{\oplus}$	3.6×10^3	1.3×10^3
$OH^{\ominus}$	$H_3O^{\oplus}$	1.4×10^4	2.5×10^{-5}
$CH_3CO_2^{\ominus}$	$CH_3COCH(H)COCH_3$	2.5^c	2×10^{2c}
Imidazole	$CH_3COCH(H)COCH_3$	4.5^c	5×10^{2c}
Phenol	$CH_3COCH(H)COCH_3$	$\sim 10^c$	6×10^{2c}

[a]Note that a reaction sequence,

$$H_2O \underset{}{\overset{\text{slow}}{\rightleftharpoons}} OH^{\ominus} + S \xrightarrow{\text{fast}} P$$

must be catalyzed by imidazole or phenol (general base catalysis). Note also that condensation reactions at the central carbon of acetyl acetone which proceed via carbanion formation, namely, proteins, the imidazole group of histidine most nearly fulfills this condition. The *maximal* rate of an imidazole (or imidazolium) general base (or acid) catalyzed reaction in aqueous solution under the condition, $[S] \gg B_0$, will be of the order of the slower proton transfer rate constant.

$$(BH^{\oplus} + S \rightleftharpoons SH^{\oplus} + B)$$

On the basis of evidence thus far accumulated, this *optimal* rate of catalyzed proton transfer, near pH neutrality, is of the order of 10^3 sec^{-1} (within approximately one order of magnitude). This absolute value is strikingly similar to the turnover numbers for enzyme-catalyzed specific substrate reactions involving proton transfer. These facts argue strongly in favor of the contention that the *rate-limiting process* in enzyme-catalyzed reactions involving proton transfer is the actual proton transfer process. Since the overall reactions catalyzed by such enzymes clearly involve other processes, such as nucleophilic attack, electrophilic attack, or one electron transfer (to be discussed below), these other steps in the catalytic reaction sequence must occur at rates which are either comparable to, or faster than, the rate of proton transfer. It would thus appear that a distinction between enzyme catalysis and homogeneous aqueous solution catalysis is the more rapid rate of the bond-forming and bond-breaking processes relative to the proton transfer rate in the former case.

It should be noted that the rate constants of catalyzed reactions in homogeneous solution are virtually always far slower than the corresponding enzyme-catalyzed processes. When nucleophilic or electrophilic attack is *intermolecular*, both *entro-*

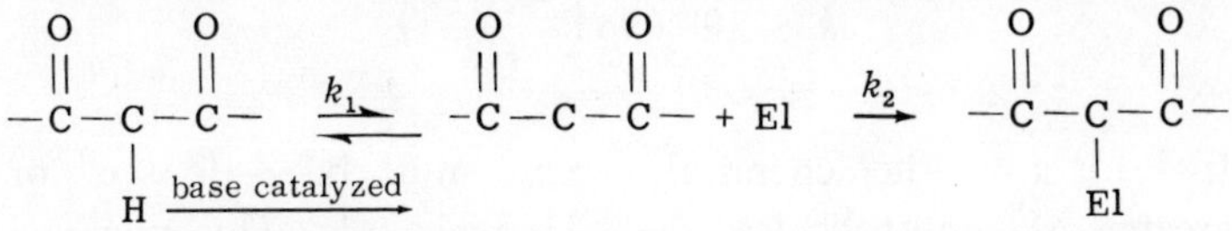

will be general base catalyzed if $k_2 \gtrsim k_1$.

[b]The rate constants are expressed as the fraction of an equivalent of the acid (A—H or B—H) which is transferred to the base per second.

[c]At total concentrations of "B" ([B] + [BH]) equal to 1 M.

pic factors (due to the translational and rotational degrees of freedom of the reactants), and interactions of the attacking reagent with the solvent, greatly hinder the approach of the attacking reagent to the reactive center. In enzyme catalysis, these inhibitory factors may be either greatly reduced or completely eliminated. If the energy barrier towards formation of the chemical bond between catalyst and substrate is not too great, the possibility that a proton transfer process becomes rate limiting is greatly enhanced.

According to the theory of absolute reaction rates (an approximate but useful theory), the absolute magnitude of the specific rate constant is given by Eq. 6–12,

$$k = \frac{RT}{Nh} \exp\left(\frac{-\Delta H^\ddagger}{RT}\right) \exp\left(\frac{\Delta S^\ddagger}{R}\right)$$

$$\sim 10^{13}(\text{sec}^{-1}) \exp\left(\frac{-\Delta H^\ddagger}{RT}\right) \exp\left(\frac{\Delta S^\ddagger}{R}\right) \quad (6\text{–}12)$$

where R is the gas constant, N is Avogadro's number, and h is Planck's constant, and $\Delta H^\ddagger$ and $\Delta S^\ddagger$ are the enthalpy and entropy differences between the initial state of the reactants and the state of highest energy in the reaction pathway.

We may inquire as to how high this energy barrier ($\Delta H^\ddagger$) can be, according to Eq. 6–12, and still allow proton transfer to be rate controlling. Since in enzymic catalysis, the proton transfer rate is of the order of 10^3 sec^{-1}, it follows that the energy barrier to nucleophilic or electrophilic attack must be such that the rate constant exceeds this value. If entropic factors are eliminated in enzyme catalysis ($\Delta S^\ddagger = 0$), Eq. 6–12 reduces to

$$k \sim 10^{13} \exp\left(\frac{-\Delta H^\ddagger}{RT}\right)$$

If k for any other chemical process must be $\sim 10^5$ sec^{-1} or greater, $\Delta H^\ddagger$ must be less than 11 kcal/mole. This *maximal* value for the energy barrier towards the transfer of an electron pair is within the range of many observed energies of activation in catalyzed homogeneous solution reactions. In nonenzymic homogeneous solution reactions, $\Delta S^\ddagger$ is invariably negative.

6–4 INTRAMOLECULAR CATALYSIS

On the basis of the previous discussion, it is clear that whatever mechanisms are rate controlling, the overall enzyme-catalytic rate constant is of the order of magnitude of 10^3 sec^{-1} (or greater). Let us consider how such rapid reaction velocities may be accomplished among the variety of chemical reactions catalyzed by enzymes. The corresponding homogeneous solution reaction rates invariably are appreciably slower. An obvious factor generally differentiating enzyme catalysis from catalysis in homogeneous solution is the rigid localization of catalytic and reactive centers, brought about by formation of the enzyme-substrate complex. One may consider this enzyme-substrate complex (kinetically) as a single molecule containing both the reactive center and the catalytic group in proximity. With this "unimolecular" model in mind, we can prepare synthetic molecules containing both the reactive center and the supposed catalytic groups, and examine these synthetic "enzyme models" for the effect of the constrained catalytic environment on the rate of reaction. Likewise, if enzyme catalysis is suspected to be the result of more than one functionally active catalytic group, we may examine the effect of a "constrained environment" on reaction rate by preparing bi- or polyfunctional catalysts. A noteworthy example of enhanced bifunctional catalysis has been found in the *mutarotation* of a glucose derivative (Fig. 6–7) in organic solvents. Such reactions are common in biochemical pathways.

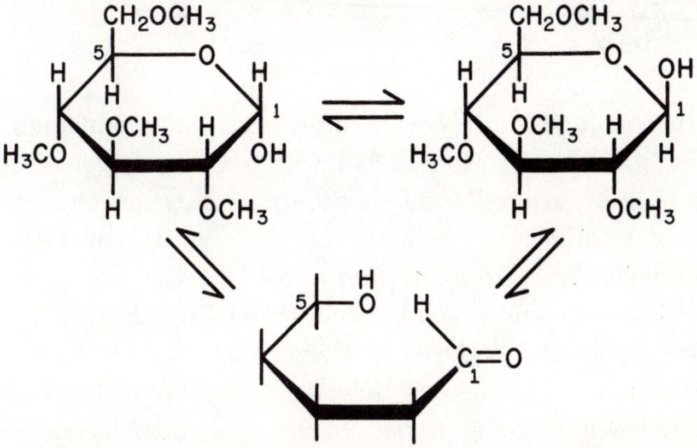

FIG. 6–7 The mutarotation of tetramethylglucose via the open-chain aldehyde.

The mutarotation of glucose in water is known to be catalyzed by both general acids and general bases. In organic solvents, however, the mutarotation of the glucose derivatives was demonstrated to be poorly catalyzed by either a base (pyridine) or an acid (cresol or phenol). When a mixture of acid and base is present, the mutarotation proceeds rapidly. In benzene solution, this reaction rate is first order in *both* the acid and the base catalyst. This result requires a "concerted" mechanism involving acid *and* base catalysis. With the bifunctional catalyst, 2-hydroxypyridine, a more rapid rate of mutarotation was observed (Eq. 6–13). Moreover, with this bifunctional catalyst, the reaction is first order in catalyst concentration. Thus a single molecule can supply both the general acid *and* the general base catalytic function. This result is all the more striking due to the fact that the hydroxyl substituent lowers the base strength of the pyridine nitrogen, and the hetero (nitrogen) atom considerably decreases the acidity of the enol. The two isomers, 3- and 4-hydroxypyridine, do not mediate rapid catalysis presumably due to the geometrical constraints to the approach of both the nitrogen base and the oxygen acid to the reactive center of the substrate.

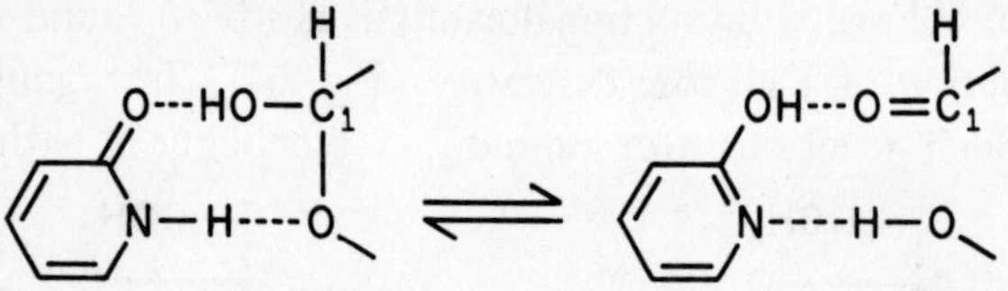

(6–13)

Intramolecular catalysis in "model enzyme" compounds has been studied in greatest detail with regard to the *solvolysis* reactions of carboxylic and phosphoric acid derivatives. A number of reactions in which the rates are greatly enhanced by intramolecular catalysis are shown in Fig. 6–8a–e. Note that in all these reactions, a combination of nucleophilic-electrophilic and (or) general acid-base catalysis is proposed in the mechanistic pathway. In some of these reactions, *intermolecular* reaction analogues can be studied. In these instances, a comparison can be made of the relative catalytic efficiencies of the intermo-

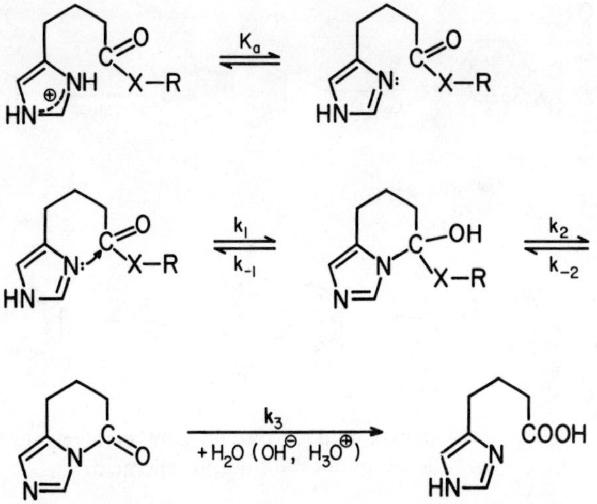

FIG. 6–8a Imidazole catalysis in γ(4′-imidazolyl)butyric acid esters.

lecular and intramolecular pathways. In other cases, intermolecular catalysis is either too slow to measure, or the reactions proceed via an alternate mechanism. Some reactions for which an intermolecular-intramolecular comparison can be made are listed in Table 6–2. The two following points should be noted concerning the rates at which such reactions occur.

FIG. 6–8b Hydrolysis of allohydroxylysylglycinamide in strong acid.

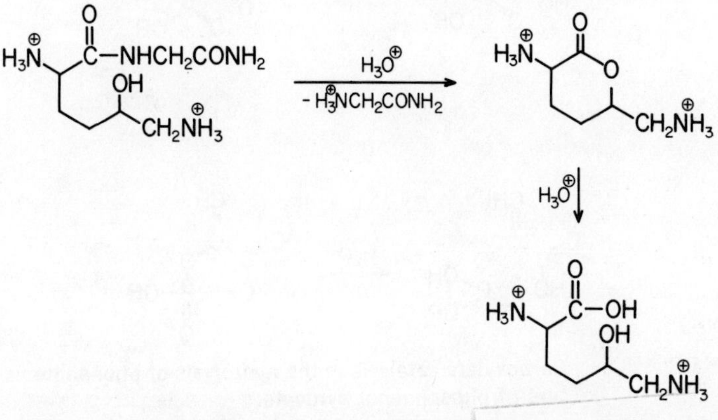

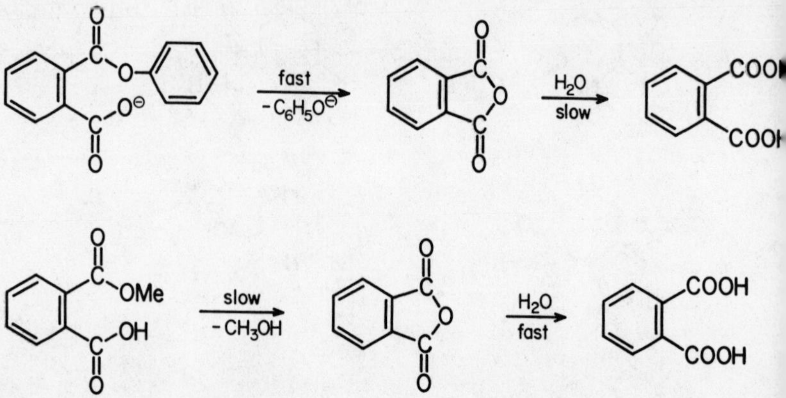

FIG. 6–8c Carboxylate, and carboxylic acid catalysis in the hydrolysis of monoesters of phthalic acid.

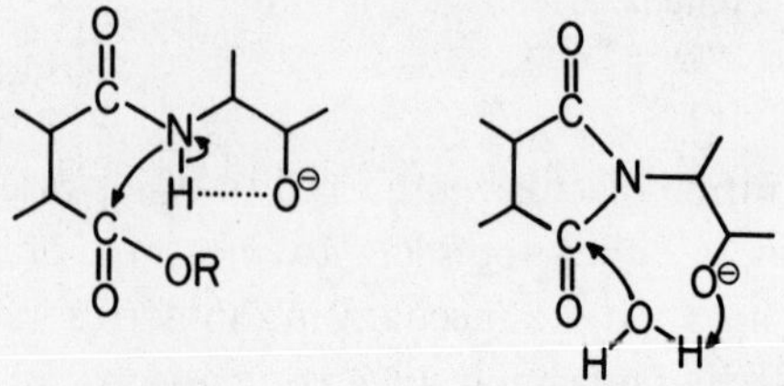

FIG. 6–8d Hydroxyl group assistance in imide formation and imide hydrolysis.

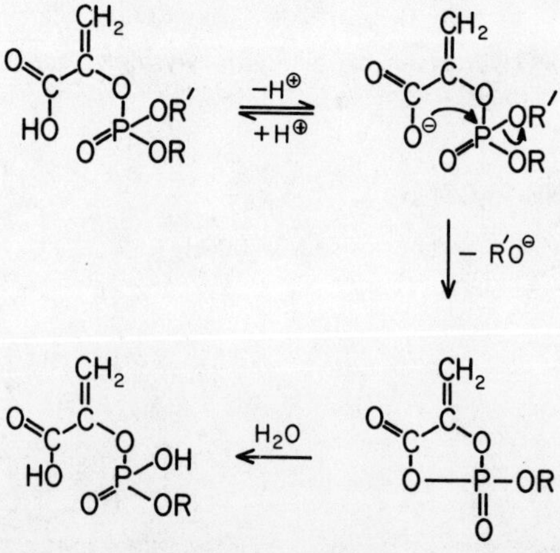

FIG. 6–8e Carboxylate catalysis in the hydrolysis of phosphate esters of phosphoenol pyruvate.

(1) The *intramolecular* rates, although markedly enhanced over the intermolecular rates, are still significantly slower than the usual rates of enzyme catalysis.

(2) The specific reactions listed in Table 6–2 are all reactions involving "activated" functional groups, that is, reactions which in homogeneous solution proceed with relative facility to completion. Those reactions which proceed much more slowly in homogeneous solution via intermolecular monofunctional catalysts, undergo apparently unique intramolecular catalysis. These apparently unique mechanistic pathways emphasize a special feature of catalytic reaction mechanisms in a constrained environment such as the enzyme-substrate complex.

Consider the reaction pathway alternatives: (a) via a single chemical intermediate with a high activation energy and (b) via a number of chemical intermediates in which the activation energy barrier for every chemical transformation is much lower than the single energy barrier in (a) (Fig. 6–9). At each step in the complex mechanism (b), the reaction will (in large part) be driven back towards reactants by the unfavorable entropic situation resulting from the constraints on the translational and rotational mobility of substrate and catalyst in the rigid complex. Although the internal (electronic) energy barriers may be lower via a complex mechanism such as (b), the highly unfavorable entropy of this path will tend to outweigh the

FIG. 6–9 Energy-coordinate diagrams (a) for a single step reaction and (b) for a multistep reaction.

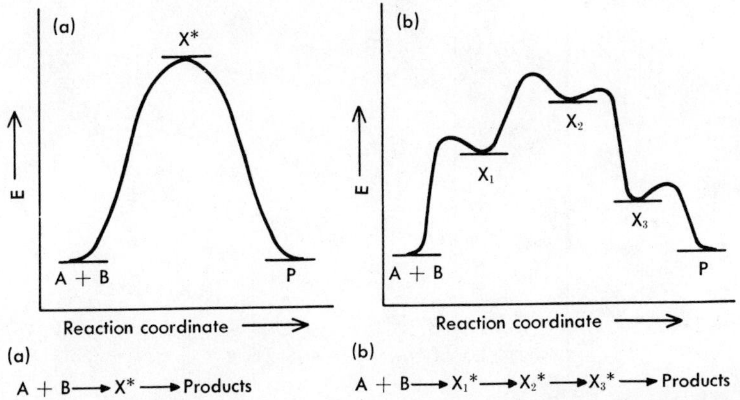

Table 6-2
A Comparison of Some Inter- and Intramolecular Catalysis of Acyl Transfer Reactions in Which the Mechanistic Pathways are Common

Reaction	Intramolecular model	$\frac{k_{intra}\text{ (first order)}}{k_{inter}\text{ (second order)}}$ $(M)^a$
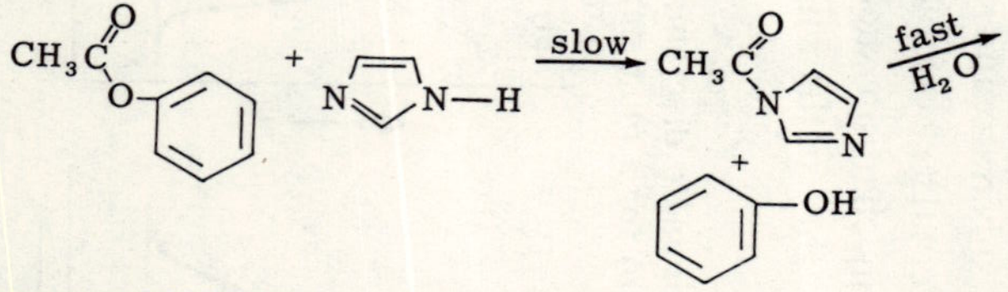	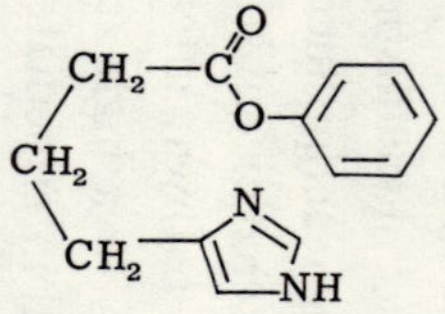	30
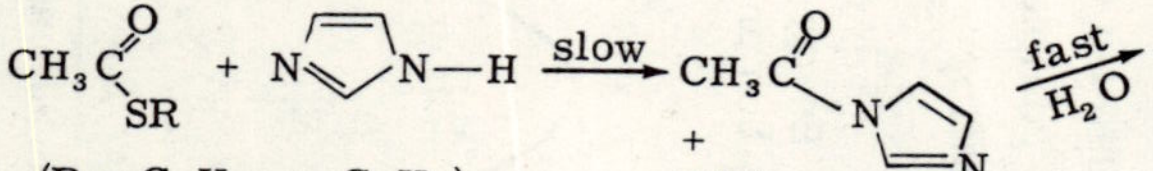 (R = C_2H_5, n-C_4H_9)	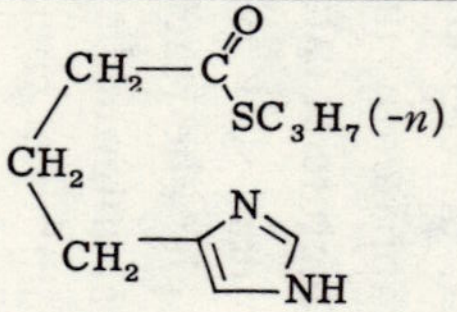	100

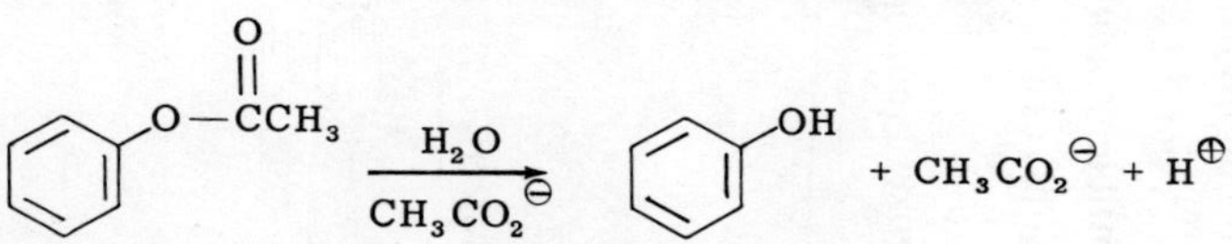	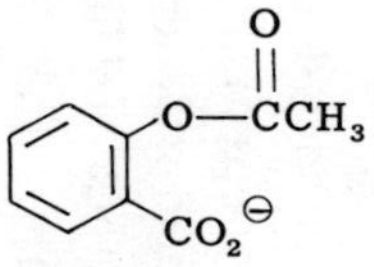	8
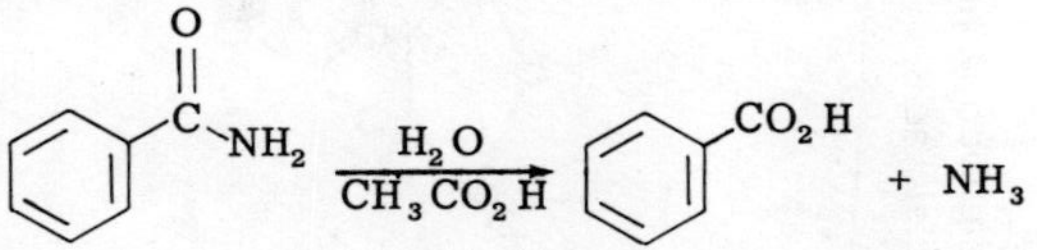	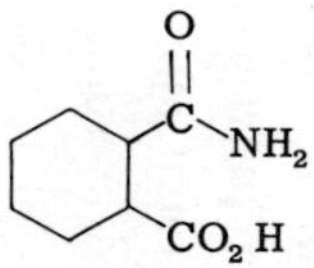	70
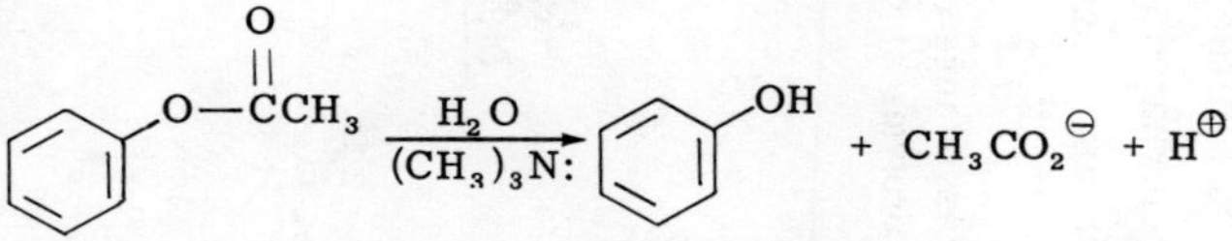	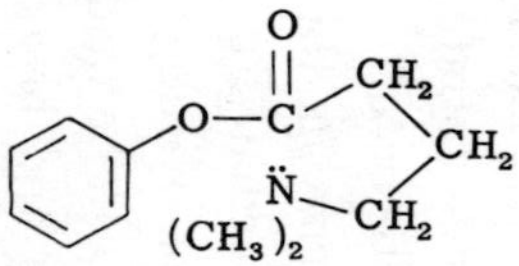	1000

[a]This ratio gives the molar concentration of intermolecular catalyst required for achieving the intramolecular reaction velocity.

advantages of the lower internal energy barriers and decrease the likelihood of the complex mechanistic pathway. In the constrained but thermodynamically stable enzyme-substrate complex, on the other hand, this multiple "entropic disadvantage" will not arise. Chemical reaction via a number of intermediate steps, each of relatively low activation energy, may become the favored pathway in intramolecular and enzymic catalysis. As we shall see when we examine the details of specific enzyme-substrate reactions, such multiple-step mechanisms frequently occur.

In our discussion thus far, we have considered catalytic mechanisms mediated only by functional groups analogous to the side chains of the constituent amino acids. General acid-base catalysis can be mediated by some of these residues. The "bond-making and breaking" and electron transfer processes which *must* occur among the variety of chemical transformations in biochemical pathways cannot *all* be mediated by these constituent functional groups alone. Very frequently, these reactions are mediated by the assistance of neighboring functional groups within the substrate molecule. A good example is the substrate-mediated hydrolysis of ribonucleic acids (Fig. 6–10), where an adjacent hydroxyl group assists in the

FIG. 6–10 Assistance of a neighboring hydroxyl group in the hydrolysis of ribonucleic acid.

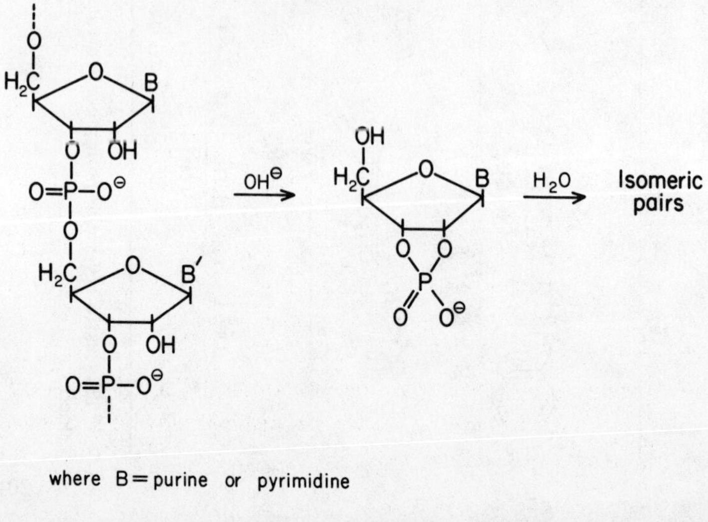

enzyme-catalyzed (and in the base- and acid-catalyzed) hydrolysis reaction. It is of doubtless significance that both *in vivo* and *in vitro* the rates of synthesis and degradation of ribonucleic acids are more rapid than the synthesis and degradation rates of deoxyribonucleic acids (in which the neighboring 2′ hydroxyl functional group is absent).

Many, perhaps most, of the chemical processes involved in metabolism require the intervention of electron donors or acceptors, other than the conventional amino acid side chains, for rapid catalysis. These catalytic species may, a priori, be imagined to be of two types: (1) New chemical compounds formed by chemical reaction between or among constituent side chains of the enzyme protein and (2) Additional molecules (not amino acid residues), firmly bound and constrained to the enzyme site. The former type of catalytic species has been frequently postulated, but never conclusively demonstrated to exist in any enzyme protein. The latter co-catalytic species are characterized by the *coenzymes, cofactors,* and *prosthetic groups* which mediate enzyme catalysis of an enormous number of distinct biochemical reactions. Catalytic mechanisms involving these cofactors are described in the following chapter.

In this chapter, as well as in Chapter 7, comparisons have been made between enzyme-catalyzed reactions and similar reactions catalyzed by small molecules. It has been shown that the *entropic* constraints which are inherent in polyfunctional intramolecular catalysts can lead to both a large enhancement of reaction rate in favorable instances, and to mechanisms which are not operable in intermolecular catalysis. It has also been shown that these intramolecular constraints make it possible for an overall reaction to proceed via a set of distinct chemical intermediates, in which the free energy barrier for transformation from one intermediate to another is always considerably smaller than the total energy barrier for direct chemical conversion in an intermolecular process. In this way it is possible to rationalize qualitatively the rapid rates of enzyme-catalyzed reactions. When specific rate constants are compared quantitatively, the model reactions catalyzed by small polyfunctional catalysts are rarely, if ever, as fast as enzyme-catalyzed specific-substrate reactions.

One difference between enzymic reactions and other catalyzed reactions, which has not been discussed in this chapter, is the potential for the enzyme protein to undergo *conformational changes.* Conformational changes in the protein concomitant with catalysis could lead to changes in the active site by bringing into proximity new constellations of functional groups, by the disruption of former catalytic functional groups, and by changing the binding specificity for substrates. It is conceivable that an "energetic" transformation at a locus far removed from the active site could result in the alteration of the three-dimensional structure at the site. The experimental identification of such conformational changes as intimate parts of the catalytic mechanism are often difficult to make. Moreover, since such changes do not occur in catalyses involving small molecules, no set of precedents and analogies can be drawn from nonenzymic systems for the purpose of discriminating one potential conformational mechanism from another. Nevertheless, an impressive body of data exists, suggesting that conformational changes in the protein molecule are involved in the detailed mechanism of at least some enzyme catalyses. A few specific examples are discussed in Chapter 8.

REFERENCES

Bruice, T. C., and Benkovic, S., *Bioorganic Mechanisms,* Benjamin, New York, Vols. 1 and 2, 1966. To date, these two admirably documented volumes contain the most detailed and comprehensive discussions of the mechanisms of catalyzed reactions of biological interest. A good background in *organic reaction mechanisms* is required for critical reading. References which start at a more elementary level are described below.

Breslow, R., *Organic Reaction Mechanisms,* Benjamin, New York, 1965. An excellent readable elementary account of the variety of mechanistic pathways by which organic molecules react.

Bender, M. L., and Breslow, R., "Mechanisms of Organic Reactions," in *Comparative Biochemistry,* Vol. 2, p. 1 (Florkin, M., and Stotz, E. H., editors), Elsevier, Amsterdam, 1962. A critical and illuminating survey of the mechanisms of organic reactions which have a counterpart in biological systems, by two of the foremost authorities.

Gould, E. S., *Mechanisms and Structure in Organic Chemistry*, Holt, Rinehart and Winston, New York, 1959.

Hine, J., *Physical Organic Chemistry*, 2nd ed., McGraw-Hill, New York, 1962. This text, as well as the one by Gould, deals with reactivity and reaction mechanisms at a more advanced level. Both books are intelligible to students with introductory courses in organic chemistry and the elements of physical chemistry (chemical principles) as background.

Bell, R. P., *The Proton in Chemistry*, Cornell Univ. Press, Ithaca, New York, 1961. The best general reference on the energetics and mechanisms of catalysis by acids and bases, as well as an excellent introduction to the applications of chemical kinetics to the determination of reaction mechanism.

Bruice, T. C., and Schmir, G., "The Influence of Mechanism on the Apparent pK_A of Participating Groups in Enzymic Reactions," *J. Am. Chem. Soc.* **81**, 4553 (1959). A critical discussion of the way in which the detailed enzyme-substrate reaction pathway may influence the pH-dependent enzyme activity. This paper presents important considerations for the correlation of the protonic equilibria of particular enzymic constituents with catalytic activity.

Eigen, M., and Hammes, G. G., "Elementary Steps in Enzyme Kinetics," *Adv. in Enzymol.* **25**, 1 (1963). The consequences of the measured rates of proton transfer to the rate of enzyme-catalyzed reactions is discussed in this article.

SEVEN — COENZYMES AND COFACTORS: ENZYME MODELS AND MECHANISMS

DESPITE THE ENORMITY OF DISTINCT BIOCHEMICAL REACTIONS, the variety of mechanistic reaction types is fairly limited. Any proposed mechanism for enzyme-catalyzed reaction must allow for the rapid turnover numbers observed experimentally. This restriction of high velocity must be responsible, in part, for the limitation on the variety of enzymic reaction mechanisms. It is well known that particular coenzymes or cofactors participate in the catalysis of many different reactions involving many different enzymes. The primary intent of this chapter is to consider the mechanisms by which a particular coenzyme or cofactor may mediate a variety of reactions.

The terms *coenzyme, cofactor,* and *prosthetic group* have been used in the biochemical literature to describe two types of enzymic functions. Often, these three terms are used indiscriminately. It is perhaps most useful to define these terms according to the catalytic functions which are mediated. Two fundamental distinctions in regard to molecular function of co-catalyst can be made:

(1) *Co-catalysts which form part of an active site, and are regenerated upon each turnover of substrate* (Eq. 7–1).

$$E + C \rightleftharpoons EC + S \rightleftharpoons X \rightleftharpoons \text{etc.} \rightleftharpoons EC + P \rightleftharpoons E + C \quad (7\text{–}1)$$

In this type of process, the co-catalyst (C) and regions of the polypeptide together form the enzyme site. The co-catalyst can usually be dissociated (reversibly) from the enzyme protein to yield catalytically inactive components (E + C). Such co-catalysts are usually referred to as *cofactors*. Good examples of this type of catalysis are the enzymic reactions involving the cofactors, pyridoxal phosphate and thiamine pyrophosphate.

(2) *Reactions in which co-catalyst and substrate form a new chemical compound.* In reactions of this type, one equivalent of the co-catalyst is converted to a new product concomitant with each turnover of substrate. The cofactor is hence not a stable constituent (*prosthetic group*) of the enzyme site but a reactant. Such cofactors are commonly referred to as *coenzymes*. The distinction between *coenzyme* and *substrate* is a subtle, and often merely semantic distinction. The coenzyme is not irreversibly exhausted; in the overall *in vivo* metabolic process, it is regenerated via other enzyme-catalyzed reactions which utilize the coenzyme-product (C*) as reactant, as illustrated by Eq. 7–2.

$$E_1 + S_1 + C \rightleftharpoons E_1 + \left\{ \begin{array}{c} C^* + P_1 \\ \text{or} \\ C\text{-}P_1 \end{array} \right.$$

$$\left. \begin{array}{c} C^* \\ \text{or} \\ C\text{-}P_1 \end{array} \right\} + E_2 + S_2 \rightleftharpoons E_2 + P_2 + C \qquad (7\text{–}2)$$

Hence, the function of the coenzyme in this latter type of reaction sequence (Eq. 7–2) is the transfer of a chemically reactive intermediate from one enzyme-catalyzed process to another. In some situations, the role of the coenzyme is the transfer of a relatively reactive intermediary metabolite to a new site of reaction (as exemplified in the reactions involving coenzyme A described below), and in others it serves alternately as an oxidant and reductant in two different catalyzed reactions (as in the reactions which utilize pyridine nucleotide and flavin coenzymes). The classification of coenzymes according to this latter criterion is obviously somewhat arbitrary; for example, the phosphorylation and dephosphorylation of adenine nucleotides (Eq. 7–3) occurs concomitantly with many enzyme-catalyzed processes.

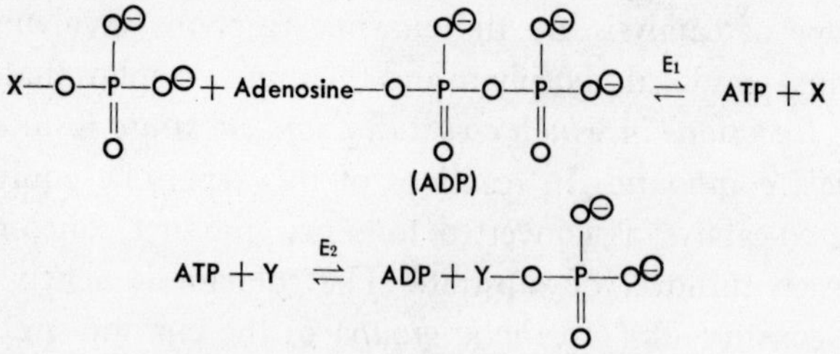

(7–3)

Adenine nucleotides can be classified as coenzymes in the phosphorylation reactions in which they participate, just as pyridine nucleotides are classified as coenzymes in biological oxidation-reduction reactions in which they participate.

A characteristic, common to all coenzyme and cofactor mediated reactions, is the high affinity of enzyme for the co-catalyst, and hence, the high degree of geometrical constraint among the various reactive centers (the catalytic groups of the enzyme, the functional groups of the co-catalyst, and the reaction center(s) of the substrate). The mechanism of action of some of the coenzymes and cofactors has been extensively investigated. In other cases, mechanistic details are virtually unknown at this time. Since our primary concern here is not in a comprehensive description of all of intermediary metabolism, we shall restrict the discussion to select examples for which mechanistic details are available. One should not draw conclusions, on the basis of the following discussion, as to the relative importance or the diversity of function of one coenzyme or cofactor relative to another.

7–1 PYRIDOXAL PHOSPHATE-MEDIATED CATALYSIS

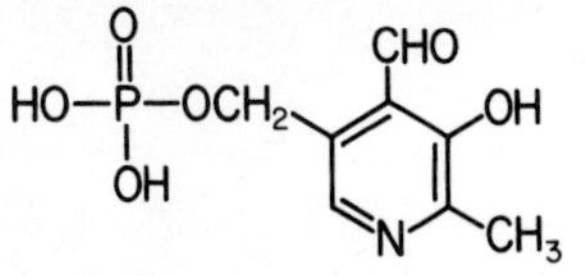

Pyridoxal phosphate

Among the most extensively investigated cofactor-mediated reactions in regard to mechanism, are those involving pyridoxal phosphate. A number of seemingly different types of catalyzed reactions require this cofactor. Some of these enzyme-catalyzed reactions are described in Table 7–1.

Interest in reactions mediated by this particular cofactor has been generated by the fact that many of the same reaction processes are catalyzed by pyridoxal phosphate in the absence of enzyme. The reaction rates in the absence of enzyme protein are, however, very much slower (by $\sim 10^6$-fold). Metastable intermediates in the nonenzyme-catalyzed reaction have been demonstrated, and the mechanistic sequence illustrated in Fig. 7–1 has been partially established. Note that this sequence of reactions involves a series of electron-pair and proton transfer processes, and is illustrative of the mechanisms discussed earlier, in which intramolecular chemical reaction occurs via a series of intermediate steps, each step involving a relatively low activation energy process. Some modification of this pyridoxal-catalyzed reaction sequence in enzyme catalysis is made necessary by the finding that pyridoxal phosphate bound to enzyme is *chemically* linked to the ε-amino group of a specific lysine residue (Eq. 7–4). In this regard, it is of interest

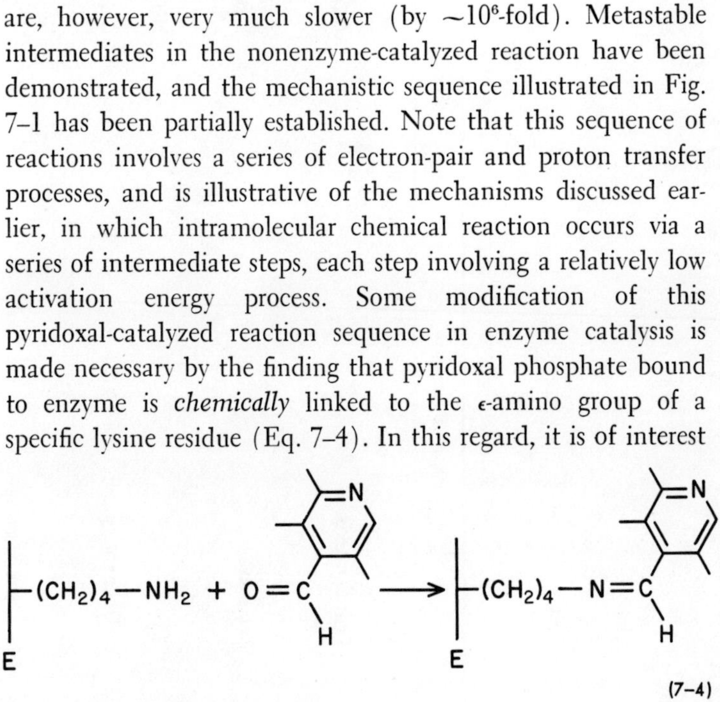

(7–4)

Table 7-1
Typical Enzyme-Catalyzed Reactions of Amino Acids Requiring Pyridoxal Phosphate

Type of enzyme	Typical reaction
Transaminase	$RC(=O)CO_2^{\ominus} + R'CH(NH_3^{\oplus})CO_2^{\ominus} \rightleftharpoons RCH(NH_3^{\oplus})CO_2^{\ominus} + R'C(=O)CO_2^{\ominus}$
Racemase	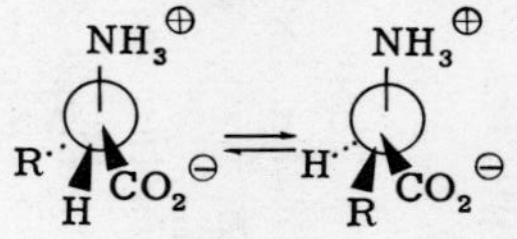
α-Decarboxylase	$RCH(NH_3^{\oplus})-CO_2^{\ominus} + H_2O \longrightarrow RCH_2NH_3^{\oplus} + HCO_3^{\ominus}$
β-Decarboxylase	${}^{\ominus}O_2C-CH_2CH(NH_3^{\oplus})-CO_2^{\ominus} + H_2O \longrightarrow CH_3CH(NH_3^{\oplus})-CO_2^{\ominus} + HCO_3^{\ominus}$

β-OH Desmolase (Aldolase)	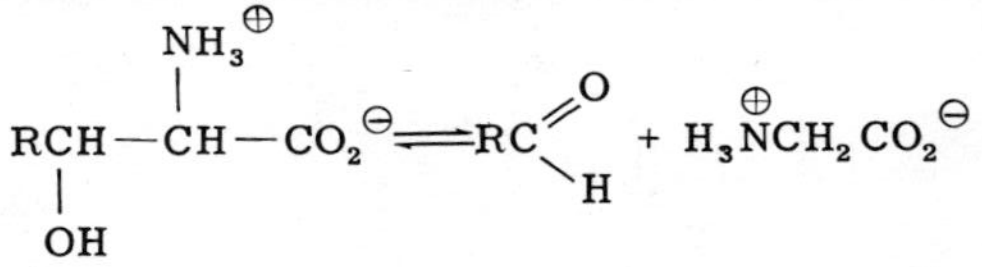
Glycine (and serine) condensation enzymes	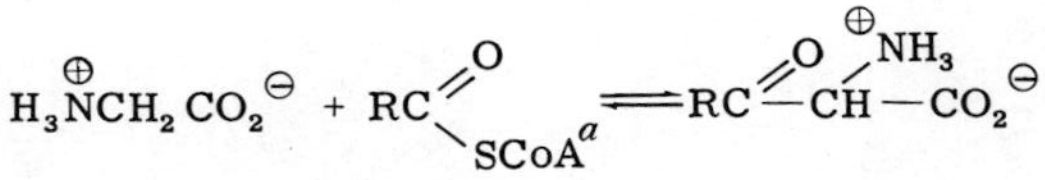
α, β-Eliminations (polar substituents)	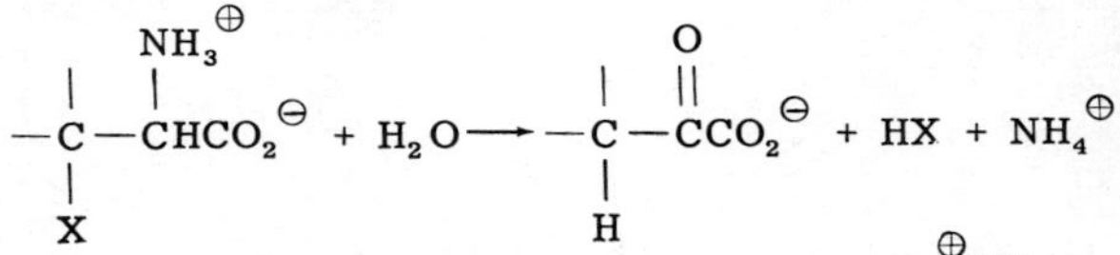
β-Condensations (synthetases)	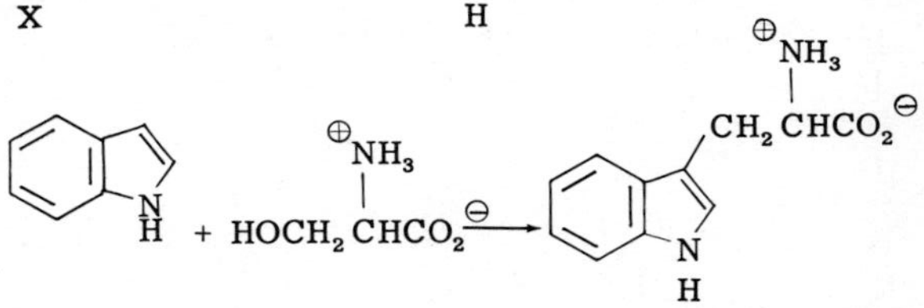

[a] Acyl coenzyme A (see Section 7-3).

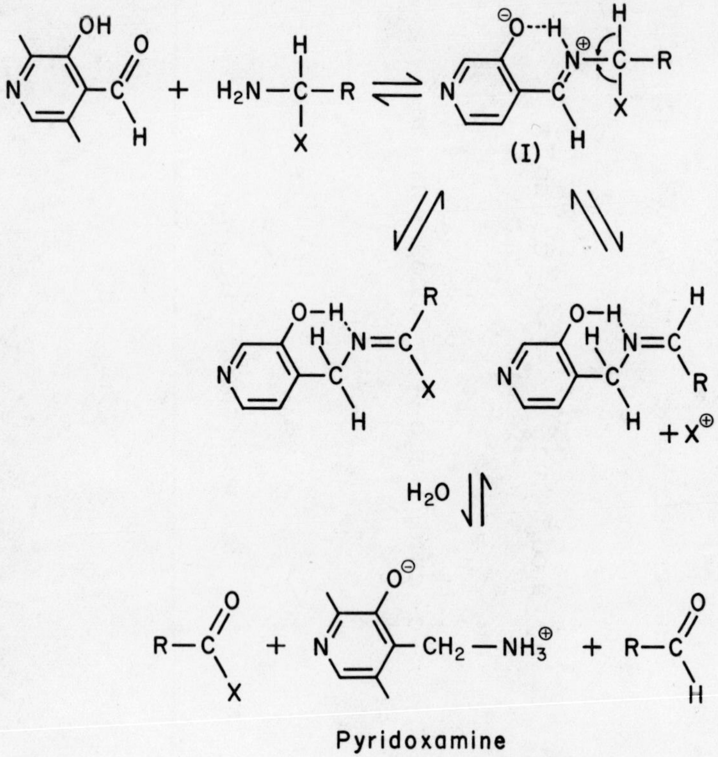

FIG. 7–1 A general mechanism for pyridoxal-catalyzed reactions.

to note that in an analogous model compound catalysis, imino derivatives of pyridoxal react with substrate (semicarbazide) more rapidly than does pyridoxal itself (Eq. 7–5).

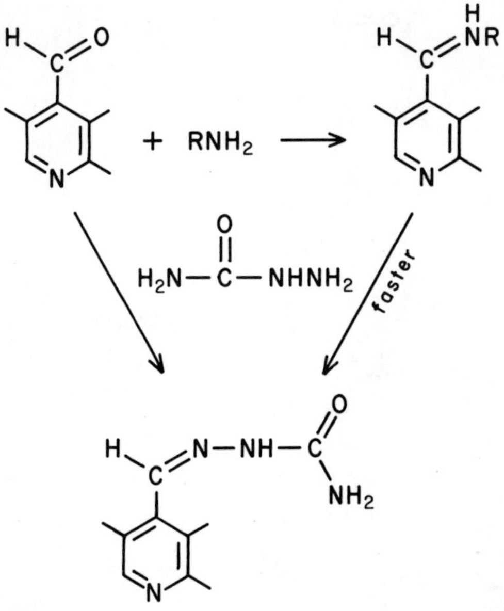

(7–5)

Hence, the coenzyme is pre-activated for reaction in its combination with the enzyme site. A plausible mechanism for the enhanced velocity of such enzyme-catalyzed reactions (other than the constraint of substrate and cofactor) is via a series of general acid-base catalyzed proton transfers. Hypothetical mechanisms illustrative of the principles, but by no means definitive models of the detailed mechanism are illustrated in Fig. 7–2. It has been well established (by techniques of relaxation kinetics) that the slowest proton transfer reactions are those involving proton abstraction from a carbon atom. The coenzyme-mediated electron-pair transfer processes facilitate general acid-base catalysis of this proton abstraction.

The "Schiff base" intermediate (I of Fig. 7–1) serves to "labilize" the hydrogen or "X" at the α-carbon atom of the substrate, by providing a facile route for the transfer of an electron pair to the coenzyme. In this way a variety of reaction pathways are facilitated, as illustrated in Fig. 7–3.

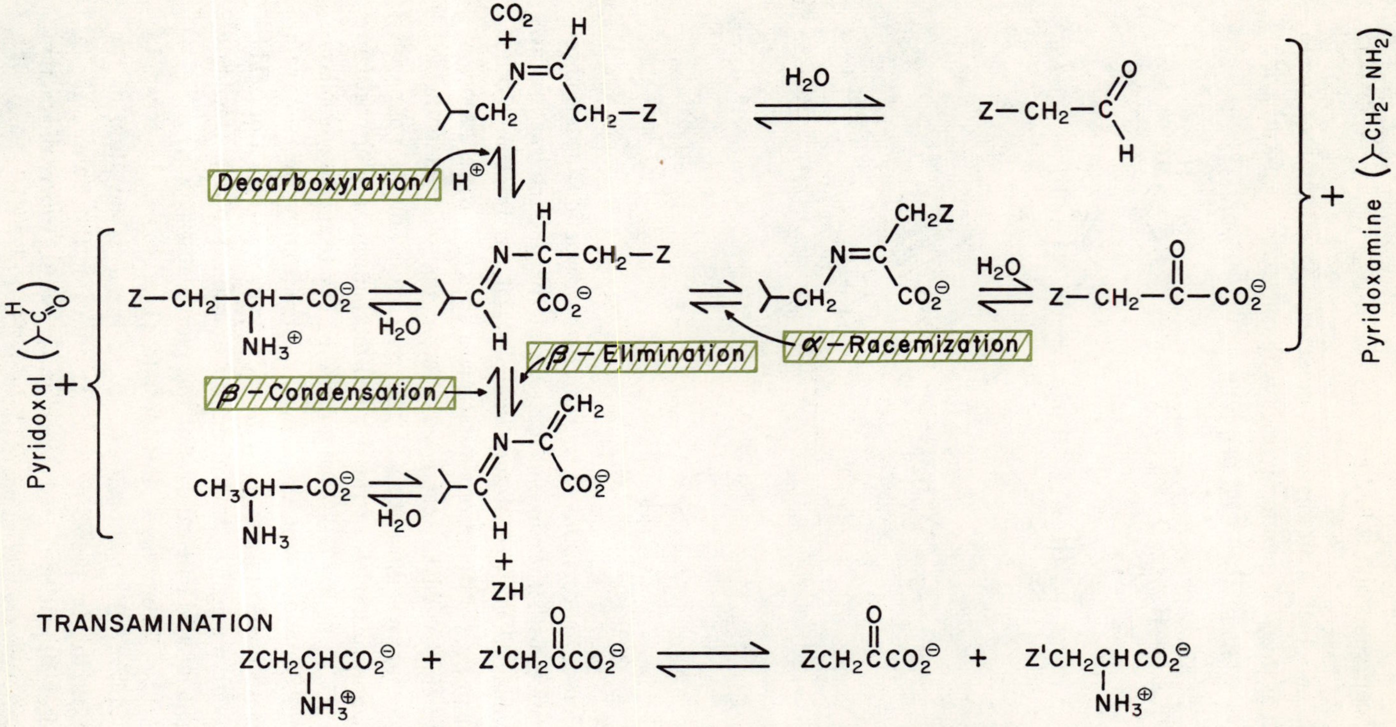

FIG. 7-3 A general mechanism for the variety of pyridoxal-catalyzed reactions.

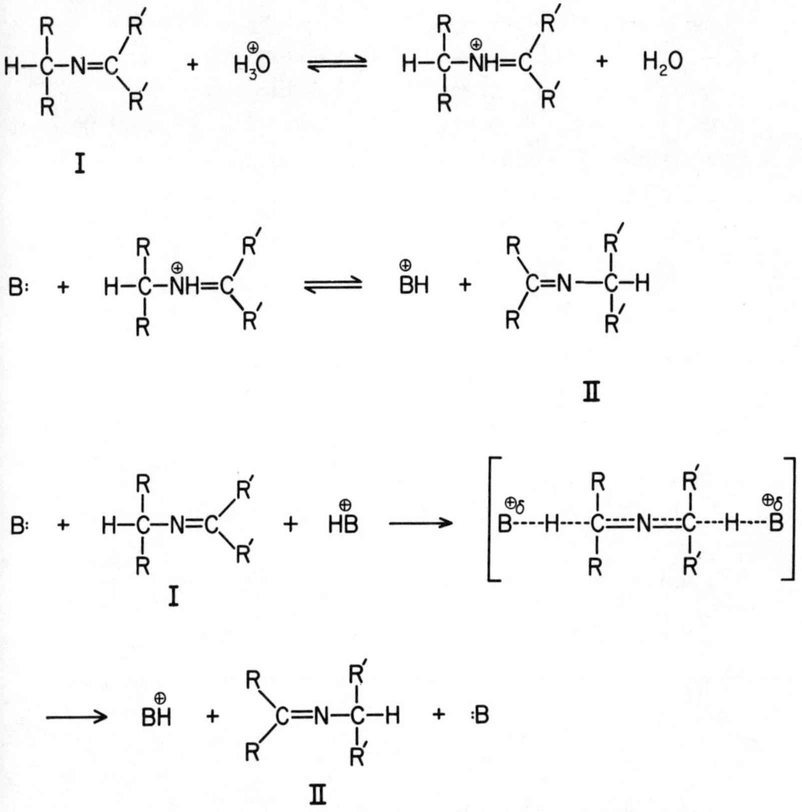

$$\underset{\text{I}}{H-\overset{R}{\underset{R}{C}}-N=\overset{R'}{\underset{R'}{C}}} \quad + \quad H_3\overset{\oplus}{O} \quad \rightleftharpoons \quad H-\overset{R}{\underset{R}{C}}-\overset{\oplus}{N}H=\overset{R'}{\underset{R'}{C}} \quad + \quad H_2O$$

$$B: \quad + \quad H-\overset{R}{\underset{R}{C}}-\overset{\oplus}{N}H=\overset{R'}{\underset{R'}{C}} \quad \rightleftharpoons \quad \overset{\oplus}{B}H \quad + \quad \underset{R}{\overset{R}{C}}=N-\overset{R'}{\underset{R'}{C}}-H$$
$$\text{II}$$

$$B: \quad + \quad \underset{\text{I}}{H-\overset{R}{\underset{R}{C}}-N=\overset{R'}{\underset{R'}{C}}} \quad + \quad H\overset{\oplus}{B} \quad \longrightarrow \quad \left[\overset{\oplus\delta}{B}\text{---}H\text{---}\overset{R}{\underset{R}{C}}{=\!=}N{=\!=}\overset{R'}{\underset{R'}{C}}\text{---}H\text{---}\overset{\oplus\delta}{B} \right]$$

$$\longrightarrow \quad \overset{\oplus}{B}H \quad + \quad \underset{R}{\overset{R}{C}}=N-\overset{R'}{\underset{R'}{C}}-H \quad + \quad :B$$
$$\text{II}$$

FIG. 7–2 General acid-base catalyzed tautomerization of the protonated Schiff's base intermediate in pyridoxal-mediated reactions; two hypotnetical mechanisms.

7–2   THIAMINE PYROPHOSPHATE-MEDIATED REACTIONS

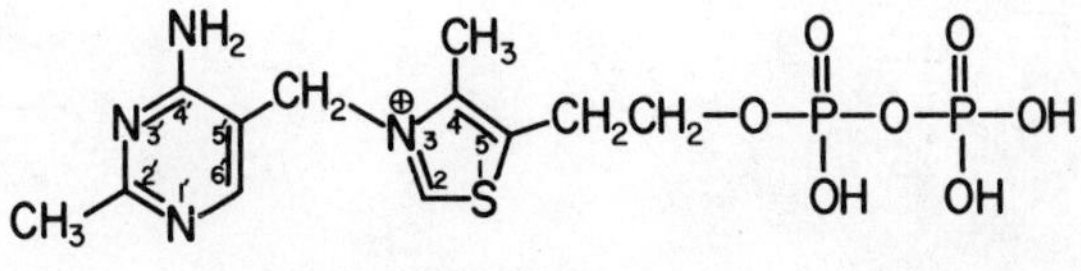

Thiamine pyrophosphate

Like pyridoxal phosphate, thiamine pyrophosphate and the free base, thiamine, can catalyze some of the reactions catalyzed by thiamine-enzyme complexes. The catalytically functional part of the molecule is the sulfur-containing five-membered thiazolium ring. The hydrogen atom at the 2-position is rapidly exchanged in aqueous solvent, and it is now well established that the catalytically active site of the molecule is at the 2-carbon. The presence of an adjacent positive charge at N-3 enhances the stability of the "ylid" structure given in Eq. 7–6.

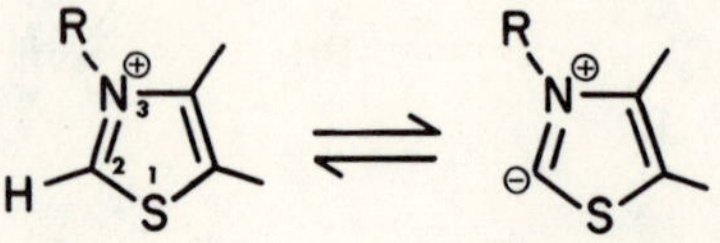

(7–6)

The probable mechanism of the nonenzymic thiamine catalysis of decarboxylation of pyruvic acid is shown in Fig. 7–4. Note the role of the coenzyme in mediating the electron-pair transfer process.

The nonenzymic mechanism proposed by Breslow (Fig. 7–4)

FIG. 7–4 A mechanism for the thiamine-catalyzed decarboxylation of pyruvic acid.

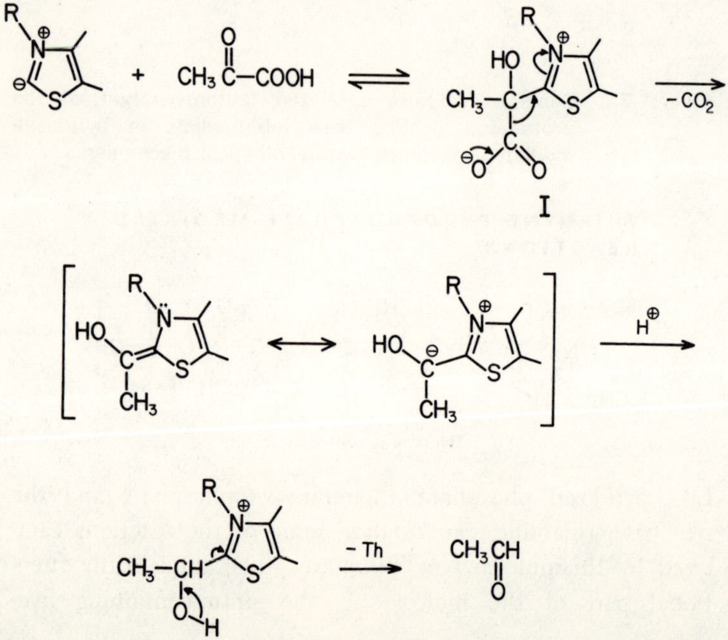

can be generalized to all thiamine-catalyzed enzymic reactions. These reactions all involve nucleophilic attack by thiamine, the cleavage of a carbonyl-X bond (where X is usually —H or $—CO_2^-$), and electrophilic attack at the α-carbon of the substrate-thiamine adduct (Eq. 7–7).

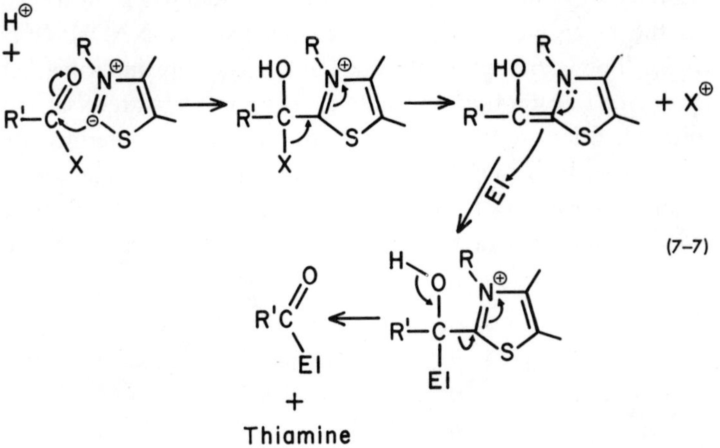

(7–7)

7–3 PYRIDINE NUCLEOTIDES (NICOTINAMIDE ADENINE DINUCLEOTIDES)

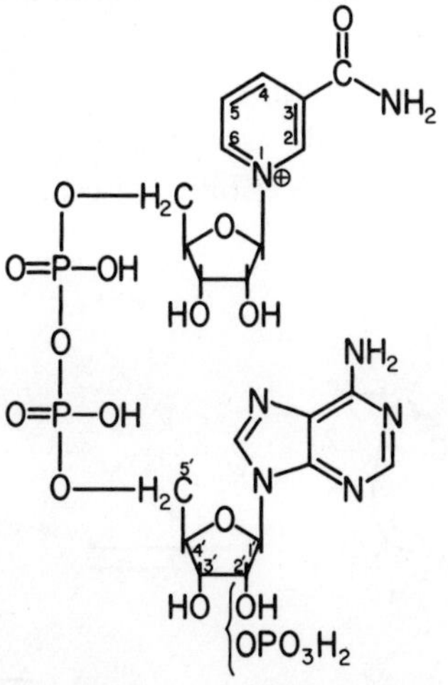

Oxidized pyridine nucleotides
(NAD^+ when —OH is present, $NADP^+$ when $—OPO_3H_2$ is present)

Pyridine nucleotides, either in their oxidized or reduced state are involved in a great variety of enzymic hydrogenation-dehydrogenation reactions. In the process, hydrogen is either added to or abstracted from the C–4 atom of the pyridine nucleus. Reduction (hydrogenation) of the pyridine nucleotides results in a tetrahedral C–4 atom (Eq. 7–8). It has been shown, in every enzyme-catalyzed process so investigated, that hydrogen transfer to this C–4 position is stereospecific, as is the hydrogen abstraction from the substrate. The stereochemistry is illustrated in the reversible reduction of acetaldehyde by NADH (Fig. 7–5). This stereospecificity does not necessarily indicate that hydrogen transfer from substrate to coenzyme, or its reversal, is a direct hydride transfer process. It may involve an intermediate transfer of hydrogen to an enzyme or solvent molecule, and a rigid arrangement of coenzyme, enzyme catalyst, and substrate which can still lead to a stereospecific reaction. Such stereo-

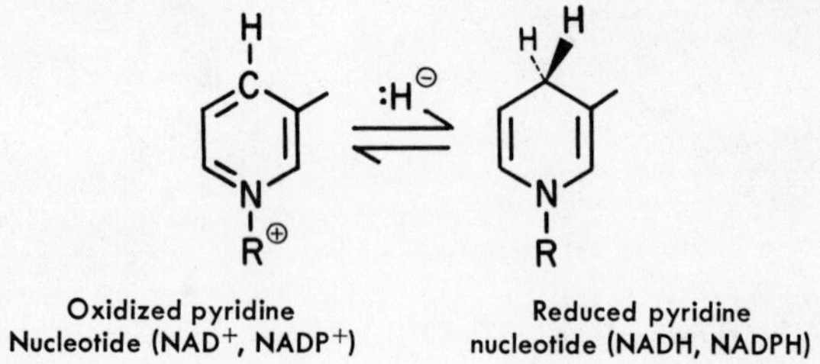

(7–8)

specificity, no matter what its mechanistic origins, is indicative of the rigid constraints on the geometry of reactive centers in the enzyme-coenzyme-substrate complex. Note once again (for example, in Fig. 7–5), the possibility of general acid-base partici-

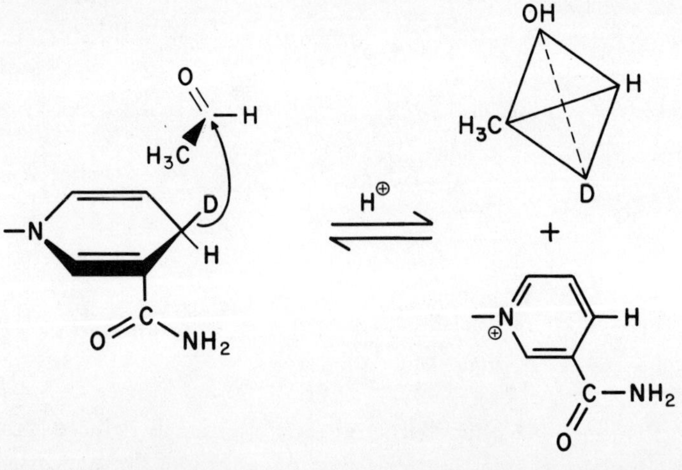

FIG. 7–5 Stereospecificity in the reversible reduction of acetaldehyde by NADH.

pation in the enzyme-catalyzed reaction. Not too much is known about the detailed mechanism of the presumed hydride transfer reaction, since in the absence of enzyme neither the coenzyme nor closely related model compounds will react with any of the extensively investigated substrates of the enzymic reactions, even when product formation is thermodynamically favored, as for example in the reduction of either acetaldehyde or pyruvic acid by NADH. The kinetics of the reactions catalyzed by the enzymes alcohol, lactic, and malic dehydrogenase have been very extensively investigated, and are discussed in the following chapter.

7−4   THIOL COENZYMES

Many of the reactions of intermediary metabolism lead to carbon compound products which are at the oxidation state of carboxylic acids. Very often, these carboxylic compounds are thermodynamically extremely stable relative to the carbon compounds from which they were formed (aldehydes, ketones, olefins, and alcohols). Indeed, the formation of carboxylate anion presents a "low energy trap," inhibiting the further utilization of the metabolite. In order to reutilize a carbon compound at this oxidation state for further biosynthesis, it is necessary to provide a "detour" between the high energy metabolic precursor (for example an aldehyde) and the stable (unreactive) carboxylate anion. One of the most common of such routes in biosynthetic pathways is via the chemical coupling of oxidation to the formation of thiol esters (Eq. 7–9).

$$
\text{RC}\!\!\begin{array}{c} \text{H} \\ \diagup \\ \diagdown \\ \text{O} \end{array} + \text{Ox} \xrightarrow{\text{R'SH}} \text{RC}\!\!\begin{array}{c} \text{SR'} \\ \diagup \\ \diagdown \\ \text{O} \end{array} + \text{OxH} \qquad \text{(7–9)}
$$

$$
\downarrow \text{H}_2\text{O}
$$

$$
\text{RCO}_2^{\ominus} + \text{H}^{\oplus} + \text{R'SH}
$$

Further biosynthesis

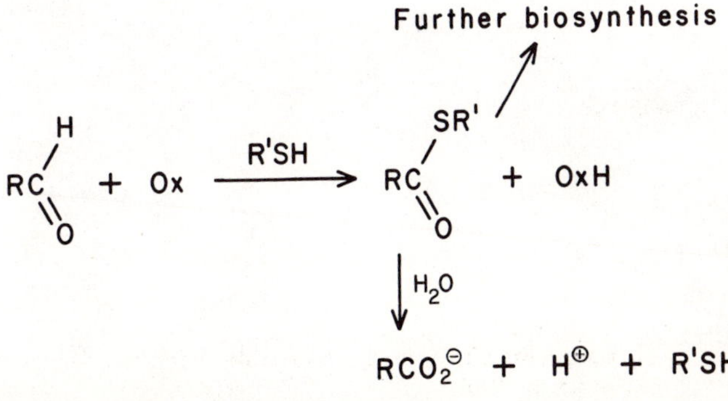

The thermodynamic stability of thiol esters has been investigated. Thiol esters are thermodynamically less stable than carboxylate anion but somewhat more stable than the usual metabolic precursors of carboxylate. In the absence of appropriate catalysts nevertheless, thiol esters are kinetically stable in neutral aqueous solution. An enzyme, by providing appropriate catalytic centers, may allow for further reaction (utilization) of a thiol ester intermediate in biosynthesis. This particular catalyzed pathway is one of the principal routes by which carbon-carbon bonds are synthesized in biological systems. Given an appropriate thiol ester, further reaction may occur via either of the two enzymic mechanisms discussed previously, namely, by the reaction of the thiol ester with a second substrate at the original enzyme (-coenzyme) site, or by the transfer of the thiol ester coenzyme to a different enzyme site where it subsequently reacts with a second substrate. In actuality, both these mechanisms have been demonstrated. The latter "transfer" mechanism appears to be much more extensively utilized, particularly with the most prevalent thiol acyl acceptor, coenzyme A.

Thiol esters are chemically differentiated from the corresponding alcohol (oxygen) esters in a number of ways:

(1) Chemical bonding, via the 3s and 3p orbitals of sulfur, results in a longer and weaker bond than with corresponding oxygen compounds. Hence, there will be a lesser interelectronic repulsion, and a more facile ester bond cleavage in nucleophilic displacement reactions at the carbonyl carbon atom (Eq. 7–10).

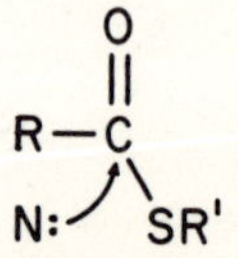

(7–10)

(2) The much smaller tendency of sulfur towards "π-bond" formation, as compared with oxygen, results in a far greater *localization of charge* in the carbonyl group (Eq. 7–11).

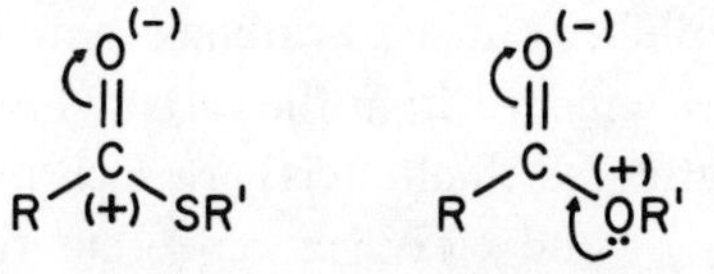

(7–11)

As a result, the carbonyl carbon atom of the thiol ester is far more electrophilic than that of the corresponding oxygen ester, and thus will tend to induce electronegativity in an adjacent α-carbon, for example, by carbanion formation (Eq. 7–12).

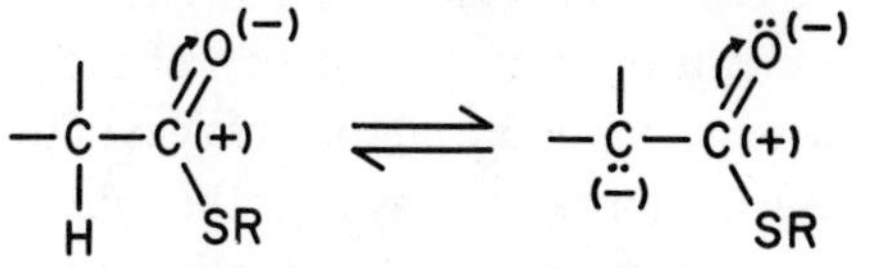

(7–12)

Thiol esters resemble carbonyl compounds (ketones and aldehydes) electronically in their reactivity at the α-carbon. Such compounds undergo facile *condensation reactions* with electrophilic reagents at the α-carbon atom. A familiar example in organic chemistry is the base-catalyzed aldol condensation, involving two moles of acetaldehyde (Eq. 7–13).

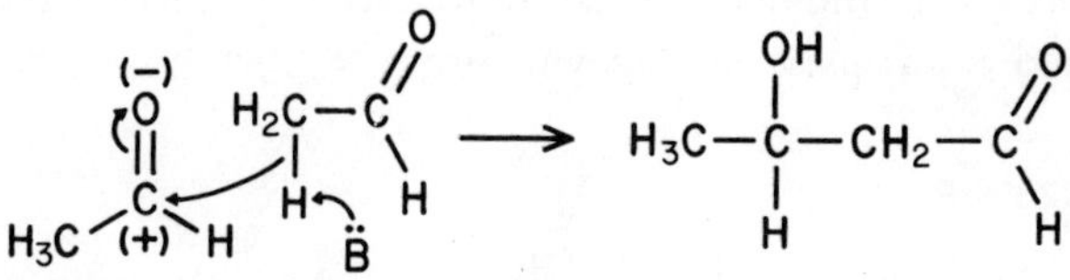

(7–13)

Similarly, thiol esters undergo facile condensation reactions with both carbonyl compounds, and with a second mole of thiol ester (*Claisen* condensations, Eq. 7–14).

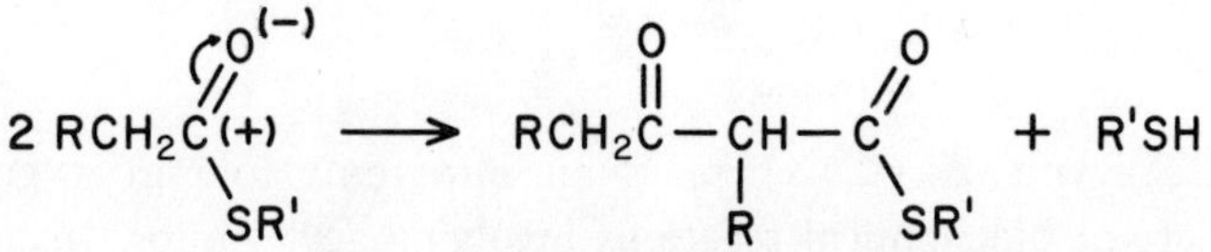

Claisen condensation of thiol esters

(7–14)

Thiol esters thus undergo two distinct types of reactions: nucleophilic displacement reactions, and reactions with electrophilic reagents at the α-carbon atom. In special situations, the induced nucleophilicity at the α-carbon atom can be transferred to atoms further removed from the carbonyl carbon (for example in α, β unsaturated thiol esters). In such instances, condensations and (or) bond cleavages can occur further down the carbon chain.

Since thiols occur in the constituent cysteine residues of enzyme proteins, and thiol ester enzyme-substrate compounds have been demonstrated (see Chapter 8), one may question the necessity for coenzymes which help carry out a function for which the pure protein enzyme is already equipped. A plausible explanation presents itself, if the function of the coenzyme is to transfer an active acyl group from one enzyme (the site of thiol ester synthesis) to another enzyme (the site of subsequent reaction). The coenzyme in this situation can perform two useful functions: It can store active acyl intermediates in concentrations stoichiometrically comparable to its own (rather than in concentrations comparable to the number of enzyme sites), and it can transfer the acyl intermediate to a new catalytic environment (the second enzyme site) which may differ markedly from the original catalytic environment. In this way the coenzyme may reduce the variety of catalytic functions required at any particular enzyme site.

COENZYME A

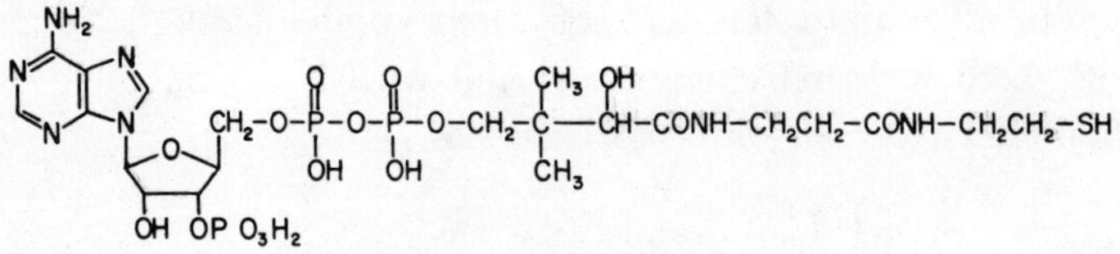

Coenzyme A (CoA) plays an important role in a great variety of biochemical reactions involving acyl groups. Invariably, the two processes, acylation of the coenzyme, and reaction of the resultant thiol ester, take place at *different* enzyme sites, for example, in the formation of acetoacetyl-CoA from acetaldehyde (Eq. 7–15).

In the course of formation of the acyl intermediate, ace-

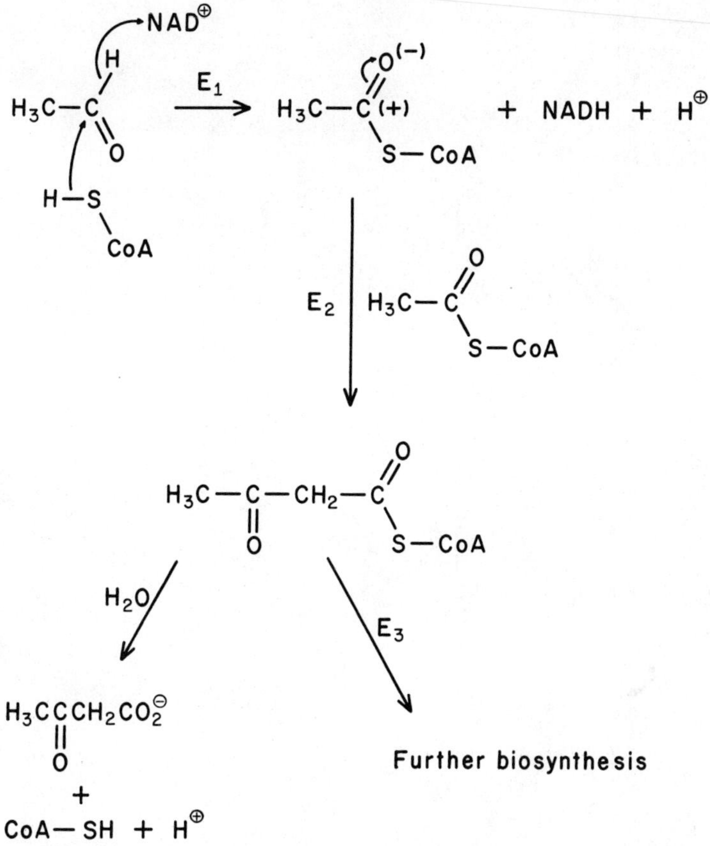

CoA-mediated utilization of acetaldehyde (7–15)

tyl-CoA (Eq. 7–15), the enzyme catalyst can enhance the rate of thiol ester formation by general base catalysis at the S—H linkage. Once the thiol ester has been formed, reactions involving the acyl group may proceed via either nucleophilic displacement at the carbonyl carbon or by electrophilic attack at the α-carbon. A partial list of reactions of acyl-CoA via these alternate pathways is summarized in Table 7–2. Note that CoA can be phosphorylated (Eq. 7–16) as well as acylated.

$$\text{CoA—SH} + \text{X—O—}\overset{\displaystyle O}{\overset{\|}{\underset{\underset{\displaystyle O^{\ominus}}{|}}{P}}}\text{—OH} \longrightarrow \text{CoA—S—}\overset{\displaystyle O}{\overset{\|}{\underset{\underset{\displaystyle O^{\ominus}}{|}}{P}}}\text{—OX} + H_2O \qquad (7\text{–}16)$$

Table 7-2
Enzyme-Catalyzed Reactions of Acyl and Phosphoryl CoA

General mechanism	Class of enzyme	Typical reaction
Nucleophilic attack on carbonyl carbon	Acylase	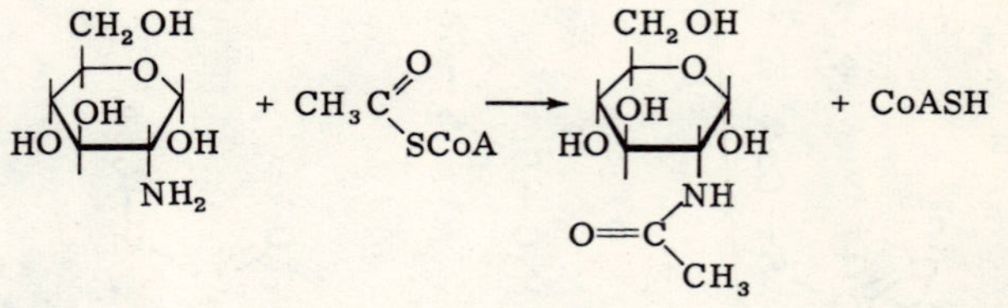 (N-acetylglucosamine)
		$(CH_3)_3\overset{\oplus}{N}CH_2CH_2OH$ + acetyl-CoA $\longrightarrow$ $(CH_3)_3\overset{\oplus}{N}CH_2CH_2O{-}C({=}O)CH_3$ (acetylcholine)
	(Deacylase)	$CH_2{-}C({=}O)SCoA$ / $CH_2{-}CO_2^{\ominus}$ (Succinyl-CoA) + H_2O $\longrightarrow$ $CH_2CO_2^{\ominus}$ / $CH_2CO_2^{\ominus}$ + CoASH

	(Transacylase)	succinyl-CoA + acetoacetate $\longrightarrow$ acetoacetyl-CoA + succinate
	(Phosphotransacetylase)	$CH_3C(=O)SCoA + H_2PO_4^{\ominus} \longrightarrow CH_3C(=O)OPO_3H^{\ominus} + CoASH$
	(Thiokinase)	acetyl–CoA + adenylic acid + $O{-}P(=O)(OH){-}O{-}P(=O)(OH){-}O \longrightarrow$ ATP + acetate + CoASH
	Acyl reductase	Acyl-SCoA + NADH + $H^{\oplus} \longrightarrow$ aldehyde + $NAD^{\oplus}$ + CoASH
Nucleophilic attack on phosphoryl	Phosphokinase	$CoA{-}S{-}P(=O)(O^{\ominus}){-}OH$ + adenosine diphosphate (ADP) $\rightleftharpoons$ CoASH + ATP (S-phospho–CoA)
	Phosphoryl transferase	Succinate + S-phospho–CoA $\longrightarrow$ succinyl-CoA + $H_2PO_4^{\ominus}$

(continued)

(Table 7-2 continued)

General mechanism	Class of enzyme	Typical reaction
Condensations at the carbonyl carbon	β-Ketoacyl thiolase	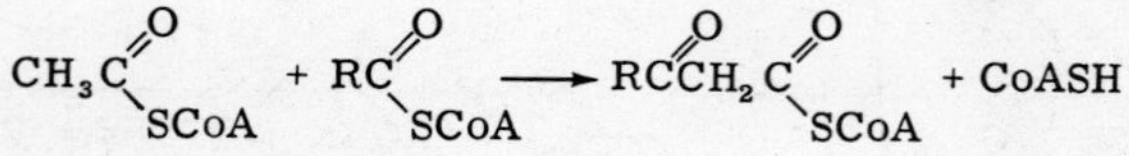
	Transacetylase	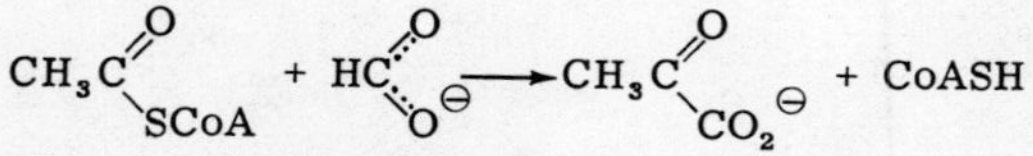
Electrophilic condensations at the α-carbon	Synthetase	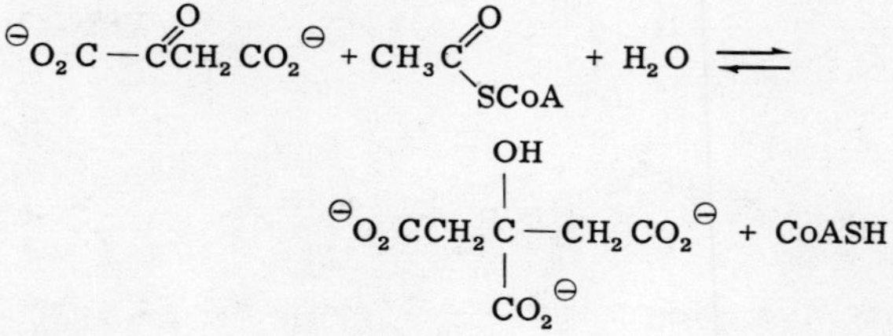

	Carboxylase	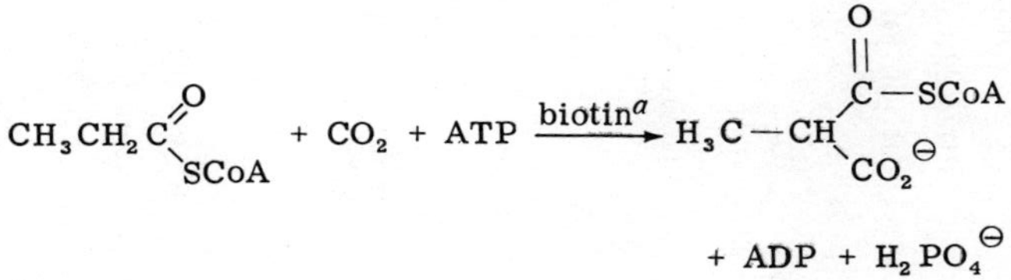
	Decarboxylase	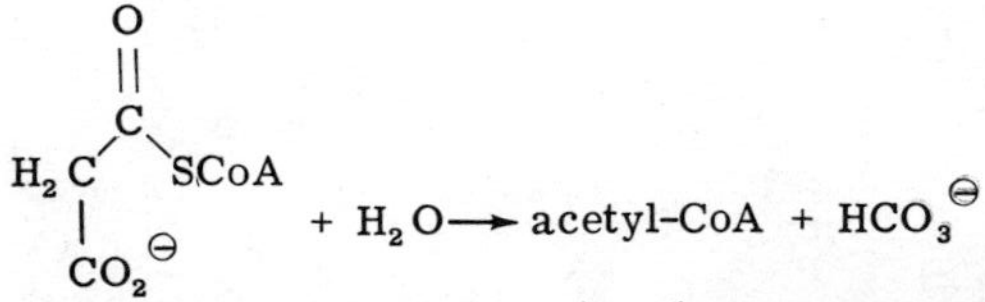
α, β-Elimination	Acyl dehydrogenase	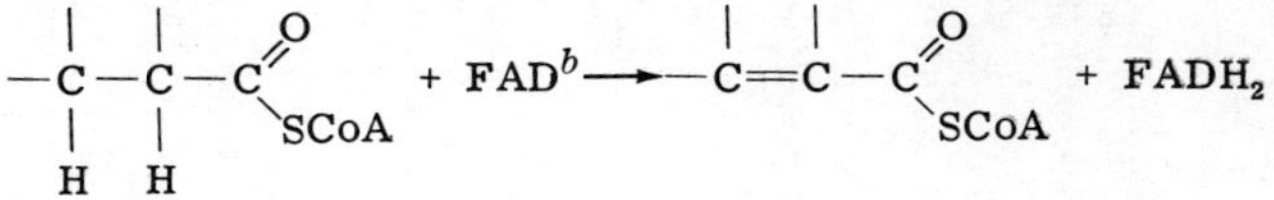
β-OH Oxidation	β-OH Acyl dehydrogenase	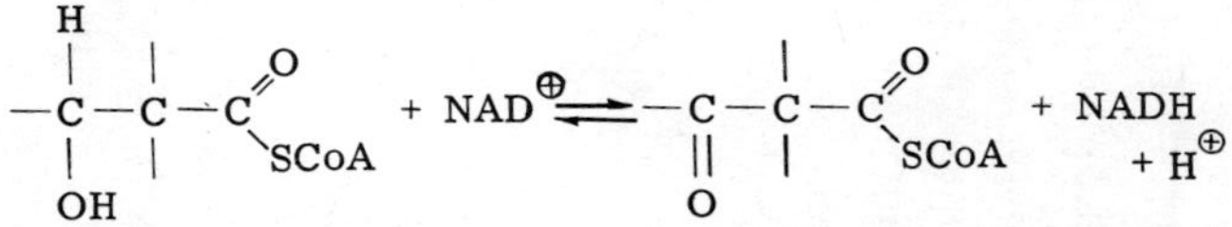

[a] A cofactor in reactions involving the activation of CO_2.
[b] Flavin adenine dinucleotide. See Section 7-4.

Such phosphorylated derivatives undergo reactions analogous to the nucleophilic displacements at the carbonyl carbon of acyl-CoA, yielding phosphoric esters and anhydrides.

In addition to its function in nucleophilic displacement and condensation reactions, CoA is involved in elimination reactions, for example, in the formation of unsaturated fatty acids from saturated fatty acids (Eq. 7–17).

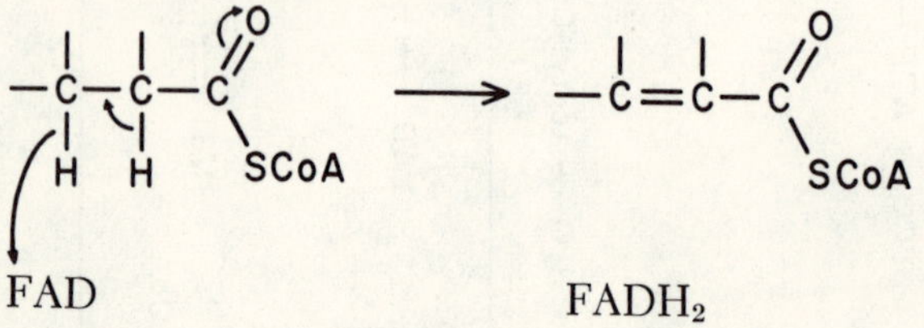

(7–17)

Enzymic catalysis in the nucleophilic displacement reactions involving acyl- or phosphoryl-CoA may be similar in mechanism to what has already been suggested for other carboxylic acid derivatives (see Section 6–3). Reactions at the α-carbon can be plausibly envisaged to involve proton abstraction from the α-carbon followed by general acid catalysis of the electrophilic attack by second substrate as in Eq. 7–18.

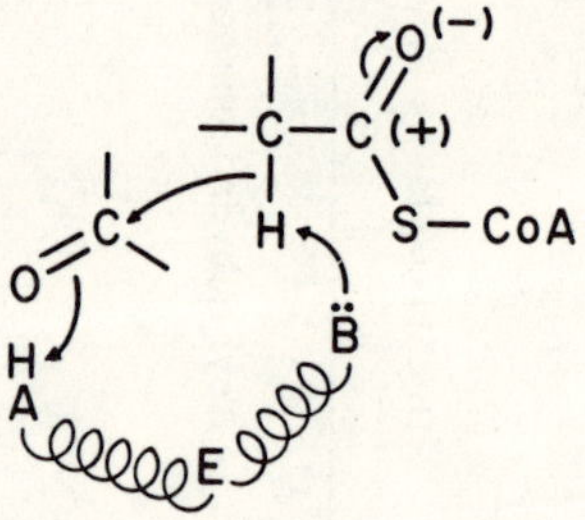

(7–18)

Thiol groups are chemically reactive in two other types of addition reactions, namely, the addition to carbonyl compounds, and the addition to double bonds (Eq. 7–19).

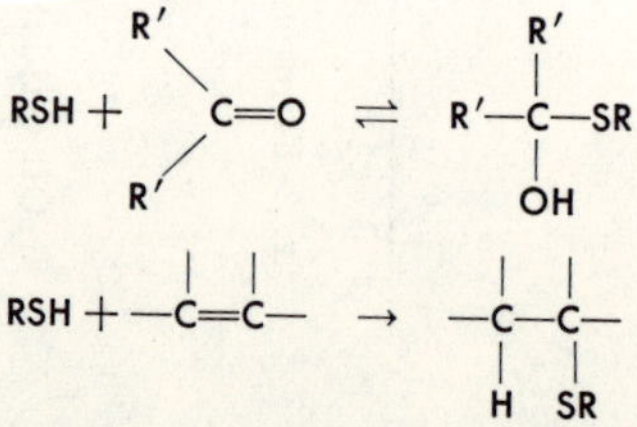

(7–19)

Neither of these additions (Eq. 7–19) are observed in enzymic reactions involving CoA. Reactions of these types are carried out either by the thiol groups of cysteine residues of the enzyme protein, or by still other thiol coenzymes.

GLUTATHIONE

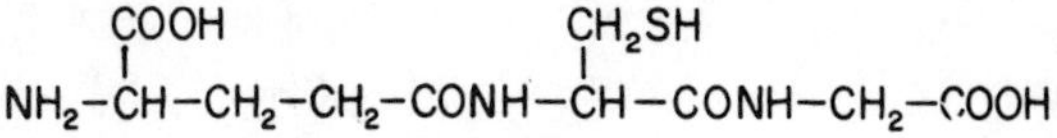

$$\underset{\text{Glutathione}}{\overset{\displaystyle \underset{|}{\text{COOH}} \qquad\qquad \overset{\text{CH}_2\text{SH}}{\underset{|}{}}}{\text{NH}_2-\text{CH}-\text{CH}_2-\text{CH}_2-\text{CONH}-\text{CH}-\text{CONH}-\text{CH}_2-\text{COOH}}}$$

The *addition* reactions of thiols with aldehydes (described above) are sometimes mediated by this coenzyme. When glutathione adds to an aldehydic carbonyl compound, the corresponding hemithioacetal (Eq. 7–20) is formed. Enzymes which utilize glutathione as coenzyme catalyze further reaction to form the corresponding thiol ester.

$$\text{R}'-\overset{\displaystyle O}{\underset{\displaystyle H}{C}} + \text{RSH} \;\rightarrow\; \text{R}'-\overset{\text{OH}}{\underset{\text{H}}{C}}-\text{SR} + \text{NAD}^\oplus \;\rightarrow\; \text{R}'-\overset{\displaystyle O}{\underset{\displaystyle SR}{C}} + \text{NADH} + \text{H}^\oplus$$

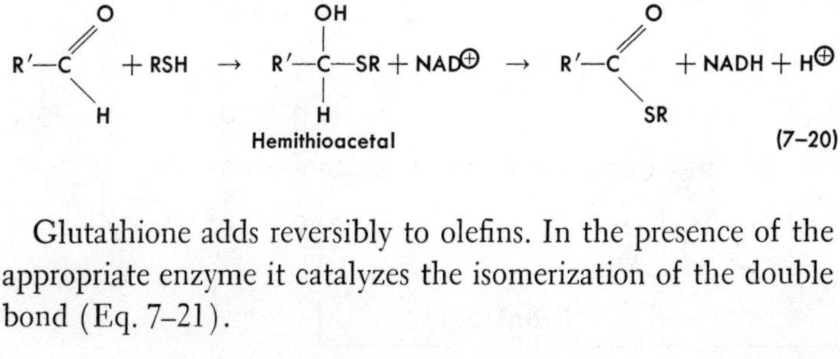

$$\text{Hemithioacetal} \qquad\qquad\qquad\qquad (7\text{–}20)$$

Glutathione adds reversibly to olefins. In the presence of the appropriate enzyme it catalyzes the isomerization of the double bond (Eq. 7–21).

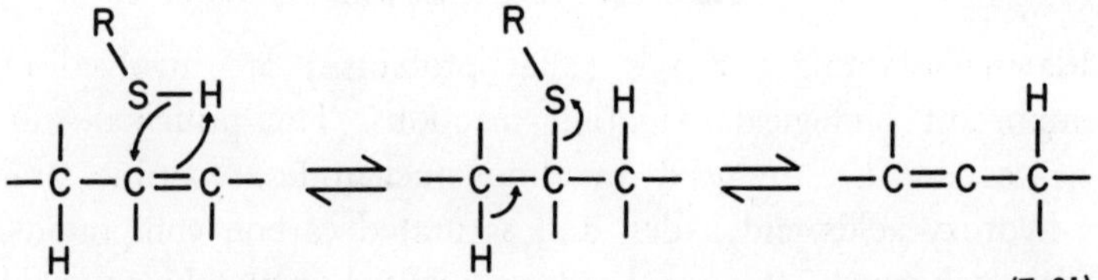

$$(7\text{–}21)$$

Unlike coenzyme A, which is involved in the *transfer* of acyl metabolite between two enzyme sites, gluthathione action is restricted to a single enzyme site.

Surprisingly, no enzymic process is known which utilizes the facile oxidation-reduction properties (Eq. 7–22) of either of these two thiol coenzymes.

$$2RSH \rightleftharpoons RS - SR \tag{7–22}$$

A still more facile disulfide formation occurs in the coenzyme *lipoic acid.*

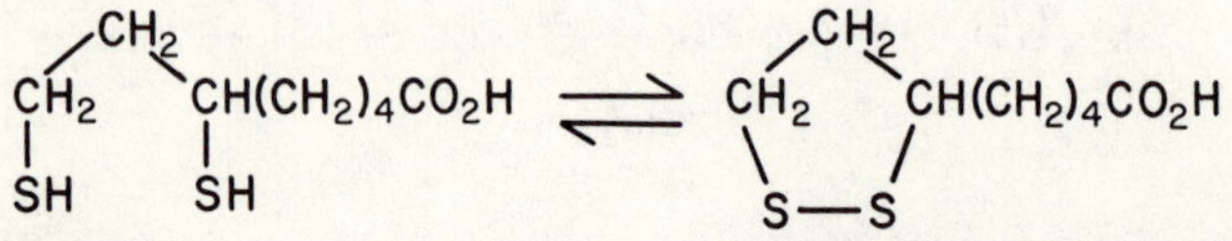

Lipoic acid

It is with this coenzyme that thiol-disulfide coupled oxidation-reduction is carried out enzymically. The molecular details of the enzymic mechanism, however, are still quite obscure.

7–5 FLAVINS AND HEME COMPOUNDS

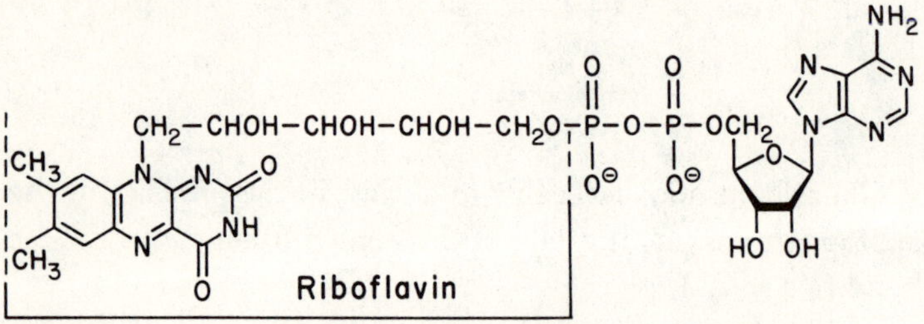

Flavin-adenine dinucleotide (FAD)

Flavin-containing enzymes ("flavoproteins") are involved in important biological oxidation reactions. The principal substrates are the reduced pyridine nucleotides, α-amino and α-hydroxy acids, aldehydes, and saturated carbon compounds. Although much attention has been focused on the dynamics of these biological oxidations, the detailed molecular mechanisms are obscure. Model studies with flavins and related structures have established that reduction of the flavin proceeds via two one-electron transfers (Eq. 7–23).

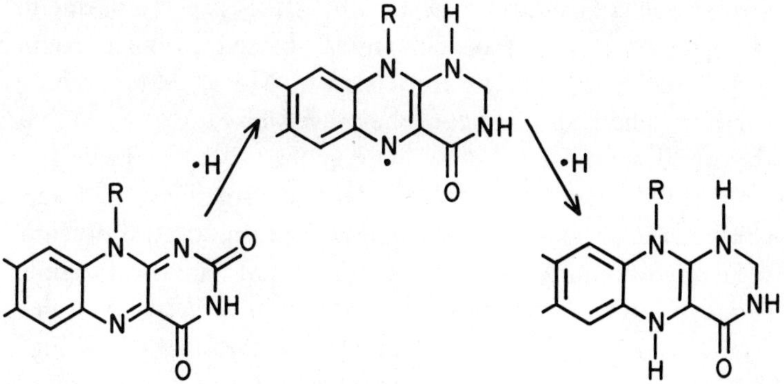

(7–23)

The intermediate free radical (Eq. 7–23) is referred to as a *semiquinone,* by analogy with the semiquinone formed from *p*-benzoquinone (Eq. 7–24).

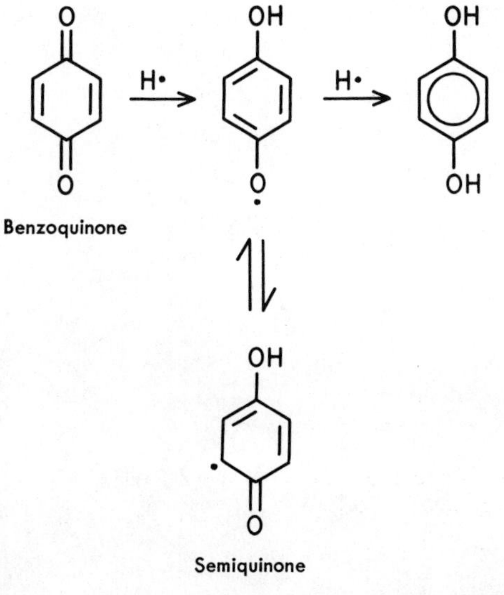

(7–24)

Free radicals can be detected by measuring the absorption of microwave radiation in a strong magnetic field. This technique is known as an *electron spin resonance* (ESR). By means of this technique, free radicals have been detected in flavoprotein reactions. In the enzyme, *xanthine oxidase* (a flavoprotein), a second free radical can be detected by its ESR spectrum due to the presence of a transition metal cofactor, molybdenum (Mo), and the transient conversion of Mo^{6+} to Mo^{5+} during catalysis. The ESR kinetics of the conversion of Mo^{6+} to Mo^{5+} is the same as the kinetics of appearance of a riboflavin free radical, indicative of a concerted mechanism. Details of substrate participation in the reaction mechanism are still obscure.

Heme-containing enzymes are of crucial biological significance in primary oxidation processes and in *oxidative phosphorylation* (the coupling of oxidation to the formation of high energy ATP). In some heme proteins, the site and nature of the interaction of heme with substrate has been identified, for example, in hemoglobin, myoglobin, catalase, and peroxidase. In a large number of processes, however, the detailed interrelationship between heme group, substrates, and products of reaction is totally obscure. The general structure of the porphyrin-iron complex (heme) is illustrated in Fig. 7–6. Note that the porphyrin provides four of the six coordination linkages to the

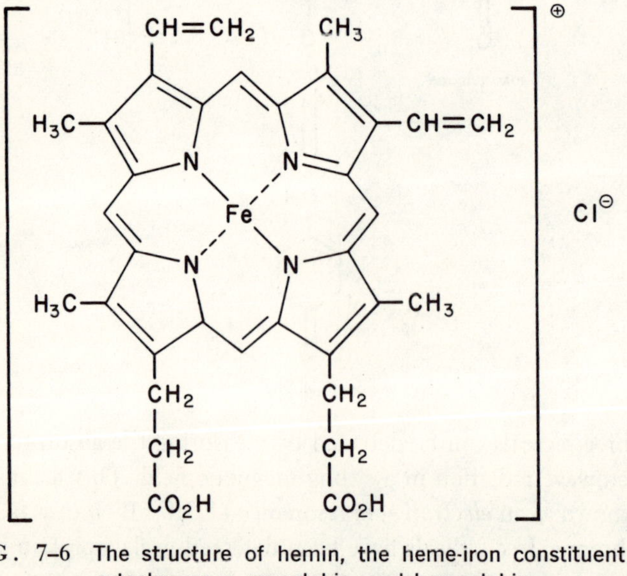

FIG. 7–6 The structure of hemin, the heme-iron constituent of cytochrome c, myoglobin, and hemoglobin.

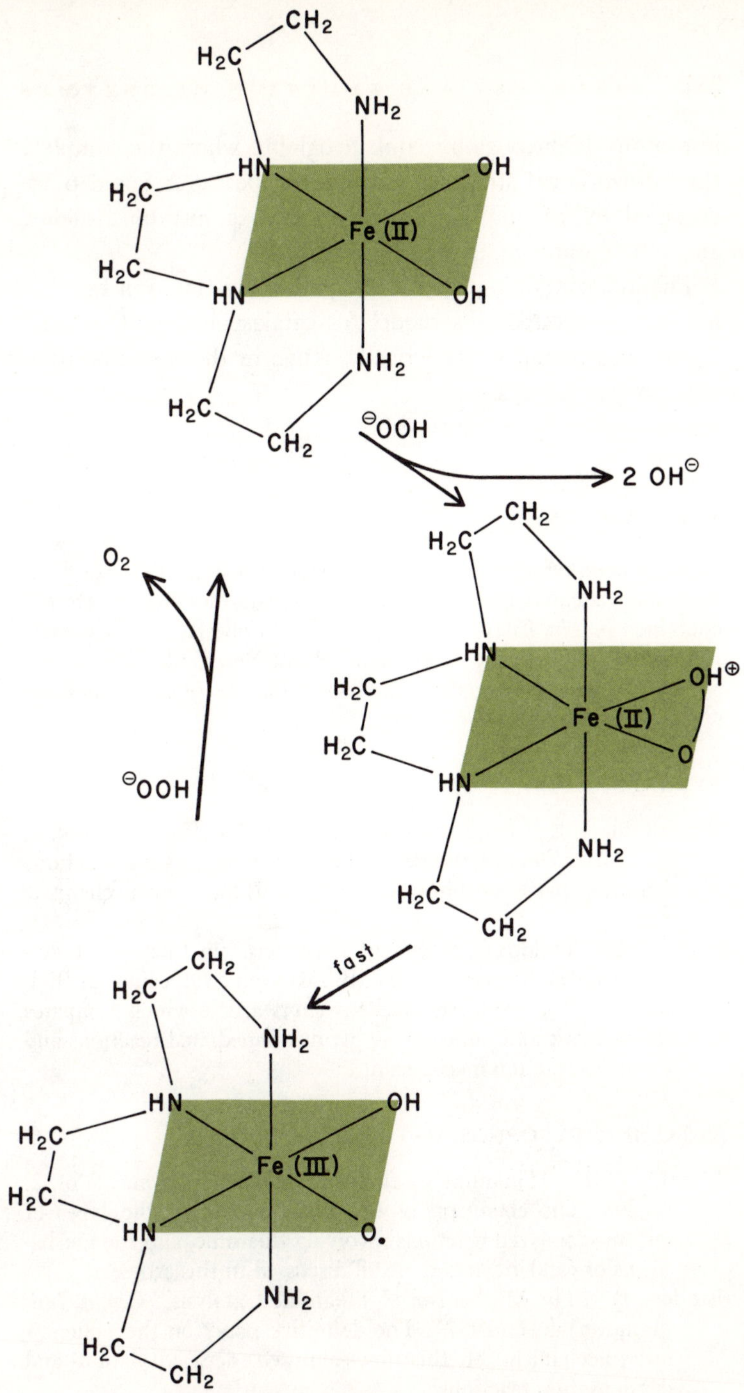

FIG. 7–7 A proposed mechanism of catalysis, by an iron chelate, in the cleavage of the O—O in hydrogen peroxide and the formation of oxygen. [Redrawn from J. H. Wang, *J. Am. Chem. Soc.* **77**, 4715 (1955).]

iron atom. In hemoglobin and myoglobin where the complete three-dimensional structures are known, positions 5 and 6 are occupied by an imidazole nitrogen from a histidine residue, and either water or oxygen.

The iron-containing chelate compound illustrated in Fig. 7–7 has been suggested as a model for catalase action. This iron-chelate compound is catalytically active in the decomposition of hydrogen peroxide.

REFERENCES

For general references on reaction mechanisms, see Chapter 6.

A comprehensive general reference on coenzymes and cofactors is contained in *The Enzymes* (Boyer, P. D., Lardy, H., and Myrbäck, K., editors), Vols. 2 and 3, Academic Press, New York, 1960.

The following are specific references to the role of coenzymes and cofactors in catalysis.

PYRIDOXAL PHOSPHATE

Braunstein, A. E., "Pyridoxal Phosphate," in *The Enzymes*, Vol. 2, p. 113. A 70-page review of the variety of enzymic reactions requiring pyridoxal phosphate and their common mechanistic basis.

Snell, E. E., "Pyridoxal-Dependent Reactions," in *Enzyme Models and Enzyme Structure*, p. 32, Brookhaven Symposium in Biology, No. 15, 1962. An excellent short article which compares the enzymic and nonenzymic pyridoxal-mediated reactions and discusses reaction mechanism.

THIAMINE PYROPHOSPHATE

Metzler, D. E., "Thiamine Coenzymes," in *The Enzymes*. Vol. 2, p. 295. The chemistry of the thiazolium ring, the types of enzyme-catalyzed reactions involving thiamine, and the mechanisms of catalytic action are all discussed in this article.

Breslow, R., "The Mechanism of Thiamine Catalysis," *Chem. Ind.* (London) 893 (1957). The definitive paper on the nonenzymic mechanism of thiamine-catalyzed decarboxylation and condensation reactions.

Comprehensive discussions of the chemistry of, and catalysis by, pyridoxal and thiamine are to be found in Bruice, T. C., and Benkovic, S., *Bioorganic Mechanisms*, Vol. 2, p. 181, Benjamin, New York, 1966.

PYRIDINE NUCLEOTIDES

Kaplan, N. O., "The Pyridine Coenzymes," in *The Enzymes*, Vol. 3, p. 105.

Kosower, E. M., "Charge-Transfer Complexing of Pyridinium Rings," in *The Enzymes*, Vol. 3, p. 171.

Bruice, T. C., and Benkovic, S., *Bioorganic Mechanisms*, Vol. 2, p. 301.

All three references discuss the chemistry of nicotinamide and reduced nicotinamide derivatives extensively and well, and indicate how little is understood about the actual mechanistic details of enzymic hydride transfer.

THIOL ESTERS AND COENZYME A

Jaenicke, L. and Lynen, F., "Coenzyme A," in *The Enzymes*, Vol. 3, p. 3. A comprehensive review of the chemistry of coenzyme A, its function in metabolism, mechanism of action and the enzymes which require it. Highly recommended.

Some further details of chemical mechanisms involving thiol esters are discussed in Bruice, T. C., and Benkovic, S., *Bioorganic Mechanisms*, Vol. 1, p. 259.

GLUTATHIONE

Knox, W. E., "Glutathione" in *The Enzymes*, Vol. 2, p. 253. A concise review of the known metabolic pathways requiring glutathione, with a discussion of catalytic mechanisms involving hemithioacetal formation.

FLAVIN COENZYMES

Beinert, H., "Flavin Coenzymes," in *The Enzymes*, Vol. 2, p. 340. This article contains a good deal of information on the oxidation-reduction properties of flavin compounds and on systems requiring *flavin adenine dinucleotide*. As indicated, little is known about the mechanistic pathway of hydrogen transfer.

HEME PROSTHETIC GROUP MODELS

Wang, J. H., "On the Detailed Mechanism of a New Type of Catalase-like Action," *J. Am. Chem. Soc.* 77, 4715 (1955). An elegant example of the testing of a postulated mechanism by the synthesis of an appropriate enzyme model.

Wang, J. H., "The Molecular Mechanism of Oxidative Phosphorylation," *Proc. Natl. Acad. Sci., U.S.* 58, 37 (1967). An illuminating discussion of the mechanisms by which the oxygen-controlled oxidation state of iron in heme can in turn control the formation of N-phosphoimidazole from imidazole and inorganic phosphate. As is discussed, N-phosphoimidazole reacts with AMP and ADP to form ATP.

EIGHT ~ ENZYME CATALYSIS; SPECIFIC EXAMPLES

IN THIS CHAPTER WE SHALL CONSIDER IN DETAIL, THE PROPERties of a select group of enzyme-substrate systems in an attempt to relate the principles and methods thus far discussed to specific mechanisms of enzymic catalysis. In order to keep the discussion within reasonable dimensions, a somewhat arbitrary selection of enzymes has been made. The selection is based primarily on the amount of information available, the variety of the information (structural, chemical, and mechanistic), and to a lesser extent, on the significance and diversity of the biological reactions represented. Presumably, conclusions and generalizations drawn from these particular systems will be applicable elsewhere; however, this particular selection of topics should not necessarily be construed as representative of all the diverse types of enzyme-substrate systems.

In order to carry out chemical and structural investigations, a reasonably large sample of enzyme must be available. Measurements of enzyme activity, on the other hand, require minute amounts of enzyme. In the past, this consideration of quantity placed a severe restriction on the variety of enzymes for which a detailed chemical investigation was feasible. Some of the enzymes to be discussed in this chapter are readily available in gram quantities. Extensive investigations of chemical and physical properties have centered on these abundantly available enzymes.

Other factors governing the particular directions of research on enzyme catalysis are not without significance. Highly pertinent factors are (1) the ability to label, chemically, a component (or components) of the active site; (2) the presence, and the availability of techniques for measurement, of a tightly bound enzyme cofactor; (3) catalysis of a reaction in which a corresponding homogeneous solution reaction mechanism has been rather thoroughly investigated; (4) the availability or the ability to prepare crystalline enzyme samples; (5) the availability of a variety of structurally related substrates, from which a correlation between structure and reactivity may hopefully be drawn; and (6) the relevance of a specific enzyme function to a biological principle or to a physiological process.

8–1 α-CHYMOTRYPSIN AND OTHER ACTIVE SERINE ENZYMES

The two proteolytic enzymes *chymotrypsin* and *trypsin* are obtainable in large quantities from beef pancreas. Both are readily available (in gram quantities) in the crystalline state, and in fairly high purity at a relatively low cost. It is therefore not surprising that so much attention has been focused on these two enzymes. Both enzymes are involved, physiologically, in the hydrolysis of peptide bonds. A very large variety of model substrates have been prepared, and from this the substrate specificity of the two enzymes has been established. The specificity and site of reaction is shown in Fig. 8–1.

Apparently, there is no requirement of stereochemical specificity (Fig. 8–1) for the "X" substituent (the site of hydrolysis). Both enzymes will catalyze the hydrolysis of substrates other than peptides, but containing the specific structure indicated. For this reason, a variety of carboxyl derivatives of an

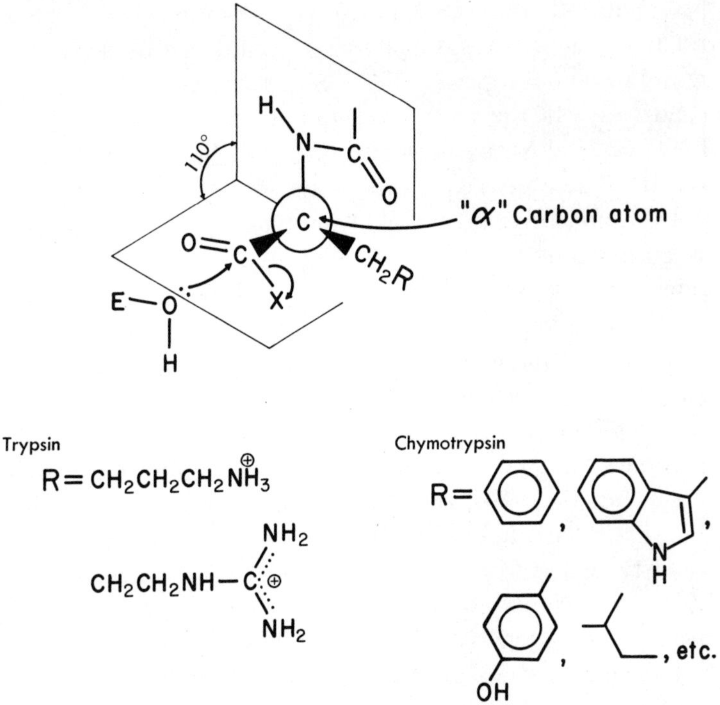

FIG. 8–1 Details of the substrate stereospecificity for trypsin and chymotrypsin catalyzed reactions.

N-acylamino acid can be examined as substrates and information can thereby be obtained on the effect of electronic structure of the X substituent on reactivity. Peptides, amides, anilides, alkyl and aryl esters, thiol esters, anhydrides, and acyl imidazoles have all been investigated and can all serve as substrates. The relative rate of hydrolysis of a given acyl derivative correlates qualitatively with its rate of hydroxide ion catalysis in homogeneous solution. Hence, there is some reason to believe that the catalytic mechanism can be interpreted with the aid of "mechanistic precedents," established on the basis of countless studies of catalysis of the solvolysis of acyl derivatives in homogeneous solution. Moreover, the structural specificities illustrated in Fig. 8–1 are not absolute. Substrates lacking one or another structural feature still are often sufficiently reactive to be studied. In this way, the effect of structure on catalyzed reactivity (as well as on binding specificity) can be studied in regard to various regions of the substrate molecule. Virtually all substrates with the general structure indicated in Fig. 8–1 exhibit "classical Michaelis-Menten kinetics"; linear plots of $[S]^{-1}$ versus v^{-1} are observed, as well as substrate saturation (Section 4–1). Competitive inhibitors of substrate reactions are known in great number. These inhibitors contain some of the structural features of substrates (Fig. 8–1) but lack a reactive center. A notable group of competitive inhibitors are the D isomers of specific L acylamino acid substrates. Whereas other features of structural specificity are not stringent, there is virtually absolute selection for L isomers in catalyzed hydrolysis. Nevertheless, the D isomers can bind firmly to the enzyme site and competitively inhibit reaction. This absolute selection of α-carbon configuration in catalysis must be indicative of a critical disposition of catalytic centers within the enzyme site.

Irreversible inhibitors of the proteolytic enzymes are well known. An important discovery in regard to enzymic mechanism is the mode of action of organophosphate esters of the type (I)

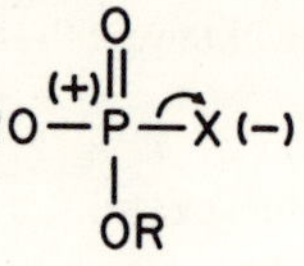

(I)

on proteolytic enzymes and esterases. It was first discovered that such compounds react specifically and irreversibly with the physiologically important enzyme *acetylcholinesterase*. Later, it was demonstrated that the same mode of irreversible inhibition occurred with chymotrypsin and trypsin. Radioactive labeling of the phosphorus atom of such compounds (with ^{32}P) can be used to follow the fate of the phosphoryl group after reaction with the enzyme. After reaction, one equivalent of radioactive phosphorus is found to be firmly bound to the enzyme. Denaturation and degradation of the ^{32}P-containing enzyme, as described in Chapter 5, leads to a stable ^{32}P-phosphoserine peptide (II). This type of experiment was the first in which an

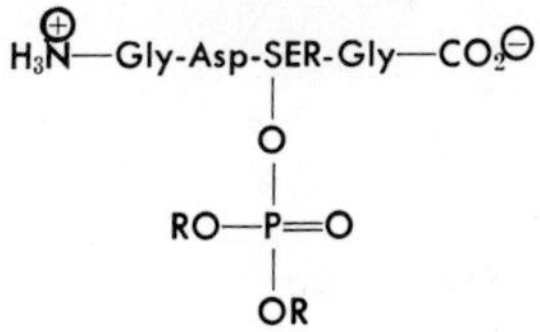

(II)

amino acid side chain involved in covalent interaction with substrate was chemically identified. The occurrence of a serine-phosphate ester came somewhat as a surprise to "enzyme mechanists" since, on the basis of both the pH dependence of enzyme activity and model studies of nucleophilic catalysts (see Fig. 6–1), a plausible mechanism of catalysis was assumed to involve nucleophilic attack by an imidazole group of histidine. The chemical similarity between the phosphate esters (II) and carboxylate esters is noteworthy. It has been postulated that the phosphate ester formed in the irreversible inhibition reaction is a stable analogue of a reactive acyl enzyme intermediate in the catalytic reaction with substrate. Strong suggestive evidence for the validity of this hypothesis comes from experiments on the reaction of nitrophenyl acetate with α-chymotrypsin. By studying the reaction with this pseudosubstrate at nearly comparable concentrations of enzyme and pseudosubstrate (a few-fold excess of pseudosubstrate), it can be demonstrated that an initial "burst" of nitrophenol occurs in amounts corresponding roughly to the number of equivalents of enzyme. This burst is followed by a slower "turnover" of the residual nitrophenyl ester. Such experiments are consistent with the rapid formation of an acyl enzyme intermediate, followed by a slower hydrolysis of the acyl enzyme to regenerate catalyst (Eq. 8–1). Indeed, by utilizing an excess of nitrophenyl ^{14}C-acetate as substrate, and by rapid denaturation of the protein after the above-noted burst reaction, a covalent ^{14}C-labeled acetyl enzyme can be recovered. The same amino acid sequence noted above (II) has been established about the site of the radioactive acetyl label.

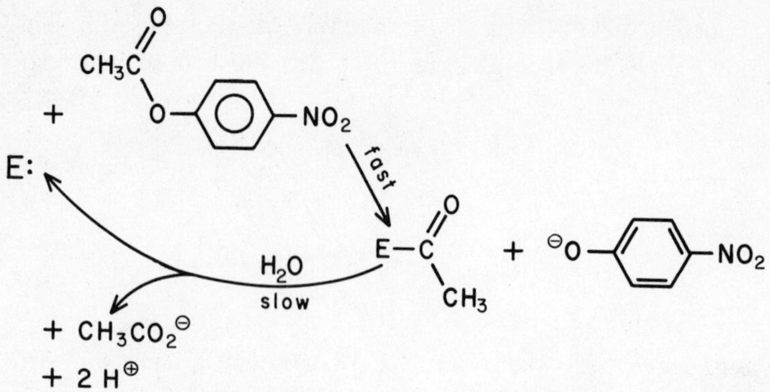

(8–1)

Such experiments demonstrate the presence of acyl ester after denaturation and degradation. They do not, however, necessarily establish the uniqueness (or the actual existence) of such acyl intermediates in the course of catalysis by native enzyme. In an attempt to examine this point, active acylating compounds of the type (III) have been prepared. These com-

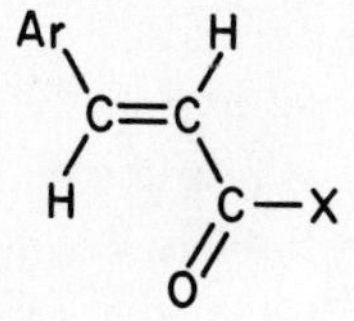

(III)

Ar = aromatic residue, e.g.,

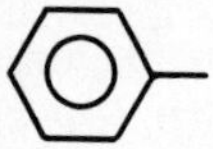

pounds exhibit strong absorption of near ultraviolet radiation due to their extended conjugation. Moreover, the wavelength position of the ultraviolet absorption bands are dependent on the precise electronic nature of the reactive acyl substituent (X). If acyl enzyme intermediates are formed during the course of catalysis, they should become apparent by examination of the ultraviolet spectrum during the catalyzed reaction process. Indeed, such acyl chymotrypsin intermediates are observable, by virtue of their intense and characteristic ultraviolet spectra. These intermediates can be trapped by denaturation of the acyl enzyme protein. Unlike the native acyl enzyme, the denatured acyl enzyme does not undergo subsequent hydrolysis at neutral or moderately acidic pH. Both from the characteristic ultraviolet spectra, and by direct chemical analysis, these denatured acyl intermediates have been shown to be acyl serine derivatives. A peculiarity in this otherwise consistent acyl enzyme intermediate picture is that the native (active) chromophoric acyl enzyme intermediates do not correspond in spectra with the corresponding denatured or the degraded acyl derivatives, nor with the spectra of small synthetic model O-acyl serine esters. Perturbations of the spectra due to the "solvent effect" of the native enzyme site environment in the native acyl enzyme is apparently ruled out as an explanation on the basis of the magnitude of the spectral shift.

Similar acylation reactions can be carried out with the proteolytic enzymes trypsin and bacterial subtilisin. In all of these enzymes, a particularly reactive serine residue has been shown to be the site of acylation. The spectra of the native acyl enzymes of the type III are essentially the same as in the corresponding native acyl chymotrypsins.

The complete amino acid sequences of the catalytically inactive precursors chymotrypsinogen and trypsinogen (i.e., *zymogens*) are illustrated in Fig. 8–2. In each case, activation of the zymogen is achieved by proteolytic (tryptic) hydrolysis at a particular peptide bond as indicated in the figure. Further rapid proteolytic hydrolysis of chymotrypsin at a few specific peptide linkages leads to different multichain structures. These various forms of chymotrypsin differ only trivially both in activity and in substrate specificity. Completion of this rapid proteolysis leads to the three-chain active enzyme structure commonly referred to as α-chymotrypsin. Recent evidence indicates that in the primary tryptic activation process (Fig. 8–2), an *essential* $^{+}H_3N$-terminal group is created which is in some way responsible for maintaining the structural integrity of the catalytic site. The reversible pH-dependent formation of active site is correlated with the pK_a of this terminal α-ammonium group. Due to

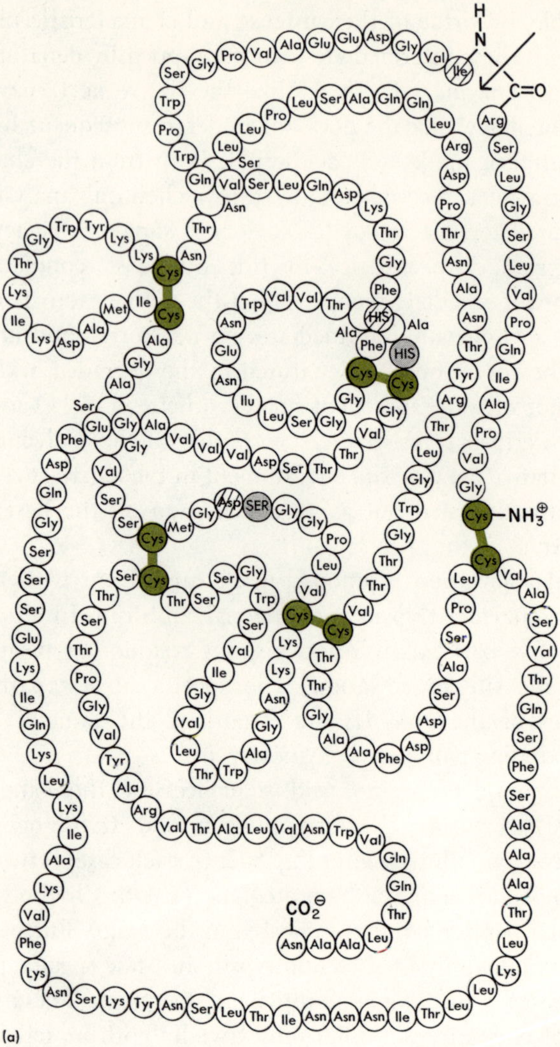

(a)

FIG. 8–2 The primary structures of chymotrypsinogen (a) and trypsinogen (b). (Courtesy of Dr. J. Brown.) The arrow indicates the position of primary cleavage (in each structure) for activation of the zymogen to the functional enzyme. Residues shown in gray are known to play a crucial role in the catalysis. Residues shown in lined area are presumed to be of significance to the catalytic mechanism.

(b)

this pK_a, the dissociation constant for competitive inhibitors varies with pH (Fig. 8–3, colored curve), whereas the V_{max} versus pH profile for specific substrates of chymotrypsin is a sigmoid-shaped curve (Fig. 8–3, black curve). At any particular concentration of substrate, the pH-activity profile will be either bell-shaped, or sigmoidal, depending on the extent of saturation of enzyme with substrate (a pH variable). Since the dependen-

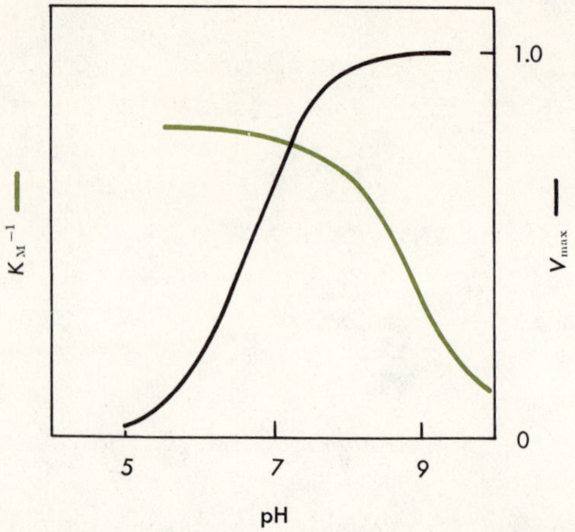

FIG. 8–3 Dependence of binding affinity (K_M^{-1} or K_I^{-1}) and V_{max} on pH for α-chymotrypsin.

cies on substrate concentration and pH have not usually been examined in such kinetic detail with most enzyme-substrate systems, one may wonder to what extent the commonly observed bell-shaped pH-activity profiles in other systems are due to a superposition of two sigmoidal curves, one for binding specificity and one for catalytic reactivity.

Chymotrypsin and trypsin, like most other enzyme proteins, are catalytically active in "heavy water" (D_2O). Deuterium solvent slows down the rate of catalysis by a factor of about two. The binding of competitive inhibitors is not significantly affected by the transfer from H_2O to D_2O. These facts again argue strongly for the involvement of general acid-base catalyzed steps in the rate controlling processes, since the rate of transfer of a deuteron is slower than the rate of transfer of a proton. In nonenzymic reactions of related carboxylic acid derivatives, the same effect of deuterium isotope on rate is observed, whenever general acid-base catalysis can be shown to be a rate-limiting process. An important distinction between imidazole catalysis via a *nucleophilic* pathway and imidazole catalysis via a *general acid-base* pathway is the predicted effect of deuterium isotope on rate. In nucleophilic catalysis, where the free base attacks at the electrophilic carbonyl carbon, there should be no effect of the deuterium solvent on the reaction rate (to a first approximation), whereas in general acid-base catalysis, the rate of transfer of a deuteron will be measurably slower than the corresponding proton transfer rate.

That an imidazole residue is functioning catalytically in the proteolytic enzymes is suggested by the fact that V_{max} is invariably correlated with a reversible enzyme —H^+ dissociation, consistent with that predicted for a single residue with a pK_a of approximately 7 (close to that of the imidazolium cation). Direct evidence for the involvement of imidazole in chymotrypsin catalysis has been obtained utilizing the bifunctional reagent, tosyl-L-phenylalanylchloromethyl ketone described in Chapter 5. The above discussion suggests that an imidazole residue (of histidine) participates in a general acid-base catalyzed process, but does not exclude the additional possibility of imidazole participation via a nucleophilic pathway.

Thus far we have presented direct evidence in chymotrypsin for the catalytic necessity of the $^+NH_3$-terminal isoleucine, of at least one of the two histidine residues (residue 57), and of one serine residue (residue 195). The residues known to participate in the catalytic process are located in very different parts of the covalent amino acid sequence. There has been some suggestive evidence for the involvement of a carboxylate residue based on the catalytic and binding specificity behavior at lower pH's (where there is an apparent dependence on the ionization of a carboxylic acid residue). The emphasis on aspartate 194 is based merely on its obvious proximity to the active serine hydroxyl (Fig. 8–2).[1] Although model studies have shown unusual reactivity for aspartyl-seryl peptides (see Fig. 6–3), the correlation of such intramolecular models with proteolytic enzyme catalysis is questionable in light of the very similar catalysis observed with bacterial proteolytic enzymes, *subtilisins* (see later).

Rapid reaction techniques have been applied to the study of specific substrate reactions with both α-chymotrypsin and trypsin. Since, at neutral pH, these reactions are driven to comple-

[1] See Appendix 8–I.

tion by ionization of the carboxylic acid product, the relaxation techniques described in Chapter 4 are not usually applicable. Such techniques require significant concentrations of both reactants and products at equilibrium. The flow methods described in Chapter 4 are, on the other hand, readily applicable to these reactions. Four methods for following such reactions have been utilized.

(1) *The use of a chromophoric competitive inhibitor as an indicator of otherwise available (unoccupied) sites.* If the binding of a chromophoric competitive inhibitor to enzyme results in a change in the spectrum of the competitive inhibitor (as is the case with both chymotrypsin and trypsin and the acridine dye, proflavin), temporal changes in the *dissociability* of the enzyme-substrate complex due to chemical reaction will be reflected in a concomitant change in the extent of bound competitive chromophore. This technique is illustrated in Fig. 8–4.

(2) *The use of a chromophoric substrate to measure changes in enzyme-substrate affinity concomitant with chemical reaction.* It is possible to covalently attach a strong chromophore to regions of the substrate outside of the specificity region indicated in Fig. 8–1, for example at the N-acylamino substituent. If the spectra of chromophores are sensitive to the immediate environment (as strong chromophores usually are), the temporal transitions from chromophore in aqueous solution to physically adsorbed enzyme-substrate complex to enzyme-substrate chemical intermediates, and thence to free enzyme and chromophoric products in aqueous solution, can be fol-

FIG. 8–4 The transient fate of the trypsin-proflavin (enzyme-competitive inhibitor) complex during the reaction of enzyme with substrate (carbobenzyloxy-L-lysine nitrophenyl ester) at pH 5.3, 25°C; (a) 100 msec per major division and (b) 10 msec per major division. The spectral changes (at 470 mμ) can be directly interpreted in terms of an initial desorbtion of proflavin due to the formation of an enzyme-substrate covalent intermediate, and a subsequent reappearance of the enzyme-proflavin complex as the enzyme-substrate intermediate is converted to a product of low binding affinity.

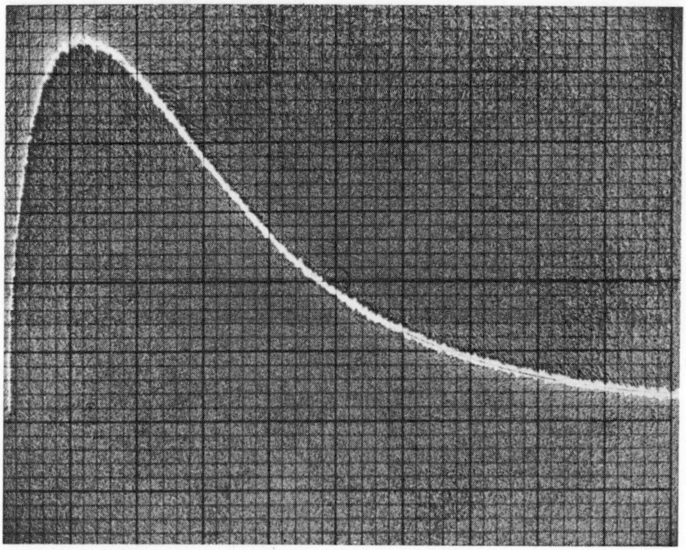

(a)

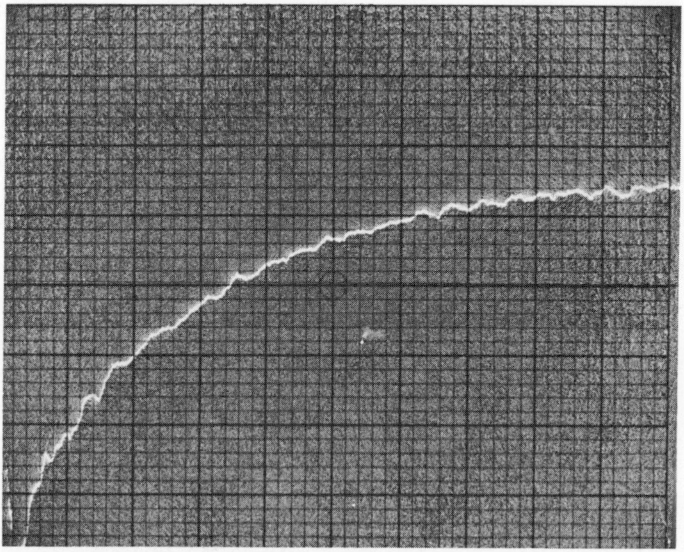

(b)

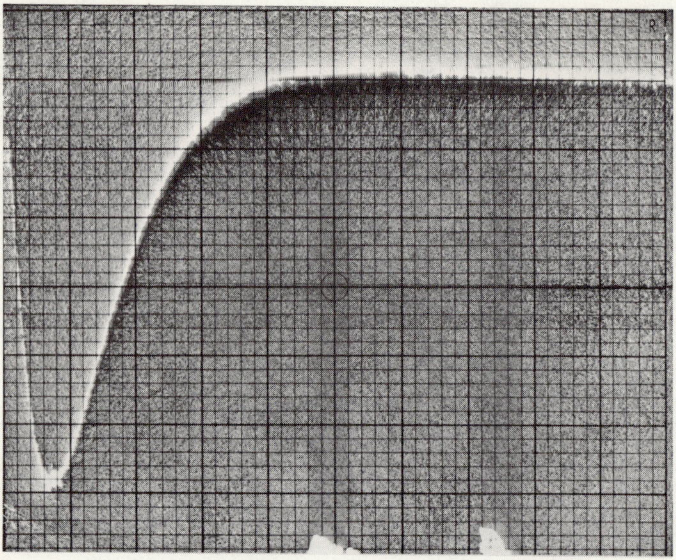

FIG. 8–5 Changes in the absorption spectrum of the β-(2-furyl)-acryloyl chromophore in the specific substrate N-furyl-acryloyl-L-arginine methyl ester, during its transient reaction with an excess of trypsin; pH 5.3, 25°C, 330 mμ; 500 msec per major division. Notice (as in Fig. 8–4) the occurrence of a *minimum* of two consecutive transient chemical reactions.

lowed spectrophotometrically. An example is shown in Fig. 8–5.

(3) *The detection of one of the products of reaction by spectral or chemical properties.* In the overall reaction (Eq. 8–2), if the product (HX) is independently measurable, the

$$\mathrm{RC(=O)X} + \mathrm{H_2O} \rightarrow \mathrm{RCO_2^{\ominus}} + \mathrm{H^{\oplus}} + \mathrm{HX} \qquad (8\text{–}2)$$

rate of release of X relative to the overall turnover of substrate will give information regarding potential intermediates in the chemical catalytic process. For example, according to the acyl enzyme hypothesis, the product (X) should appear concomitant with formation of the acyl enzyme, the overall turnover of substrate, however, being dependent on the hydrolysis as well as on the formation of the acyl enzyme intermediate. This method has been utilized by taking advantage of the spectrophotometric properties of particular products, and by direct chemical analysis of X, as is illustrated in Figs. 8–6 and 8–7.

Although mechanistic conclusions based on rapid flow experiments are far from complete, it is apparent (even from the illustrative examples in the figures) that more steps in the reaction sequence are observed than would have been predicted from the straightforward acyl enzyme hypothesis. Rapid flow kinetic studies have thus far led to the conclusion that at least one other enzyme-substrate intermediate is formed reversibly, and with measurable kinetics, from the physically bound enzyme substrate (ES) complex (Eq. 8–3).

$$\mathrm{E+S} \rightleftharpoons \mathrm{ES} \rightleftharpoons \mathrm{ES'} \rightleftharpoons \begin{array}{c}\mathrm{X}\\ + \\ \mathrm{P_1}\end{array} \rightleftharpoons \mathrm{EP'} \rightleftharpoons \mathrm{EP} \rightleftharpoons \mathrm{E+P} \qquad (8\text{–}3)$$

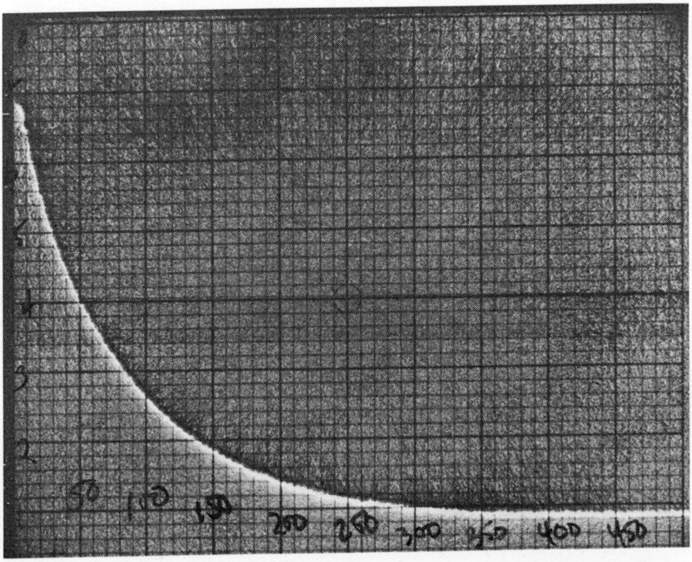

FIG. 8–6 The release of *p*-nitrophenol from the identical *p*-nitrophenyl ester-enzyme system described in Fig. 8–4. Transient changes in the ester-nitrophenol spectrum (50 msec per major division); pH 5.30, 25°C, 340 mμ. Notice the difference in time constant in comparison with Fig. 8–4a, b.

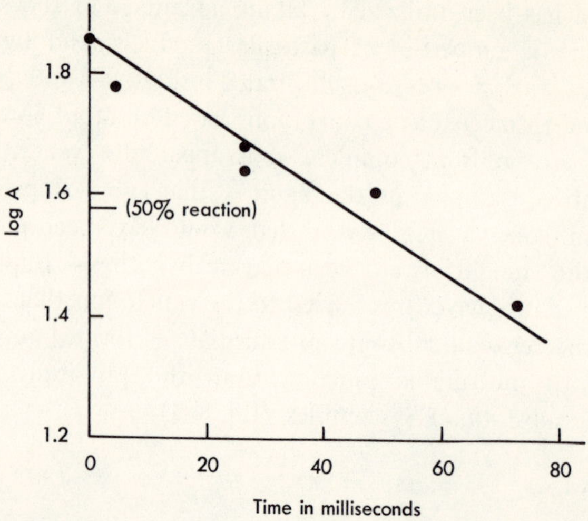

FIG. 8–7 The liberation of ethanol from benzoyl-L-arginine ethyl ester. [From T. E. Barman and H. Gutfreund, *Proc. Natl. Acad. Sci. U.S.* **53**, 1245 (1965).]

The necessity for postulating a second intermediate (ES′) arises from the dependence of reaction rate (as determined by one or another of the indicator processes discussed above) on enzyme and substrate concentration, and from the fact that the rate of P_1 liberation (Fig. 8–6) can be much slower than the fastest *measurable* rate utilizing either of the above-mentioned chromophoric indicator techniques. The possibility that this second intermediate (ES′) is in actuality a composite of a number of intermediates, for example, as in Eq. 8–4,

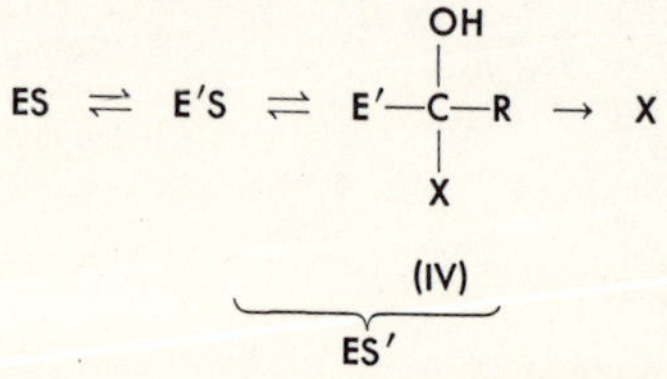

(8–4)

can in no way be excluded; it is merely the simplest mechanism consistent with available experimental facts. Two molecular interpretations of this second intermediate have been postulated. One possibility, for which there is good presumptive evidence is that the transition ES $\rightarrow$ ES′ represents a conformational change in the enzyme protein, or more precisely, ES $\rightleftharpoons$ E′S. This possibility is strongly suggested by the observable changes in spectral characteristics of the constituent tyrosine and tryptophan chromophores of the protein, which occur concomitantly with the early phase of substrate reaction. A second possible explanation involves the formation of a chemical intermediate in which none of the original substrate bonds have been broken (loss of P_1 is irreversible, whereas the transition ES $\rightleftharpoons$ ES′ must, on the basis of experimental evidence, be reversible). A plausible intermediate is the tetrahedral intermediate, IV, of Eq. 8–4. A comparison of these two alternative suggestions is interesting as a reflection of current thinking in enzymology. The first alternative suggests a dynamic role for the enzyme protein in mediating catalysis, whereas the latter relies on previous precedents for organic reaction mechanisms, and implicitly assumes a stereochemically *invariant* and catalytically complementary relationship between enzyme and substrate throughout the course of catalysis.

The *bacterial subtilisins* are known to contain a similar total number of amino acid residues per active site. Unlike trypsin and chymotrypsin (Fig. 8–2), they contain no disulfide linkages. Hence, the three-dimensional structure of these molecules must be governed by specific *intrachain* hydrogen bonds, and by van der Waals and coulombic forces exclusively. These enzymes contain a similarly and uniquely reactive serine residue, but the covalent sequence about the serine is markedly different (Eq. 8–5).

Thr-Ser*-Met-Ala (subtilisin)
Asp-Ser*-Gly-Gly (trypsin and chymotrypsin) (8–5)

Since there is no aspartate residue near the reactive serine in the covalent sequence of subtilisin, the necessity of this residue for chymotryptic and tryptic catalysis is questionable.

In addition to the proteolytic enzymes (trypsin, chymotrypsin, thrombin, elastase, and the subtilisins), a number of other enzymes involved in the hydrolysis of carboxylic acid derivatives have been shown to contain a similarly (and uniquely) reactive serine residue. The primary function of such enzymes is the hydrolysis of naturally occurring esters. Acetylcholinesterase, liver esterases, and serum esterases apparently function via the same catalytic mechanisms. Phosphate esters have many chemical properties similar to those of carboxylic acid esters. Enzyme proteins with a single uniquely reactive serine hydroxyl residue are known to participate both in the hydrolysis of phosphate esters (alkaline phosphatase) and in the transfer of phosphate from one phosphate ester linkage to another (phosphoglucomutase and phosphofructomutase). In these enzymes, the phosphoryl group can be considered analogous to the acyl group. It has been demonstrated that in the reaction of alkaline phosphatase with substrate, an intermediate *phosphoserine* enzyme (Eq. 8–6) is formed. An extensive sequence

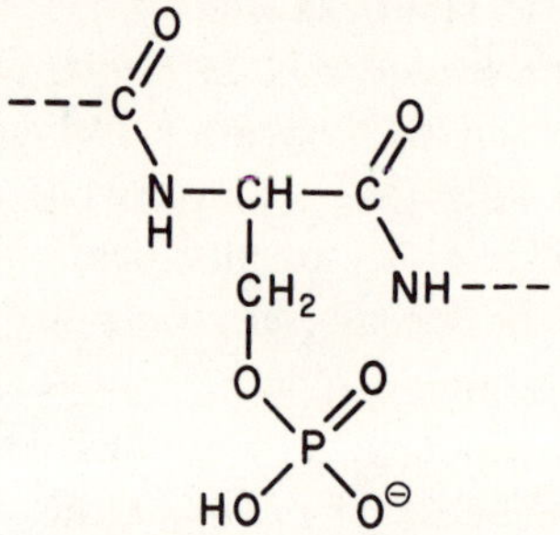

(8–6)

Phosphoserine peptide

of amino acids about this reactive serine has been determined in alkaline phosphatase (see Table 5–2) using radioactive ^{32}P-phosphate ester substrate, or ^{32}P-phosphate itself, for the formation of the phosphoserine ester linkage. The fact that appreciable amounts of phosphoserine ester are formed from free phosphate and enzyme at neutral pH is somewhat surprising, in light of the known free energy of hydrolysis of phosphate esters in neutral aqueous solution (Eq. 8–7).

$$ROH + HO{-}\overset{\displaystyle O}{\overset{\|}{\underset{\underset{\displaystyle O^{\ominus}}{|}}{P}}}{-}O^{\ominus} \rightleftharpoons RO{-}\overset{\displaystyle O}{\overset{\|}{\underset{\underset{\displaystyle O^{\ominus}}{|}}{P}}}{-}O^{\ominus} + H_2O \qquad (8\text{–}7)$$

The active serine phosphate ester of the enzyme is thermodynamically more stable than ordinary phosphate esters (and ordinary serine peptide phosphate esters). Nevertheless, in neutral aqueous solution, it is far *more reactive* than the thermodynamically less stable ordinary serine phosphate esters.

8–2 RIBONUCLEASE

The primary structure of a protein, and its effect on enzyme activity, has been most extensively investigated for the enzyme ribonuclease. It is therefore appropriate to begin a discussion of ribonuclease by considering its primary structure. This is shown schematically in Fig. 8–8. The rather odd shape of the (hypothetical) two-dimensional structure shown in the figure is based on a large body of experimental fact relevant to the identification of specific amino acid residues involved in, and located at the catalytic site, as summarized in Fig. 8–9. Other residues cannot necessarily be excluded from involvement in the catalytic mechanism.

A discovery useful for correlating structure with function is the specific cleavage of ribonuclease between residues 20 and 21 by the bacterial proteolytic enzyme subtilisin. The resultant two peptides can be separated chromatographically in denaturing solvents. These peptides are catalytically inactive individually after removal of the denaturant. On reincubation of the two peptide fragments in the absence of denaturant, a specific noncovalent recombination takes place to yield an essentially fully active ribonuclease. Aside from the important bearing of this result on the existence of complementary intrapolypeptide chain interactions in enzyme proteins, it presents a method for specific chemical modification of amino acid residues on each

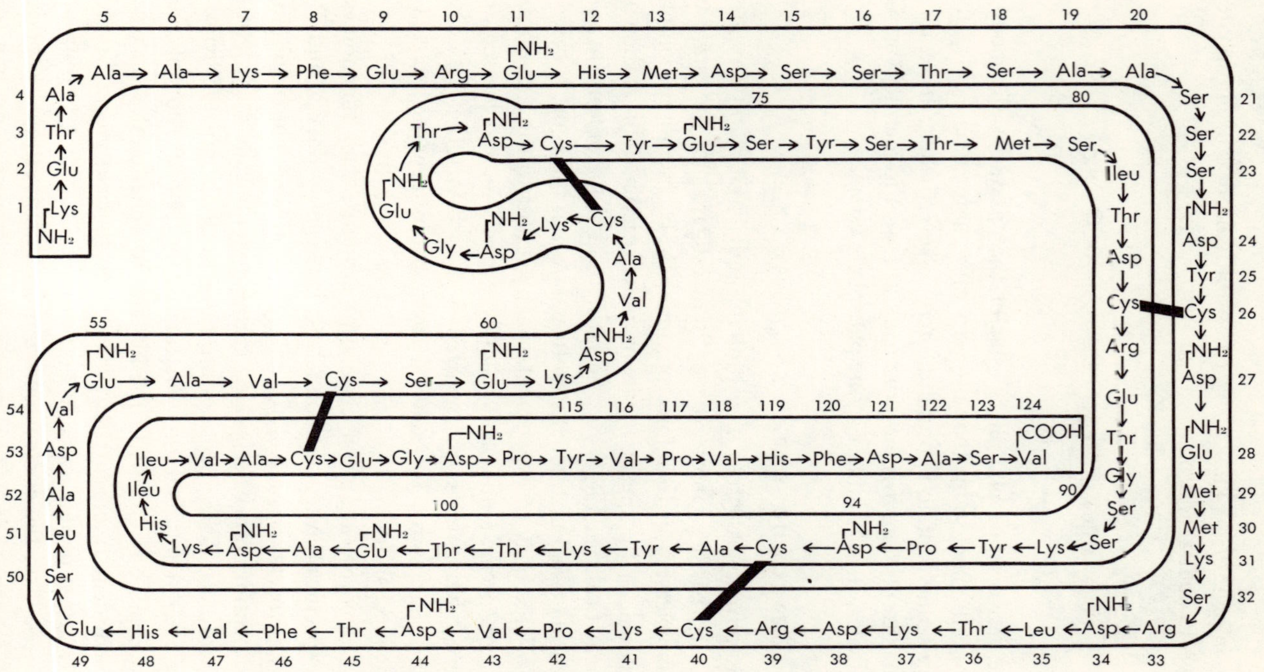

FIG. 8–8 The primary sequence of amino acids in pancreatic ribonuclease.

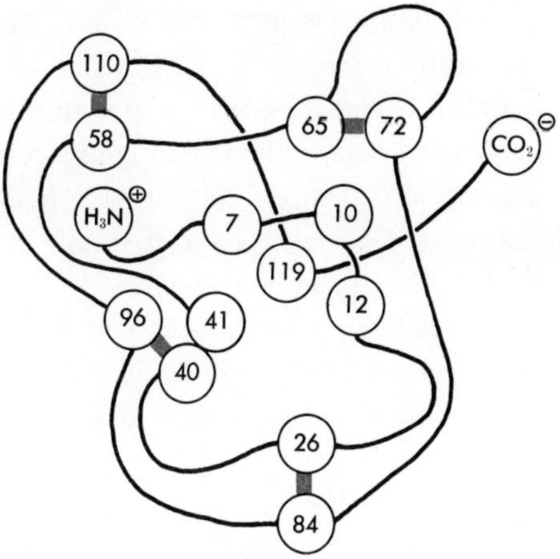

FIG. 8–9 A schematic spatial arrangement of amino acid side chains in ribonuclease (indicating the disulfide linkages) and the amino acid residues which are known to be involved in catalytic function.

of the two polypeptide chains independently, and thereby allows a study of the effect of such individual modifications on the catalytic activity of the reconstituted active enzyme protein.

Upon treatment of *native* ribonuclease with iodoacetate, histidine (12) and histidine (119) can be chemically modified independently. Thus each of these residues has been shown to be essential for catalytic activity. Note that these two histidine residues are at nearly opposite (covalent) ends of the polypeptide chain. Likewise, the two lysine residues, 41 and 7 (shown in proximity in the figure), have been modified independently. Each has been shown to be involved in the catalytic process. By inference, the two lysine residues were assumed to be in proximity. In striking confirmation, the bifunctional reagent (V), a

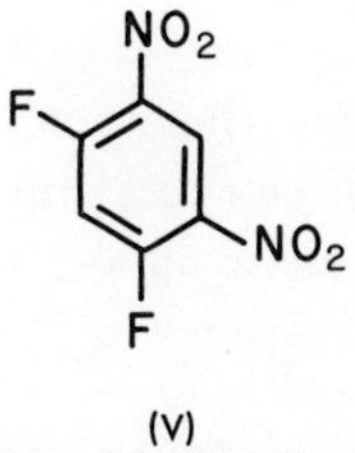

(V)
2,4-Dinitro-1,5-difluorobenzene

reagent specific for amino groups, forms a chemical bridge between the two ϵ-amino groups of these particular lysine residues. This resultant bridged derivative is catalytically inert. Phosphate, in concentrations sufficient to competitively inhibit ribonuclease action, very strongly inhibits the inactivation by this bifunctional reagent (as well as inactivation by the monofunctional derivative, 2,4-dinitrofluorobenzene).

The smaller ribonuclease peptide fragment (residues 1–20) obtainable from the specific cleavage with subtilisin has been synthesized. In the course of chemical synthesis, many shorter peptide sequences were prepared as well. The potential for reformation of active enzyme has been studied as a function of the length and composition of these shorter peptide chains. Such a synthetic approach permits more straightforward modification of specific amino acid residues.

An interesting result, bearing on the *specific* recombination of peptide residues, is the reformation of active enzyme from fully "reduced" ribonuclease, that is, ribonuclease in which all of the disulfide linkages have been reduced to (cysteine) sulfhydryls. The reduced enzyme is catalytically inert. In solvents in which the native disulfide-bridged enzyme is catalytically active, virtually complete activity can be eventually restored upon reoxidation of the sulfhydryl groups to disulfides. Under these particular conditions, all of the original specific S—S bridges are reformed. In the presence of solvent denaturants such as concentrated urea solutions, reoxidation of the reduced sulfhydryl groups to intrachain disulfides yields virtually no active enzyme (after removal of the denaturant). The kinetics of reformation of the "correct" disulfide linkages (for restoration of activity) indicate that "wrong" disulfide linkages are made during the process, but that these wrong linkages are readily reduced and reoxidized until virtually all the correct linkages have been made. Hence, *in this case,* the properly folded polypeptide chain (for catalytic activity) appears to be the thermodynamically most stable conformation. These results are indicative of very extensive complementary interactions between different regions of the polypeptide chain.

A clue to the mechanism of action of ribonuclease is the finding of 2′,3′-cyclic phosphates (VI) in ribonuclease digests of ribonucleic acid. Simple 2′,3′-nucleotide cyclic phosphates can serve as substrates of ribonuclease. The chemical fate of ribonucleic acid substrate appears to be given straightforwardly by Eq. 8–8.

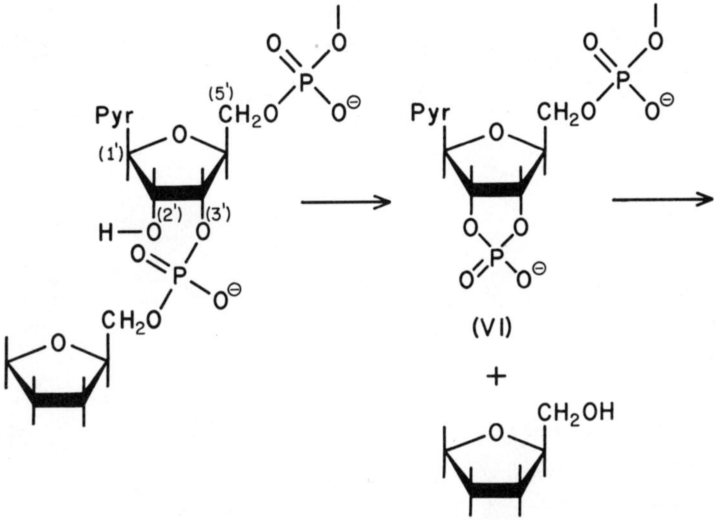

Pyr = pyrimidine ring (uracil or cytosine) (8–8)

The pH dependence of the hydrolysis of these synthetic nucleotide substrates as well as of ribonucleic acid is as shown in Fig. 8–10. This bell-shaped profile is suggestive of a concerted general acid-base catalyzed mechanism, and the apparent pK_a's are suggestive of two histidine residues, one functioning as a general acid, the other as a general base. The demonstration that the two histidine residues of the ribonuclease molecule are essential to catalysis greatly strengthens this supposition. A mechanism involving concerted acid-base catalysis

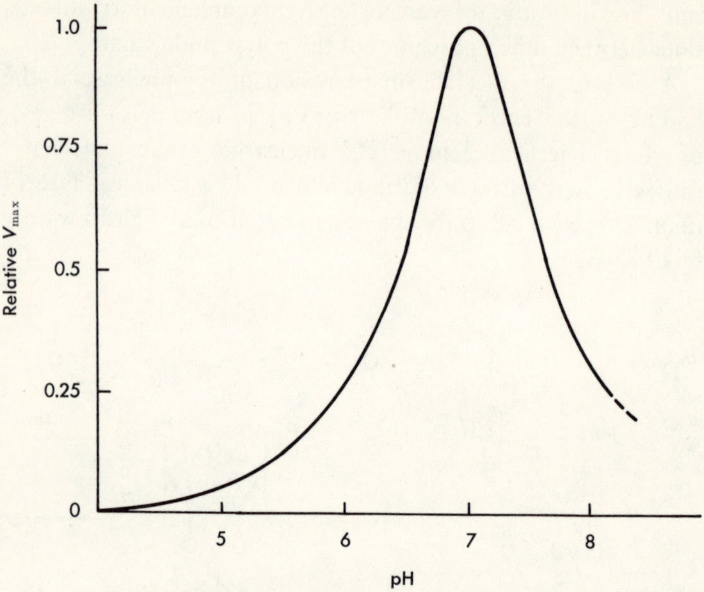

FIG. 8–10 The dependence of V_{max} on pH in the ribonuclease catalyzed hydrolysis of nucleoside 2′,3′-cyclic phosphate.

is shown in Fig. 8–11. Note the complementary functions of the two imidazole residues in the two analogous steps of the reaction mechanism.

8–3 ENZYMES REQUIRING PYRIDOXAL PHOSPHATE

Many enzymes involved in the metabolism of amino acids are known to contain firmly bound pyridoxal phosphate. It has been established, in every case thus far examined, that this firm attachment of the prosthetic group to the enzyme is via a covalent linkage to a specific lysine residue. An obvious interpreta-

FIG. 8–11 A proposed mechanism for the ribonuclease catalyzed hydrolysis of 3′,5′-phosphodiester involving two imidazole (histidine) residues, in accord with the experimental data of Fig. 8–10.

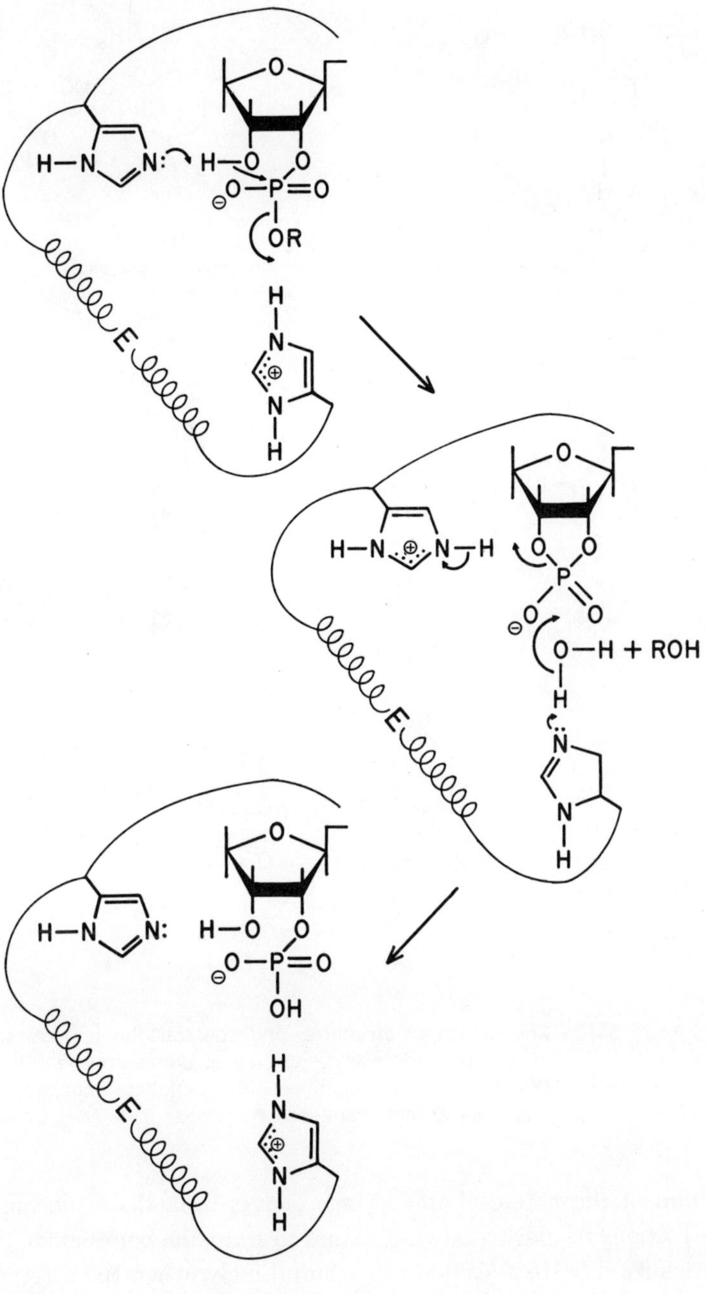
H—N
N:
H—O
O
⊖O—P=O
OR
H
N
⊕
N
H
E
H—N
⊕
N—H
O
O
P
⊖O
O
O—H + ROH
H
N
N
H
H—N
N:
H—O
O
⊖O—P=O
OH
H
N
⊕
N
H
E

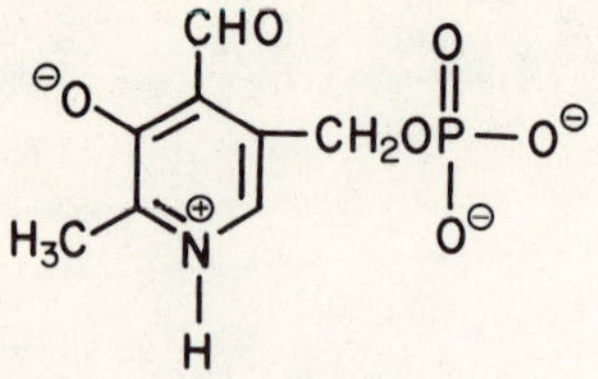

VII
Pyridoxal phosphate
(λ_{max} 388 mμ)

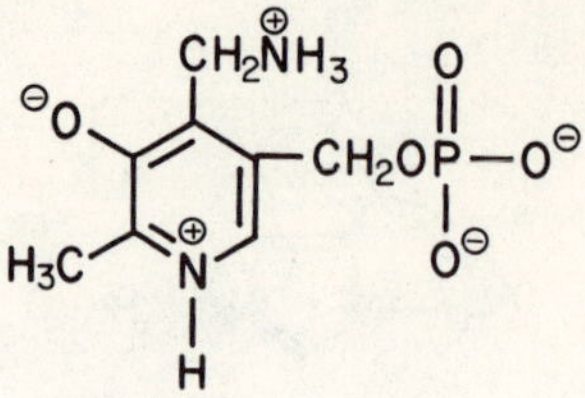

VIII
Pyridoxamine phosphate
(λ_{max} 325 mμ)

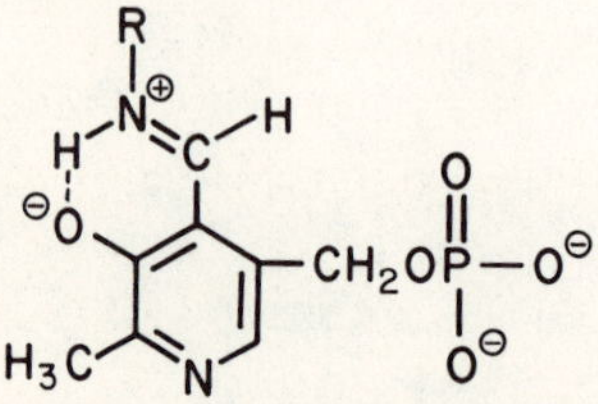

IX
(λ_{max} 400–430 mμ)

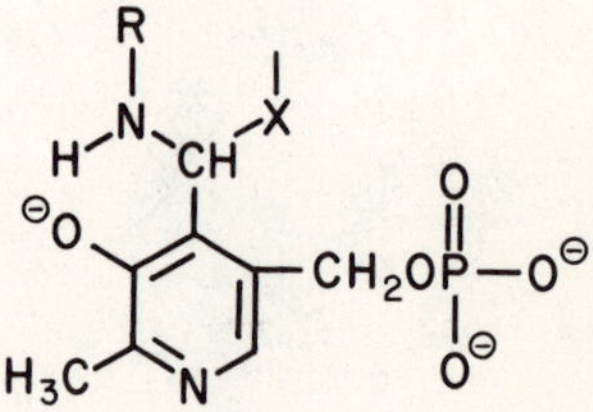

X
(λ_{max} 325–335 mμ)

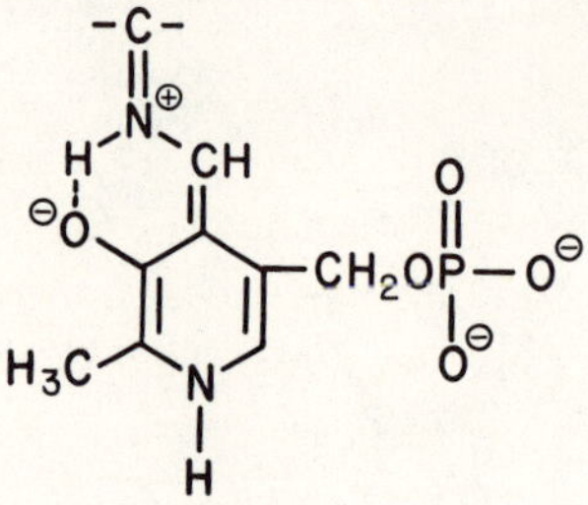

XI
(λ_{max} 500 mμ)

FIG. 8–12 The molecular structure, and spectral characteristics, of various pyridoxal derivatives of demonstrated (VII–IX) and presumed significance in pyridoxal phosphate-mediated enzyme-catalyzed reactions.

tion of the nature of this linkage derives from the analogous reactions of aldehydes with amines to form the corresponding Schiff's base (see Section 6–3). Surprisingly, when the spectra of these chromophoric enzyme-pyridoxal compounds are examined (Table 8–1), a variety of spectral characteristics are observed, dependent on both the particular enzyme-pyridoxal complex, and on the pH. The spectra of pyridoxal phosphate and various of its derivatives have been investigated in some detail. The wavelength maxima of some structurally defined pyridoxal phosphate derivatives are shown in Fig. 8–12.

These particular pyridoxal derivatives (VII–XI) are relevant to pyridoxal-enzyme-catalyzed transamination reactions (see Fig. 7–3 for the corresponding nonenzymic pyridoxal-catalyzed reactions). Note that all of these derivatives are capable of either binding or dissociating a proton at nitrogen atoms, which are part of a conjugated system, or at the phenoxide oxygen. For this reason, all of these derivatives exhibit pH-sensitive spectra. It is noteworthy that some, but not all, of the enzyme-pyridoxal compounds listed in Table 8–1 exhibit pH-sensitive spectra.

It has been established that the firm linkage between pyridoxal phosphate and enzyme is via an ϵ-amino group of a specific lysine residue (see Chapter 5). Straightforward reaction of the aldehyde with the primary amine (of lysine) should lead to the Schiff's base (IX), as is indeed the case in model reactions of pyridoxal with amines. At moderately acidic pH, an enzyme pyridoxal complex should exhibit the same spectral characteristics as compound IX. An examination of Table 8–1 reveals that this is sometimes but by no means always the case. The observed shifts in spectral maxima to lower wavelengths may arise from a variety of causes: for example, (1) by the addition of a nucleophile to the imine double bond, as in compound X, (2) by the deformation of the planar conjugated structure of compound IX, or (3) by a tautomeric rearrangement of the imine proton either to the pyridine nitrogen or to the phenoxide oxygen. The variety of spectra observed is an indication that the protein environment of the chemically linked prosthetic group exerts a profound and *enzyme-specific* effect on both the detailed chemical structure and the catalytic reactivity of the coenzyme. The reaction of amines with the enzyme complex is an amine-interchange reaction (Eq. 8–9).

Table 8-1
Spectral Properties of Transaminases and Pyridoxal Derivatives[a]

Transaminase	Source	pH	Pyridoxal enzyme λ_{max}, mμ	p*K*
Glutamic-aspartic	Pig heart	4.6	430	6.3
	Supernatant	8.4	362	
Glutamic-aspartic	Beef liver	5.0	430	6.2
	Supernatant	9.0	362	
Glutamic-aspartic	Beef liver	5.0	435	6.2
	Mitochondria	9.0	355	
Glutamic-alanine	Pig heart	4.5	426	
	Supernatant	8.5	multiple 330–450	7.3
Leucine-isoleucine	Pig heart Supernatant	4.8–10.5	414, 326	none observed
D-amino acid	*Bacillus subtilis*	5–9	330, 415	none observed

Pyridoxamine pyruvate	*Pseudomonas* sp.	—	415	(Bound pyridoxal)
Aspartic-β-decarboxylase[b]	*Achromobacter* sp. *Clostridia* sp.	3–8	360	none observed
Serine[b] transhydroxymethylase	Rabbit liver	5–9.7	430	none observed
Pyridoxal phosphate	—	7	388, 330	—
Pyridoxal	—	7	318, 390	—
Pyridoxal-valine imine	—	7 12	324, 425 361	10.9

[a]Courtesy of Prof. W. T. Jenkins.

[b]Aspartic-β-decarboxylase and serine transhydroxymethylase will both react slowly with certain amino acids to yield the corresponding keto acid.

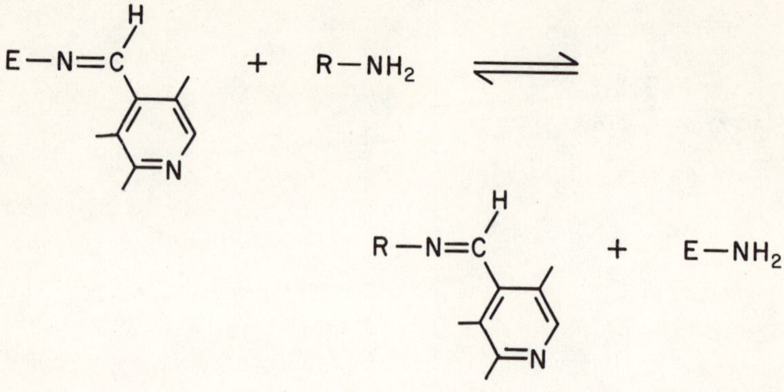

(8–9)

Thus, one might anticipate only minor changes in the spectra of enzyme-pyridoxal compounds upon addition of amine substrates, especially if the coenzyme remains rigidly fixed in the enzyme protein environment via complementary physical interactions (firm binding of pyridoxamine phosphate has been observed). Nevertheless, large changes in spectral characteristics do occur upon the addition of amine substrates. One spectral change, observable with particular amine substrates (e.g., amino acids), and similarly observable during the course of nonenzymic pyridoxal catalysis, is the appearance of a spectral band similar to that exhibited by compound XI. Note that no such spectrum is detectable in enzyme-pyridoxal complexes in the absence of added substrates (Table 8–1). This is probably due to the difficulty of breaking a C—H bond without the (electronic) assistance of other functional groups in its immediate covalent vicinity. Such electronic assistance is lacking in the lysyl-ε-imino Schiff's base (IX). When the corresponding *aldimine* is formed from an α-amino acid, rather than an ε-amino derivative, the α-carboxyl group enhances the lability of the C—H bond, allowing for tautomerization and the formation of the *ketimine* (XI). Note that the addition of one equivalent of water to compound XI can lead to either of two reactions, depending on which of the two double bonds (external to the ring) is being attacked (see Fig. 7–3). Hydration of the substrate C—N bond leads ultimately to the formation of a keto acid and pyridoxamine phosphate (VIII), but addition to the other N—C bond leads to the re-formation of pyridoxal phosphate and the original amino acid. Since, with amino acids as substrate, the reaction sequence in forming compound XI involves the removal of the α-C—H hydrogen, racemization may occur concomitantly with rehydration. Indeed, there exist specific, pyridoxal phosphate-linked enzymes, the function of which appears to be either racemization, or the specific intraconversion of L to D amino acids. Note that if reaction proceeds in the direction of formation of ketoacid and pyridoxamine phosphate, the predicted spectral characteristics at equilibrium will depend on the magnitude of the equilibrium constant for the prior hydration-dehydration reaction (XI → VIII). Amino acid analogues with particularly labile α-C—H bonds, in which this equilibrium greatly favors the dehydrated ketimine structure (XI) are known; for example, α-hydroxy aspartate. Such compounds are potent inhibitors of enzyme activity, and have been used to examine the spectral characteristics of the ketimine.

Pyridoxal enzymes are nearly ideal enzymic systems for the study of the intermediate details of catalysis; the strong chromophoric properties of the conjugated pyridinium system can be used as an indicator of transient chemical events involved in these catalytic processes. Moreover, the potential "pH-indicator" properties of both the enzyme-coenzyme complex and the ternary (enzyme-coenzyme-substrate) chemical complexes may be of aid in the chemical identification of the intermediates. A further advantage of pyridoxal-enzyme systems for the study of detailed catalytic mechanisms is that the reactions are readily reversible, and hence it is possible to vary the equilibrium distribution of reactants and products (and thereby, the concentration of intermediates) over very wide ranges of concentration. This ability to control the position of equilibrium is particularly noteworthy in pyridoxal phosphate-enzyme-catalyzed transamination reactions. One such transaminase, glutamate-aspartate transaminase, has been extensively investigated.

Glutamate-aspartate transaminase catalyzes reversibly the overall reaction given in Eq. 8–10. Each of the substrate re-

$$\overset{\oplus}{\underset{\text{L-Glutamate}}{}}$$

$$\ominus O_2C-CH_2-CH_2-\underset{\text{L-Glutamate}}{\overset{\overset{\oplus}{NH_3}}{CH}}-CO_2^{\ominus} + \ominus O_2C-CH_2-\underset{\text{Oxaloacetate}}{\overset{O}{C}}-CO_2^{\ominus} \rightleftharpoons$$

$$\ominus O_2C-CH_2-CH_2-\underset{\alpha\text{-Ketoglutarate}}{\overset{O}{C}}-CO_2^{\ominus} + \ominus O_2C-CH_2-\underset{\text{L-Aspartate}}{\overset{\overset{\oplus}{NH_3}}{CH}}-CO_2^{\ominus}$$

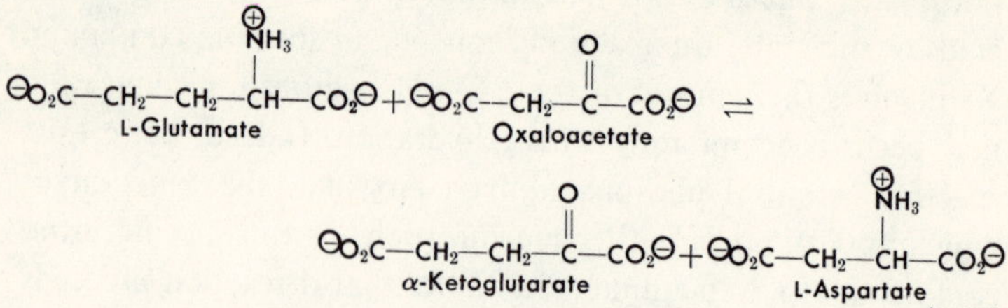

(8–10)

actions with the enzymic cofactor can be broken down into two steps (Eq. 8–11). These steps can occur independently

$$E-\overset{\oplus}{N}H_3 + \quad \rightleftharpoons \quad$$

(Enzyme–pyridoxal)

(a)

$$+ \ R_1\overset{\overset{\oplus}{NH_3}}{\underset{CO_2^{\ominus}}{CH}} \ \overset{K_1}{\rightleftharpoons} \ E\overset{\oplus}{N}H_3 \left(H_2N-CH_2- \right)$$

$$+ \ H_2O$$

$$+ \ R_1\overset{O}{\underset{CO_2^{\ominus}}{C}}$$

(b)

$$E\overset{\oplus}{N}H_3 \left(H_2N-CH_2- \right) + R_2\overset{O}{\underset{CO_2^{\ominus}}{C}} \ \overset{K_2}{\rightleftharpoons} \ E$$

(Enzyme–pyridoxamine)

$$+ \ R_2\overset{\overset{\oplus}{NH_3}}{\underset{CO_2^{\ominus}}{CH}}$$

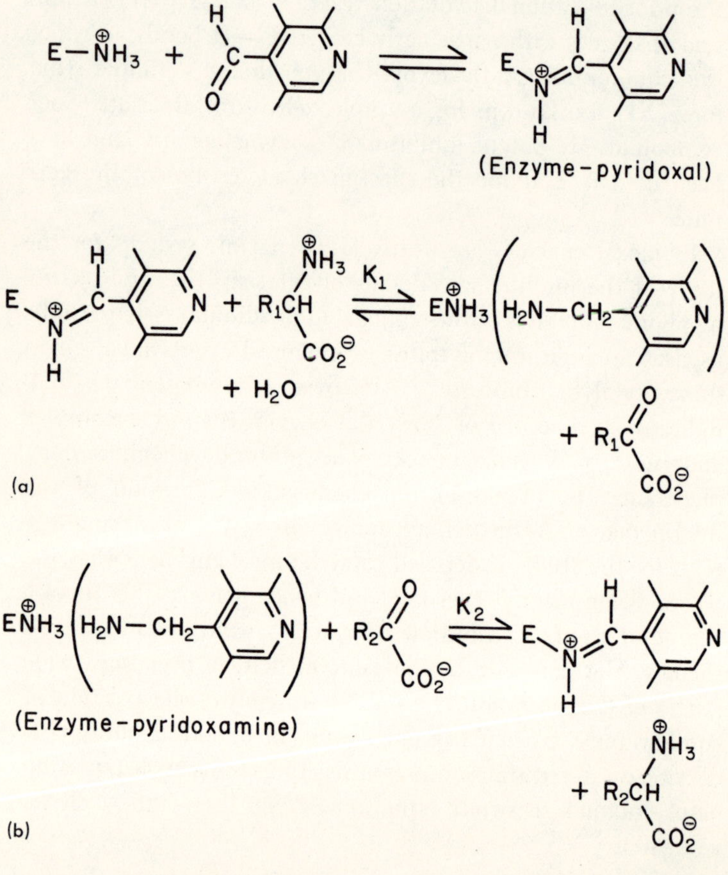

(8–11)

within stoichiometric limits. Since the reactions are appreciably reversible, the concentration ratios of keto to amino acids at equilibrium will depend on the starting ratios of amino and keto acids. Note that the concentration of enzyme-coenzyme species (as, for example, structures VII–XI) will be determined

$$K_{eq} = K_1K_2 = \frac{[R_1COCO_2^{\ominus}]\quad[R_2CH(\overset{\oplus}{N}H_3)CO_2^{\ominus}]}{[R_2COCO_2^{\ominus}]\quad[R_1CH(\overset{\oplus}{N}H_3)CO_2^{\ominus}]}$$

by the ratio of oxidants (keto acids) to reductants (amino acids).

If, at equilibrium, all substrates and products are in great excess over the total enzyme-cofactor concentration, the condition of Eq. 8–12 will obtain. Interconversion of keto and

$$\frac{[\text{E-pyridoxal}]}{[\text{E-pyridoxamine}]} = \frac{[\text{keto acids}]}{[\text{amino acids}]} \qquad (8\text{–}12)$$

amino acids (*transamination*) as in Eq. 8–11, however, must proceed by transfer of amino nitrogen via the cofactor. Such a system is particularly suitable to the study of rapid reaction velocities by the *relaxation* methods discussed previously (Chapter 4). The substrates and products in this reaction do not absorb visible or near-ultraviolet light. The enzyme-coenzyme chemical intermediates as well as the ternary enzyme-coenzyme-substrate intermediates, on the other hand, have characteristic ultraviolet and (or) visible spectra (dependent on the particular chemical structures, as for example, in Fig. 8–12). A rapid perturbation of a "condition of state" (e.g., a change in temperature) will result in a change in the equilibrium constant (K) for the reaction, as described by Eq. 8–12. Moreover, the individual equilibria, as described in Eq. 8–11, will likewise be perturbed, and hence lead to changes in spectra. The rapid spectrophotometric measurements which can be carried out concomitant with an abrupt change in temperature (as described in Section 4–9), should hence be indicative of the chemical pathway of catalysis (by molecular interpretation of the specific spectral changes), and should serve as "kinetic indicators" (concentration variables) for determining specific reaction rates.

Aspartate-glutamate transaminase has a broad pH optimum (over the region pH 5–9). In view of our discussion (in Section 6–2), it might have been anticipated that proton transfer processes involving the participation of the imidazole residue(s) of histidine would contribute to rate control. Such processes should be dependent on the pK_a of imidazolium cation (approximately 7). The lack of either a sharp pH optimum (as was observed in Fig. 8–10), or of a sigmoidal pH dependence (as observed in Fig. 8–3) in the pH-rate profile might be interpreted in a number of different ways: (1) Imidazole residues are not involved in rate control; (2) proton transfer processes, although essential for catalysis, are not rate-controlling; and (3) general acid and general base catalysis involving both imidazolium$^+$ and imidazole residues occur in a number of different rate-governing steps in the catalytic pathway in such a way that a virtually constant activity is observed over the neutral pH range. It is interesting that photo-oxidation of the enzyme (utilizing the photo-oxidant methylene blue), a process believed to result in the destruction of imidazole residues of histidine, results in complete inactivation of the enzyme.

Both the binary enzyme-pyridoxal phosphate compounds (Table 8–1) and the ternary enzyme-pyridoxal phosphate-substrate compounds give rise to spectral bands with wavelength maxima distinctly lower than the Schiff's base (compound IX), but distinctly higher than the pyridoxamine (VIII). These spectral bands possibly may arise by the addition of a nucleophilic residue of the enzyme protein to the Schiff's base, as in compound X. Such nucleophilic addition could be effected via free sulfhydryls on the enzyme protein. It has been found that the titration of particular cysteine residues of aspartate-glutamate transaminase with mercurial compounds results in the loss of enzyme activity. Nucleophilic addition reactions could assist in the rapid chemical interchange of amines between the ϵ-imino Schiff's base of lysine-pyridoxal and the substrate-pyridoxal Schiff's base (Eq. 8–9).

An interesting observation with aspartate-glutamate transaminase, which appears to be of more general significance, is the phenomenon of stabilization of the enzyme protein structure upon formation of a chemically linked enzyme-cofactor (or, in general, a chemically linked enzyme-substrate) compound. A similar and striking structural stabilization of the "active-serine" containing proteins described in Section 8–1 has also been noted upon formation of the acyl (serine hydroxyl substituted) enzyme. In the present instance, the coenzyme plays a functional chemical role in the catalytic process, and exerts a profound influence (via chemical reaction) on the enzyme protein structure.

This phenomenon of a chemically induced protein structural change (or structural stabilization) is noteworthy in the enzyme, *phosphorylase*, an enzyme which does not utilize any of the potential catalytic functions of pyridoxal, but which nevertheless requires the formation of an enzyme (ϵ-amino)-pyridoxal phosphate Schiff's base for the maintenance of its specific catalytically active conformation.

8–4 DEHYDROGENASES REQUIRING PYRIDINE NUCLEOTIDES

An enormous number of different enzymes involved in oxidation-reduction reactions require the direct and specific participation of the pyridine nucleotides (see Section 7–3). In all of these reactions, one proton and two electrons are transferred from the substrate to the (oxidized) coenzyme (NAD^+ or $NADP^+$); in the reverse direction, these are transferred from the (reduced) coenzyme (NADH or NADPH) to the substrate (Eq. 8–13).

$$SH_2 + NAD^{\oplus} \rightleftharpoons NADH + S + H^{\oplus} \qquad (8\text{–}13)$$

With every enzyme investigated, an absolute stereospecificity has been observed in this hydrogen transfer process. For example, in the conversion of ethanol to acetaldehyde, catalyzed by alcohol dehydrogenase, there is *both* stereospecific abstraction of a hydrogen atom from the alcohol and stereospecific addition of a hydrogen atom to the pyridine ring, as illustrated in Fig. 7–5. These stereospecific processes have been identified by the synthesis of the two monodeuterated derivatives of ethanol, XII and XIII. The stereospecific abstraction and addition of a deuterium atom is demonstrated by the isolation of only one of the two potential isomeric monodeuterated reduced coenzyme derivatives (XIV and XV), and by the fact that only one of the two isomeric deutero ethanols can lose deuterium in the oxidation process. In surveying this stereospecific phenomenon over a large variety of dehydrogenases, it has been found that only one of the two isomers (XIV and XV) is ever found in any single enzyme-catalyzed process. However,

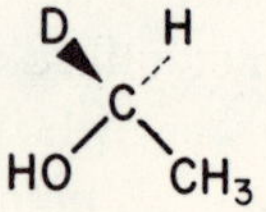

XII

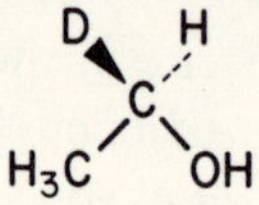

XIII

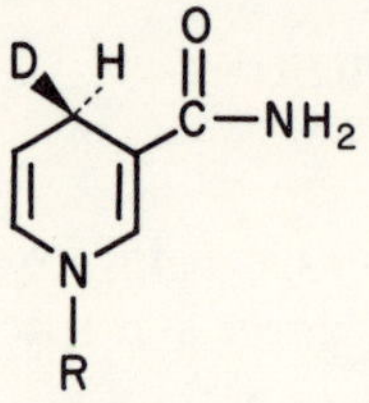

A
XIV

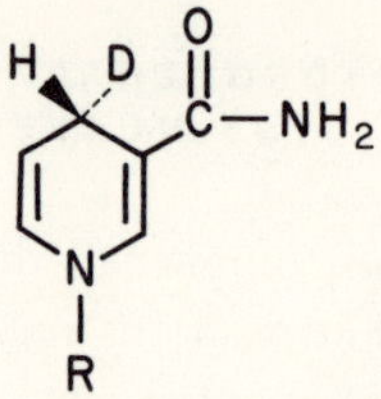

B
XV

which of the two isomers is found depends on the particular enzyme involved. These stereospecific processes once again point out the detailed and multisited interactions which must occur among enzyme protein, substrate, and coenzyme at the active site. Detailed stereospecific interrelationships are suggestive of the existence of discrete chemical intermediates among enzyme, coenzyme, and substrate (since chemical bonds exert the most stringent constraints on geometry). Numerous experiments have been designed to investigate the existence of such potential intermediates. A number of techniques can be used in this search:

(1) The binding of coenzyme to the enzyme can be studied in the absence of substrate by measurement of concomitant changes in either the fluorescence emission or the absorption spectrum of the coenzyme. If the equilibrium dissociation constant differs from the steady state Michaelis constant, K_M (see Section 4–4), there is suggestive evidence for the formation of a stoichiometrically significant quantity of enzyme-coenzyme-substrate *ordered* chemical intermediate.

(2) If an ordered mechanism (Section 4–4, Eq. 4–22) can be demonstrated by kinetic analysis, there is suggestive evidence for discrete chemical intermediates.

Experiments along these lines have provided evidence for the existence of discrete chemical intermediates, under particular conditions and with particular dehydrogenases. In other cases, no evidence for chemical intermediates has been found. The inability to demonstrate an ordered mechanism need not necessarily exclude such order. A variety of happenstances regarding the magnitudes of specific rate constants in the reaction sequence can readily obscure the experimental detection of intermediates. With yeast alcohol dehydrogenase, the equilibrium dissociation constant of the enzyme-coenzyme complex agrees fairly well with the K_M determined from steady state kinetics in the presence of substrate. With liver alcohol dehydrogenase, on the other hand, the K_M for NADH is markedly different from its equilibrium dissociation constant. This difference between the two alcohol dehydrogenases is more plausibly explained by differences in *magnitude* of particular specific rates, rather than by differences in the enzyme-catalyzed reaction pathways.

The absorption spectra and fluorescence emission spectra of the coenzymes can be utilized to follow the transient reactions of enzyme, substrate, and coenzyme. The reduced coenzyme (NADH) exhibits a change in absorption spectrum upon binding to the enzyme, liver alcohol dehydrogenase, and another spectral change upon oxidation. Utilizing rapid flow transient techniques, the kinetics of the liver alcohol dehydrogenase-catalyzed intraconversion of alcohol and acetaldehyde has been shown to follow a compulsory ordered pathway (Eq. 8–14).

$$\begin{array}{ccccc} E + NAD^{\oplus} & \rightleftharpoons & E(NAD^{\oplus}) & \xrightleftharpoons{C_2H_5OH} & (C_2H_5OH)E(NAD^{\oplus}) \\ & & & & \downarrow\uparrow \\ E + NADH & \rightleftharpoons & E(NADH) & \xrightleftharpoons[CH_3CHO]{} & (CH_3CHO)E(NADH) + H^{\oplus} \end{array} \quad (8\text{–}14)$$

By combining the information derived from these transient experiments with an analysis of the detailed steady state kinetics, it can be shown that the rate-determining process in either direction (oxidation of alcohol or reduction of acetaldehyde) is the rate of dissociation of the coenzyme (for example, NAD^+ or NADH) from the enzyme. The observed difference between the equilibrium dissociation constant, in the absence of substrate, and K_M (for coenzyme dissociation) is a necessary consequence of this rate-limiting dissociation. In the case of alcohol dehydrogenase derived from yeast, where $K_{M(NADH)}$ and the dissociation constant are equal, a random order mechanism (Eq. 8–15)

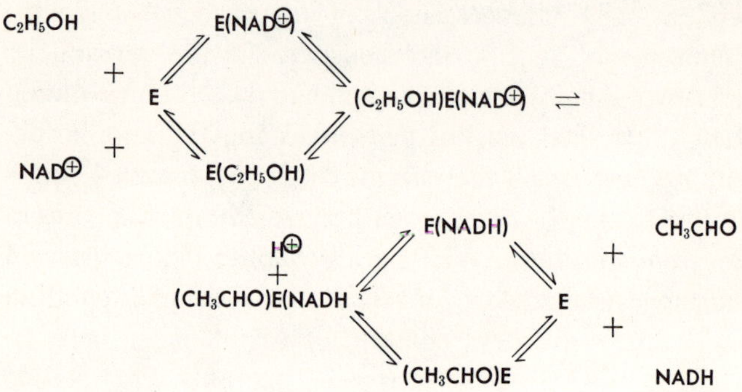

(8–15)

has been inferred from steady state analysis. The reversible dehydrogenation of malic acid (Eq. 8–16) has been studied by

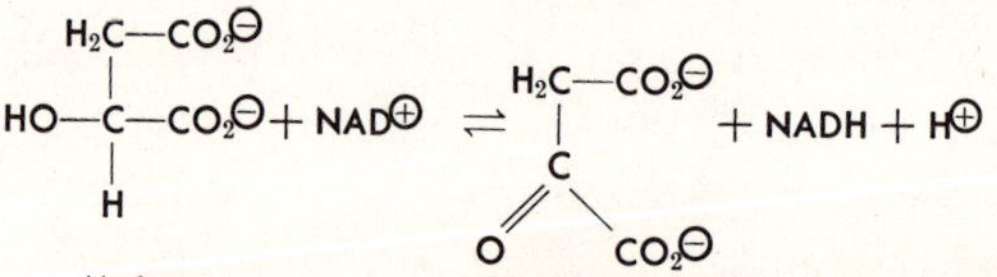

L-Malate Oxaloacetate (8–16)

steady state techniques at various concentrations of reactants and products. On the basis of this study a compulsory order mechanism has been postulated. Once again, a slow dissociation of enzyme-NADH complex has been inferred to be rate controlling. The magnitude of this rate constant can be evaluated from the steady state data, and the value thus obained has been compared with the more directly measured dissociation of the enzyme-reduced coenzyme complex (in the absence of substrate) as obtained by relaxation techniques (temperature jump measurements of changes in the intensity of fluorescent emission of the reduced coenzyme). The steady state and chemical relaxation results are in fair agreement on the value of this specific rate constant.

Unlike the pyridoxal phosphate-containing enzymes, there is no obvious chemical mechanism for the covalent binding of either the reduced or oxidized coenzyme to any of the specific amino acid side chains of the constituent protein.

It has been suggested that charge-transfer complexes are formed between the enzyme and the oxidized pyridine nucleotide. The origin of such charge-transfer interactions is illustrated schematically in Fig. 8–13. Charge-transfer interactions have not been discussed previously in this text because of the lack of any definitive information establishing the existence of such complexes in biological systems. Nevertheless, the conclusive demonstration of charge-transfer in model systems, particularly in model systems involving pyridinium rings, presents a plausible explanation of the unusually firm binding of pyridine

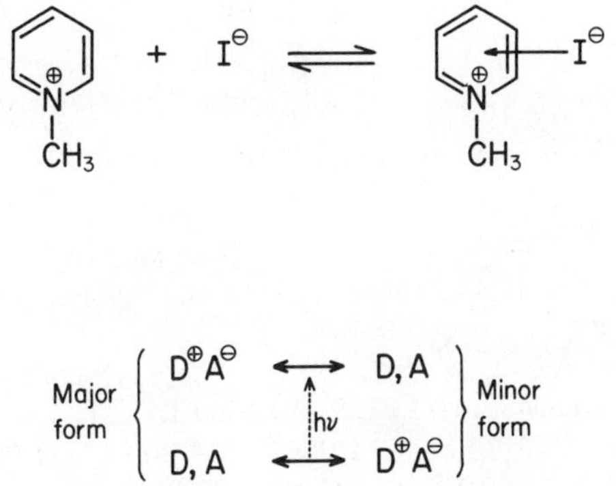

FIG. 8–13 Charge transfer complex in N-methylpyridinium iodide.

nucleotides to enzyme sites. It is entirely possible that charge-transfer complexes play a role in the intermediate catalytic steps of the enzymic pathway. Note that in the charge-transfer complex, the pyridinium ring is intermediate in oxidation state between the oxidized and the reduced form of the coenzyme. Charge-transfer complexes are usually detectable via a change in the electronic spectrum. The appearance of a broad band of low intensity, as for example, in the charge-transfer spectrum of 1-ethyl-4-carbomethoxypyridinium iodide (Fig. 8–14), is typical. Small spectral perturbations have been noted both in the equilibrium interactions of pyridine nucleotides with enzymes in the absence of substrates, and during the catalytic process (by the utilization of rapid-flow spectrophotometry). The assignment of these spectral perturbations as charge-transfer complexes is, however, uncertain. The extent of the charge-transfer interaction is determined by both "members" of the charge-transfer pair. Iodide ion is extremely effective as a charge-transfer ligand. The pyridinium iodides exhibit much more intense charge-transfer spectral bands than do other pyridinium halides. Another effective charge-transfer ligand is thiolate anion (RS^-). It should be noted that all dehydrogenases thus investigated appear to contain at least one crucial

FIG. 8–14 Absorption curves for 1-ethyl-4-carbomethoxypyridinium iodide and perchlorate in ethylene dichloride. [From T. C. Bruice and S. Benkovic, *Bioorganic Mechanisms*, Vol. 2, W. A. Benjamin, Inc., New York, 1966.]

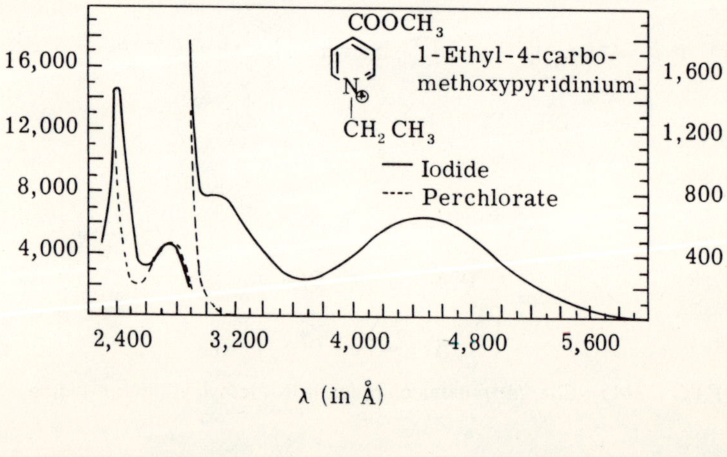

cysteine residue per catalytic site. Of special interest is the active cysteine residue in glyceraldehyde-3-phosphate dehydrogenase. This cysteine residue has been extensively investigated in regard to both the amino acid sequence at the active site and the catalytic nature of the interaction with substrate and coenzyme. The enzyme catalyzes the reaction described by Eq. 8–17. This reaction is of great importance in fermentation,

$$
{}^{\ominus}O-\overset{\overset{\displaystyle O}{\|}}{\underset{\underset{\displaystyle O_{\ominus}}{|}}{P}}-OCH_2-\underset{\underset{\displaystyle OH}{|}}{CH}-\overset{\displaystyle O}{\underset{\displaystyle H}{\diagdown}}C \quad + \quad NAD^{\oplus} \quad + \quad {}^{\ominus}O-\overset{\overset{\displaystyle O}{\|}}{\underset{\underset{\displaystyle O_{\ominus}}{|}}{P}}-OH \quad \rightleftharpoons
$$

$$
{}^{\ominus}O-\overset{\overset{\displaystyle O}{\|}}{\underset{\underset{\displaystyle O_{\ominus}}{|}}{P}}-OCH_2-\underset{\underset{\displaystyle OH}{|}}{CH}-C\overset{\diagup\!\!O}{\diagdown O-\overset{\overset{\displaystyle O}{\|}}{\underset{\underset{\displaystyle O_{\ominus}}{|}}{P}}-O^{\ominus}} \quad + \quad NADH + H^{\oplus}
$$

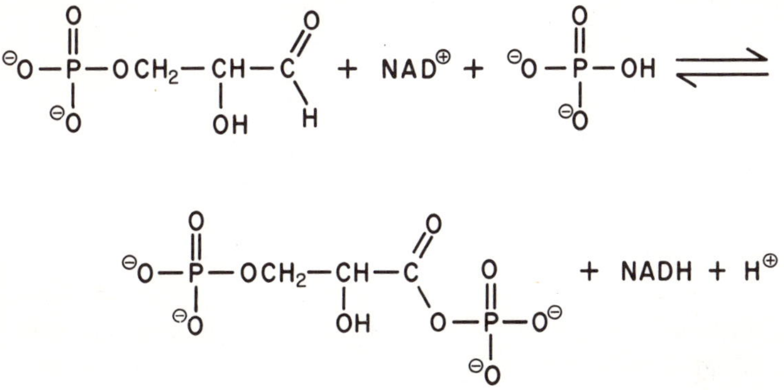

$$(8–17)$$

since in the process of oxidation of the aldehyde, one mole of inorganic phosphate is converted to a high energy acyl phosphate product. The product (3-phosphoglyceroyl phosphate) subsequently reacts with ADP to yield ATP and 3-phosphoglyceric acid (Eq. 8–18).

$$
{}^{2\ominus}O_3PO-CH_2CH-\underset{\underset{\displaystyle OPO_3^{2\ominus}}{|}}{\underset{\underset{\displaystyle OH}{|}}{C}}\overset{\diagup\!\!O}{} \quad + \quad Adenosine-O\overset{\overset{\displaystyle O}{\|}}{\underset{\underset{\displaystyle O_{\ominus}}{|}}{P}}O\overset{\overset{\displaystyle O}{\|}}{\underset{\underset{\displaystyle OH}{|}}{P}}O^{\ominus} \quad \rightleftharpoons
$$

(ADP)

$$
{}^{2\ominus}O_3PO-CH_2CH-\underset{\underset{\displaystyle OH}{|}}{C}\overset{\displaystyle O}{\underset{\displaystyle O}{\diagdown}}{}^{\ominus} \quad + \quad Adenosine-O\overset{\overset{\displaystyle O}{\|}}{\underset{\underset{\displaystyle O_{\ominus}}{|}}{P}}O\overset{\overset{\displaystyle O}{\|}}{\underset{\underset{\displaystyle OH}{|}}{P}}O\overset{\overset{\displaystyle O}{\|}}{\underset{\underset{\displaystyle O_{\ominus}}{|}}{P}}O^{\ominus}
$$

(ATP)

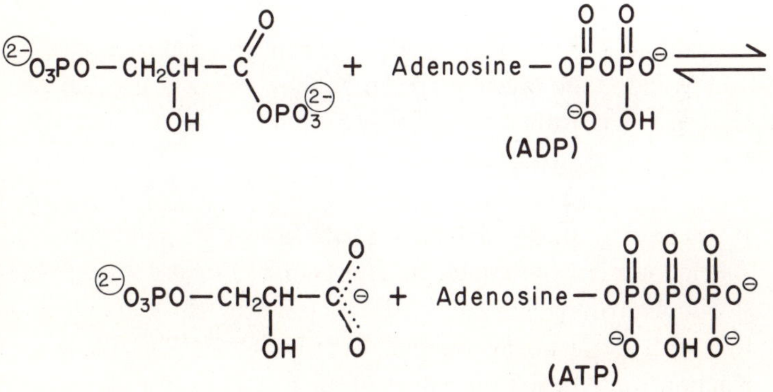

$$(8–18)$$

The sequential degradation of glyceraldehyde-3-phosphate to 3-phosphoglyceric acid is an excellent illustration of enzyme-controlled regulation of metabolic interconversions. Both reactions (Eqs. 8–17 and 8–18) have finite equilibrium constants, and one mole of ATP is produced from ADP and phosphate (an energetically unfavorable process) in the net oxidation reaction. A thermodynamically more favored reaction would be the direct oxidation of glyceraldehyde-3-phosphate by NAD^+ to 3-phosphoglyceric acid and NADH (Eq. 8–19). This direct

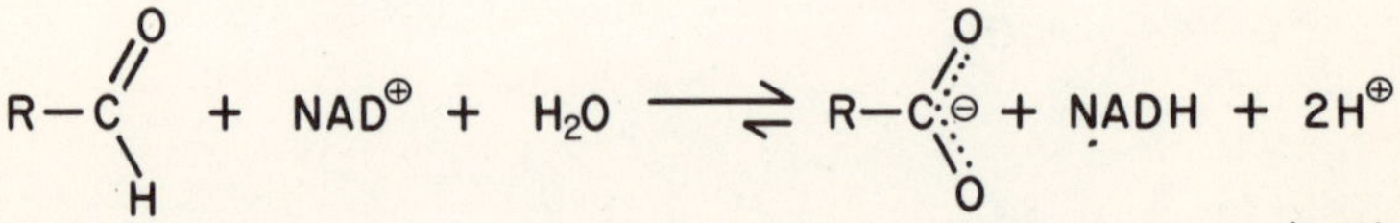

(8–19)

pathway would result in the dissipation of thermal energy in place of the synthesis of ATP. It is therefore of special significance in intermediary metabolism to examine the detailed enzyme-mediated regulatory mechanism by which glyceraldehyde-3-phosphate is converted to a "high energy" diphosphoglyceric acid intermediate.

In the absence of inorganic phosphate or of analogues of inorganic phosphate (arsenate), conversion of NAD^+ to NADH in the presence of enzyme and substrate does not take place appreciably. The enzyme-catalyzed rate of reduction of NAD^+ is dependent on the phosphate concentration. If a radioactive label is attached to the substrate, it can be shown that in the presence of NAD^+, but in the absence of phosphate, the radioactive label becomes firmly attached to the enzyme protein, and that this radioactive label can be removed by the addition of phosphate. It therefore follows that phosphate reacts directly with an enzyme-substrate intermediate compound. The enzyme can be stoichiometrically inactivated by reaction of the enzyme protein with alkylating agents such as iodoacetate. This alkylation reaction leads to the formation of an S-carboxymethylcysteinyl enzyme (an S-carboxymethylcysteinyl peptide has been isolated from the degradation products of the denatured enzyme derivative). Other reagents which react with thiols (for example, mercurial compounds) also inactivate the enzyme.

The native enzyme reacts with high energy acyl esters such as p-nitrophenyl acetate, yielding an acyl enzyme which is catalytically inert to natural substrates. When this inactive derivative is denatured, and then also degraded with proteolytic enzymes, a peptide fragment containing S-acetylcysteine can be isolated. By analysis of the adjacent amino acid sequence, it can be shown that the S-acetylation and the S-carboxymethylation reactions occur at the same cysteine residue. The acetylation experiments provide strong evidence for the supposition that the aldehyde substrate, in the presence of coenzyme, reacts directly with the enzyme yielding 3-phosphoglyceroyl thiol ester (Eq. 8–20).

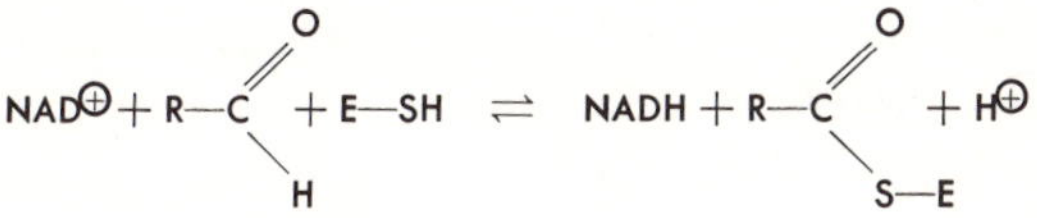

$$NAD^{\oplus} + R-\overset{\displaystyle O}{\underset{\displaystyle H}{\overset{\|}{C}}} + E-SH \;\rightleftharpoons\; NADH + R-\overset{\displaystyle O}{\underset{\displaystyle S-E}{\overset{\|}{C}}} + H^{\oplus} \qquad (8-20)$$

Thiol esters are relatively "high energy" esters. In model (nonenzymic) systems, nucleophilic attack on thiol esters by phosphate leads to appreciable formation of the corresponding phosphoryl anhydride (Eq. 8–21).

$$R\overset{\displaystyle O}{\underset{\displaystyle SR'}{\overset{\|}{C}}} + {}^{\ominus}O-\overset{\displaystyle O}{\underset{\displaystyle O^{\ominus}}{\overset{\displaystyle |}{\underset{\displaystyle |}{P}}}}-OH \;\rightleftharpoons\; R\overset{\displaystyle O}{\overset{\|}{C}}\diagdown O-\overset{\displaystyle O}{\underset{\displaystyle O^{\ominus}}{\overset{\|}{P}}}-O^{\ominus} + R'SH$$

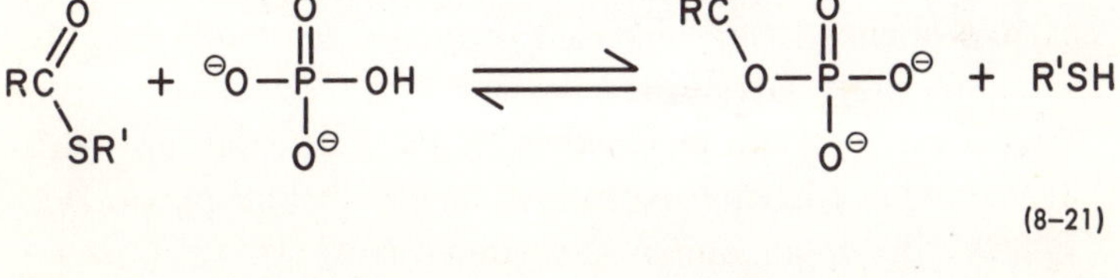

$$(8-21)$$

The effect of phosphate concentration on the overall rate of reduction of NAD$^+$ is explained by the mechanism illustrated in Eq. 8–22.

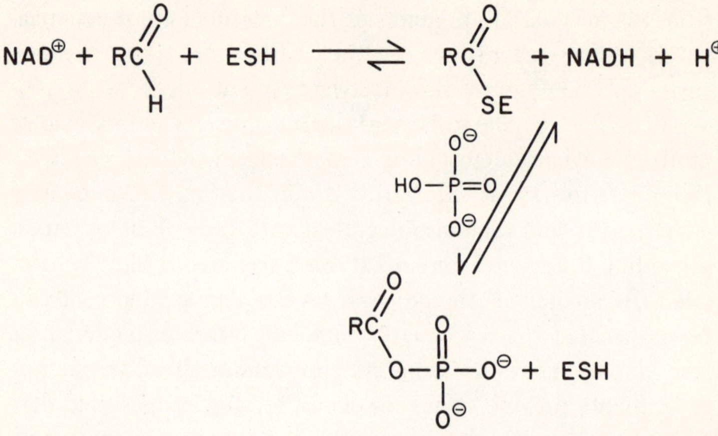

(8–22)

At low phosphate concentrations, the reaction rate is directly proportional to the phosphate concentration, whereas at very high phosphate concentrations, the overall reaction is independent of phosphate concentration. This result would be anticipated, if the second step in the mechanism of Eq. 8–22 is rate-controlling at low phosphate concentration, but not at high phosphate concentration.

It is interesting to note the common sequence of amino acids in the covalent region surrounding the reactive cysteine in glyceraldehyde-3-phosphate dehydrogenases derived from a variety of organisms (Section 3–5). This sequence is common to enzymes from microorganisms (yeast) and mammals, and very likely reflects a unique structure-function relationship in this region of the protein. Some further comparison of the enzymes obtained from different organisms is of interest. The most extensively investigated enzymes are those isolated from rabbit muscle and from yeast. Both these enzymes are fairly large proteins, having molecular weights of approximately 140,000. The total amino acid composition of each enzyme is known. From these analyses, and a knowledge of the specificity of the proteolytic enzyme trypsin (hydrolysis at lysine and arginine residues), the number of different degraded peptides which should result, on the basis of a molecular weight of 140,000, can be calculated. The actual degradation leads to only approximately one-quarter the number of different peptides. Moreover, the number of reactive cysteines as determined by titration with iodacetate is 4 per 140,000 molecular weight unit. The number of active sites per 140,000 molecular weight unit can also be estimated from the readily observable quenching of fluorescence of the reduced coenzyme (NADH) upon binding to the enzyme site. Once again, the results are consistent with a value of 4 sites per 140,000 molecular weight unit. The enzymes (from rabbit muscle, and from yeast) appear to be composed of four identical polypeptide subunits, each containing one active enzyme site. Regular aggregates of polypeptide subunits into active enzyme molecules is a common structural feature of all dehydrogenases thus far investigated. In glyceraldehyde-3-phosphate dehydrogenase, each of these four *intramolecular* sites would appear to function independently of the others under physiological conditions, where the dependence of reaction rate on substrate or coenzyme concentration is precisely that predictable from a Michaelis-Menten model (which implicitly assumes that all sites are equivalent and independent). This is not always the case with enzymes containing multiple subunits; yeast glyceraldehyde-3-phosphate dehydrogenase, for example, exhibits a *sigmoidal* dependence of reaction velocity on coenzyme concentration (Fig. 8–15) at temperatures of 40° or above. At room temperature there does not appear to be any subunit interaction.

All *dehydrogenases* thus far examined with regard to subunit structure have been shown to contain at least two (two, four, eight) subunits and a corresponding number of sites. Sometimes, as is the case with glyceraldehyde-3-phosphate dehydrogenases at room temperature, there are no apparent interactions between the multiple subunits or sites, as judged by the dependence of the reaction velocity on the concentration of substrates and effectors. In the following section we shall consider a dehydrogenase in which strongly cooperative interactions between subunits are apparent from the reaction kinetics.

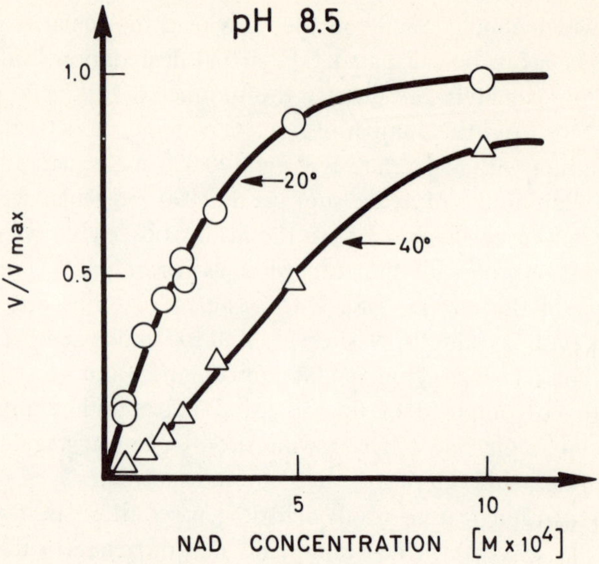

FIG. 8–15 Dependence of the velocity of yeast-GPD-catalyzed oxidation of glyceraldehyde-PO_4 on the concentration of NAD^+ at 20° and at 40°; $V_{max}(20°)/V_{max}(40°) = 2.2$. (Courtesy of Dr. K. Kirschner.)

8–5 ALLOSTERIC ENZYMES: ISOCITRATE DEHYDROGENASE AND ASPARTATE TRANSCARBAMYLASE

Enzymes in which the catalytic activity at one site is dependent on the state of binding of substrate or effector to other sites, due to protein subunit interactions, have been defined as *allosteric enzymes.* It is believed that allosteric enzyme effectors play an important role in the regulation of metabolic activity. For this reason, the molecular details of subunit interactions, and the influence of effector concentration on the detailed kinetics of the catalyzed reactions, have received considerable attention.

ISOCITRATE DEHYDROGENASE

Isocitrate dehydrogenase catalyzes the NAD^+ mediated oxidation of isocitrate to oxalosuccinate (Eq. 8–23). Under suita-

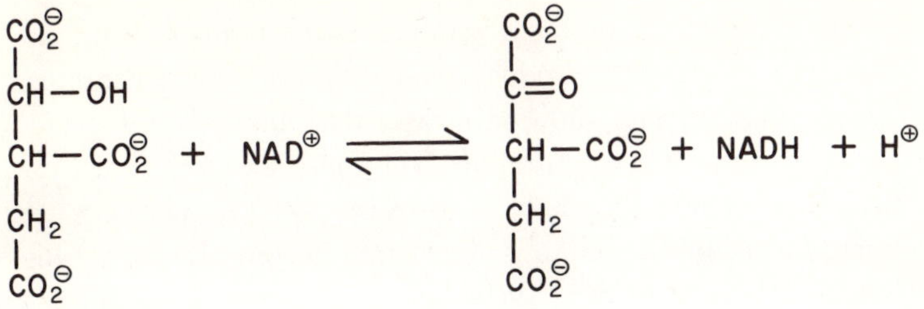

Isocitrate Oxalosuccinate

(8–23)

ble conditions of concentration, the reaction rate can be shown to be dependent upon two species playing no apparent role in the stoichiometric reaction (Eq. 8–23), Mg^{2+} and adenylic acid (AMP). The forward reaction rate can be followed via the spectrophotometrically observable appearance of the reduced coenzyme (NADH). The dependence of reaction rate on substrate (isocitrate) concentration is shown in Fig. 8–16. Notice the sigmoidal contour of the rate-concentration profile, and compare this with the hyperbolic contour anticipated for systems obeying the Michaelis-Menten model. At higher substrate

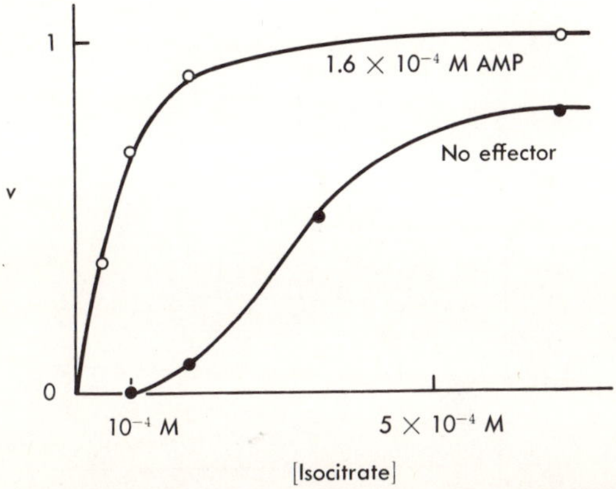

FIG. 8–16 Dependence of reaction velocity on isocitrate concentration with yeast isocitrate dehydrogenase. Lower curve, in the absence of any effector; upper curve, in the presence of the positive effector AMP. [From D. E. Atkinson *et al.*, *J. Biol. Chem.* **240**, 2682 (1965).]

concentrations the profile resembles that obtained from an effectively single-sited enzyme. At low substrate concentrations, the reaction velocity shows a greater than first-power dependence on substrate concentration. The most straightforward explanation for such kinetic behavior is that the catalytic sites are interdependent, the binding of substrate at one site increasing the catalytic efficiency at the other sites. Cooperativity among catalytic centers will be most apparent at low substrate concentration, where the fraction of sites containing bound substrate is small. Allosteric effectors, which profoundly influence the rate of reaction at low concentrations of substrate, exert little or no influence at sufficiently high substrate concentration.

The effect of AMP and Mg^{2+} on the rate of reaction is illustrated in Fig. 8–17. At sufficiently low concentrations of substrate, the rate of reaction is proportional to the square of the concentration of either of these allosteric effectors. If there is strong cooperativity among sites, the *number* of interacting sites can be estimated according to the following arguments.

Consider the general model illustrated by Eq. 8–24

$$
\begin{array}{ccccccc}
E & \underset{-S}{\overset{+S}{\rightleftharpoons}} & ES & \underset{-S}{\overset{+S}{\rightleftharpoons}} & ES_2 & \rightleftharpoons \cdots\cdots & ES_n \\
{\scriptstyle +A}\downarrow\uparrow{\scriptstyle -A} & & {\scriptstyle +A}\downarrow\uparrow{\scriptstyle -A} & & \downarrow\uparrow & & \downarrow\uparrow \\
AE & \rightleftharpoons & AES & \rightleftharpoons & AES_2 & \rightleftharpoons \cdots\cdots & AES_n \\
\downarrow\uparrow & & \downarrow\uparrow & & \downarrow\uparrow & & \downarrow\uparrow \\
A_2E & \rightleftharpoons & A_2ES & \rightleftharpoons & \cdots & \cdots\cdots & \cdots \\
\vdots & & \downarrow\uparrow & & \vdots & & \vdots \\
A_mE & \rightleftharpoons & \cdots\cdots & & \cdots & \cdots\cdots & A_mES_n
\end{array} \qquad (8\text{–}24)
$$

FIG. 8–17 Dependence of reaction velocity on effector concentration for isocitrate dehydrogenase at low (fixed) concentration of isocitrate and coenzyme; (a) under varying low concentrations of AMP; (b) under varying low concentrations of Mg^{2+}. [From D. E. Atkinson *et al.*, *J. Biol. Chem.* **240,** 2682 (1965).]

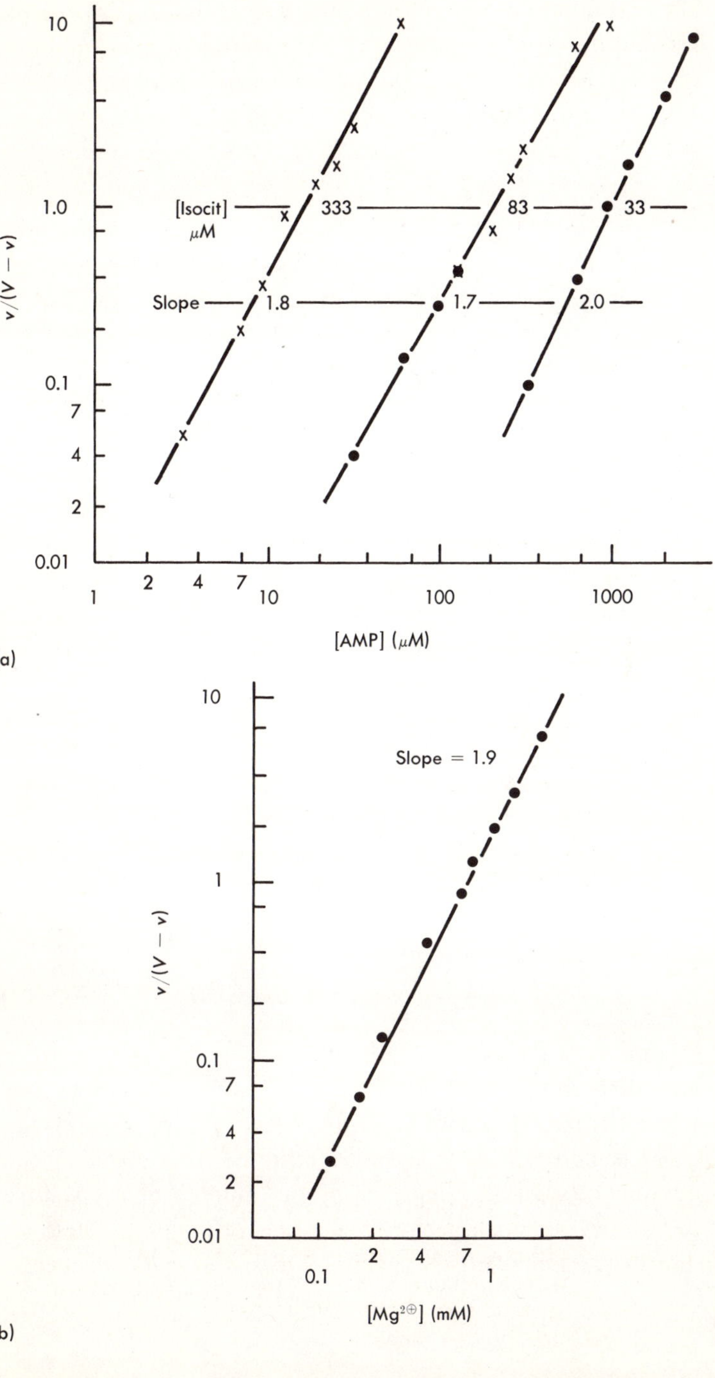

10
1.0
0.1
7
4
2
0.01
v/(V − v)
[Isocit]
μM
333
83
33
Slope
1.8
1.7
2.0
1
2
4
7
10
100
1000
[AMP] (μM)

(a)

10
1
0.1
7
4
2
0.01
v/(V − v)
Slope = 1.9
2
4
7
0.1
1
[Mg²⁺] (mM)

(b)

The overall rate will be a composite of the rates for the various distributions of filled and empty sites, namely Eq. 8–25.

$$v = \sum_{i=0}^{i=m} \sum_{j=1}^{j=n} (k_{ij}[A_i ES_j]) \qquad (8\text{–}25)$$

Let us use Y_i to denote the number of effector binding sites per molecule which are actually occupied, out of a total of m sites per molecule. For cooperativity to be observed, the *rate* of catalyzed reaction or the *extent* of substrate binding must increase more rapidly than does the experimental variable, Y_i/m. It follows from the algebraic solution for the *equilibrium* concentrations of the various complexes ($[A_i ES_j]$), that at low concentrations of effector ($Y_i/m << 1$) *and* at correspondingly low concentrations of substrate ($Y_j/n << 1$), that the variation of total rate of reaction with positive effector concentration is expressible by Eq. 8–26.

$$v = C_m[A]^m + C_{m-1}[A]^{m-1} + C_{m-2}[A]^{m-2} \cdots + C_0 \qquad (8\text{–}26)$$

where $C_m \cdots C_0$ are coefficients dependent on the magnitudes of various equilibrium constants (K_{ij}), rate constants, and the (fixed) low concentration of substrate. Under the conditions specified (low concentration), the coefficients C_{m-i} become progressively smaller as $m - i$ decreases. Hence the useful approximation of Eq. 8–27, within this concentration range,

$$v \sim C_m[A]^m$$
$$\log v \sim \text{constant} + m \log[A] \qquad (8\text{–}27)$$

Equations 8–25 and 8–26 are applicable only when many or most of both the allosteric effector sites and the substrate sites are unoccupied. Although linearity in plots of log [A] versus log v may be apparent under other conditions of concentration, the interpretation of the slope, m, as the number of *cooperative* (allosteric) sites per molecule may be incorrect in these instances (m and n are often referred to as the *Hill coefficient*).

Similar considerations in regard to the concentration-dependent behavior of an allosteric substrate (under the same conditions of effector concentration as above) lead to the same type of approximate equation (Eq. 8–28) for the dependence of rate on substrate concentration, at a low, fixed concentration of effector.

$$\log v = \text{constant} + n \log[\text{S}] \qquad (8\text{–}28)$$

Again, as noted above, the experimentally determined n may be perturbed if the fixed concentration of effector is substantial ($Y_i/m \rightarrow 1$). Whether or not n is perturbed by the allosteric effector concentration will depend on whether the effector and the substrate each function independently (albeit cooperatively, among the multiple sites for the *same* molecule), or whether there are cooperative interactions involving more than one substrate bound subunit *and* more than one effector bound subunit per molecule of enzyme.

The number of cooperative (allosteric) effector binding sites (m) and substrate binding sites (n), estimated assuming that full cooperativity is observable ($E + mA \rightleftharpoons EA_m$), are shown for isocitrate dehydrogenase in Figs. 8–17 and 8–18. Such an analysis leads to the interesting conclusion that isocitrate dehydrogenase contains four allosteric isocitrate binding sites, but only two binding sites for NAD^+, Mg^{2+}, and AMP.

The catalyzed reaction (Eq. 8–23) is inhibited by citrate, a molecule resembling isocitrate in structure. At higher concentrations of isocitrate (and of citrate), citrate inhibition is of the conventional competitive type. At very low concentrations of isocitrate, however (as for example, at concentrations which lie below the point of inflection in Fig. 8–16), *activation* of the catalyzed reaction by citrate is observed. Presumably, the inhibitor *citrate*, when bound at some of the *isocitrate* binding sites, increases the potential catalytic efficiency at other (unoccupied) isocitrate binding sites. Since at low concentrations of substrate and inhibitor the catalytic velocity rises more rapidly than the first power of the substrate concentration, the binding of competitive inhibitor, as well as substrate, should enhance

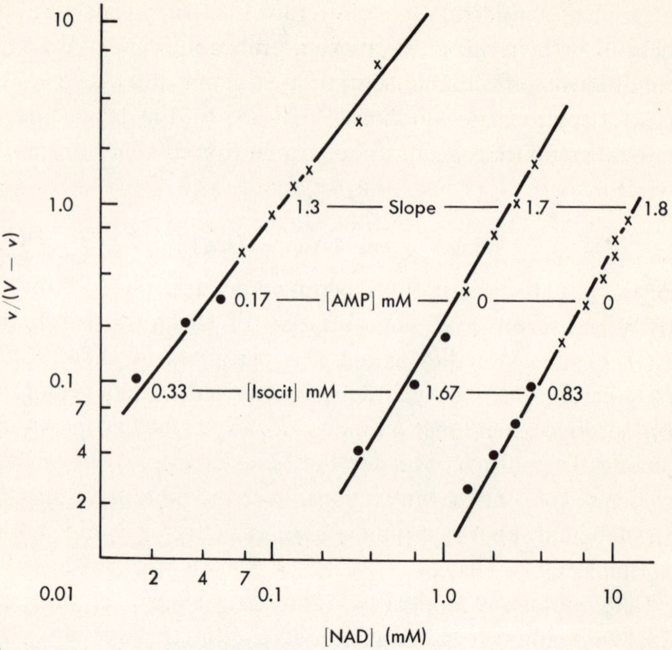
10
1.0
0.1
7
4
2
v/(V − v)
2
4
7
0.01
0.1
1.0
10
[NAD] (mM)
1.3 — Slope — 1.7 — 1.8
0.17 — [AMP] mM — 0 — 0
0.33 — [Isocit] mM — 1.67 — 0.83

(a)

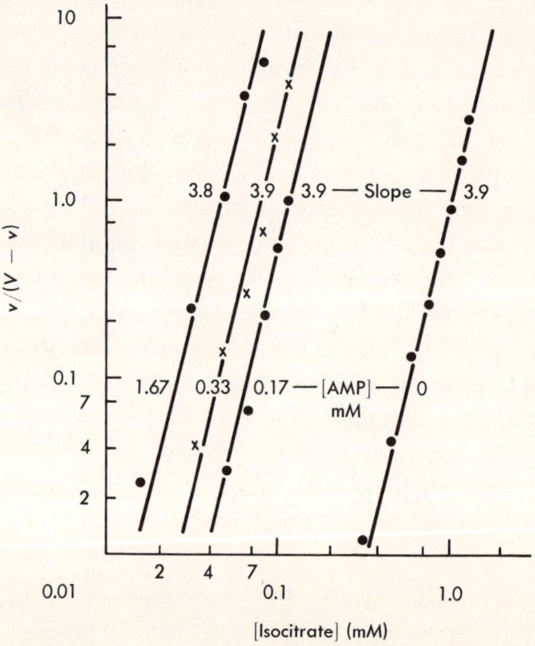
10
1.0
0.1
7
4
2
v/(V − v)
2
4
7
0.01
0.1
1.0
[Isocitrate] (mM)
3.8 3.9 3.9 — Slope — 3.9
1.67 0.33 0.17 — [AMP] — 0
mM

(b)

catalysis, provided that a substantial fraction of the isocitrate binding sites remain unoccupied. Activation, by competitive inhibitor under these conditions, is a strong indication of the cooperativity among binding sites.

One might envisage the cooperative effects described above in terms of an association and dissociation of the subunit polypeptides of isocitrate dehydrogenase which depends on the allosteric effector concentration. Were this to be the molecular origin of the regulatory mechanism, it would be anticipated that the rate of the enzyme-catalyzed reaction would likewise be dependent on a higher power (> 1) of the enzyme concentration, since the association of subunits is a bimolecular or higher order process. This is clearly not the case, since variation of enzyme concentration at fixed levels of substrate and effectors results in only a first order dependence of reaction rate on enzyme concentration. Hence, whatever is the detailed mechanism of regulation via changes in the conformations of the subunit structures of the enzyme complex, the subunits themselves must remain associated. With isocitrate dehydrogenase, all known effectors *increase* the reaction velocity (are *positive* effectors) at low substrate concentrations. *Negative* effectors have been identified in other enzyme systems, for example, in aspartate transcarbamylase which is discussed below.

When the substrate concentration is sufficiently low, so that less than a few percent of the maximal reaction velocity is realized, the rate of reaction is most sensitive to variations in the concentrations of allosteric substrates and effectors, as is dramatically demonstrated in Fig. 8–19. The effect of [isocitrate], [NAD^+], [AMP], [Mg^{2+}], and enzyme concentration on the rate of reaction are investigated simultaneously by mixing fixed amounts of enzyme, substrates, and effectors with a variable amount of solvent. The reaction velocity is depend-

FIG. 8–18 The dependence of reaction velocity on substrate concentrations according to the assumed relationship $v/(V\text{-}v) = [c]^n/K$. (a) Varying concentrations of NAD^+ at fixed low concentrations of isocitrate and AMP. (b) Varying concentrations of isocitrate at fixed low concentrations of NAD^+ and AMP. [From D. E. Atkinson *et al., J. Biol. Chem.* **240,** 2682 (1965).]

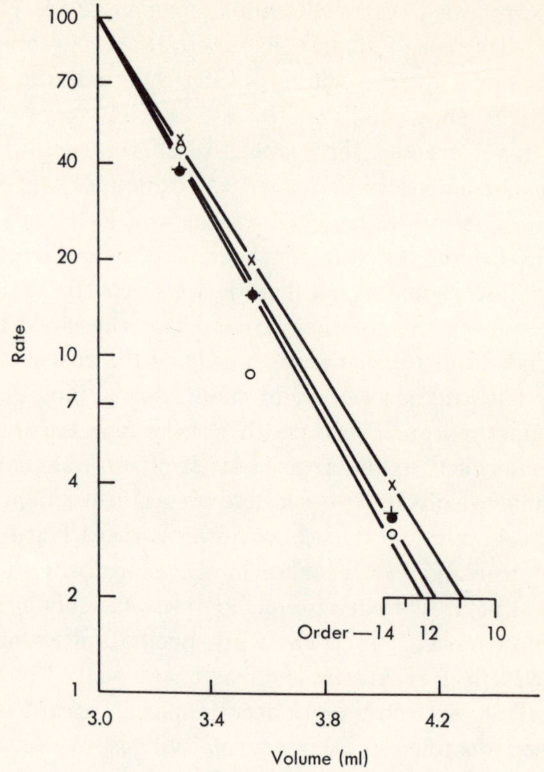

FIG. 8–19 The dependence of the rate of catalysis by isocitrate dehydrogenase on variations in the concentrations of all components of the reaction mixture. The velocity of the oxidation reaction is studied with fixed *amounts* of each of the reaction components (enzyme, isocitrate, NAD^+, AMP, and Mg^{2+}) and a varying *volume* of solvent. Substrate and effector concentrations are invariably far from saturation. [From D. E. Atkinson et *al., J. Biol. Chem.* **240,** 2682 (1965).]

ent on the *eleventh* power of the concentration of *solute* under these circumstances.

ASPARTATE TRANSCARBAMYLASE

Aspartate transcarbamylase catalyzes the reaction illustrated by Eq. 8–29. The metabolic fate of aspartate is dependent on

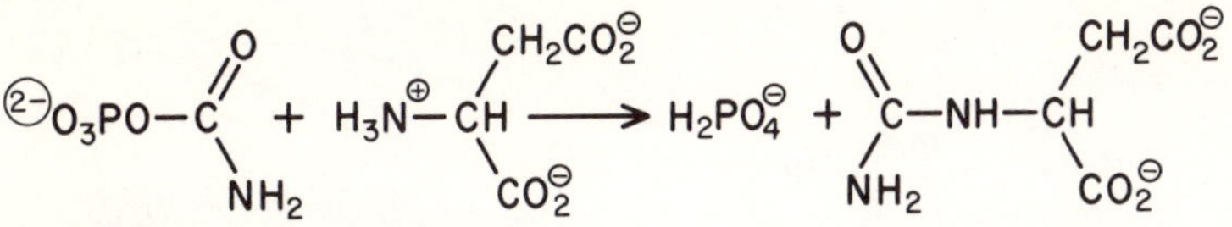

$$\overset{2\ominus}{O_3}PO-\overset{\overset{O}{\|}}{\underset{NH_2}{C}} + H_3\overset{\oplus}{N}-\overset{\overset{CH_2CO_2^\ominus}{|}}{\underset{CO_2^\ominus}{CH}} \longrightarrow H_2PO_4^\ominus + \overset{\overset{O}{\|}}{\underset{NH_2}{C}}-NH-\overset{\overset{CH_2CO_2^\ominus}{|}}{\underset{CO_2^\ominus}{CH}}$$

Carbamyl   **N–Carbamyl**
phosphate   **aspartate**

(8–29)

three different enzymatic pathways: (1) the ultimate incorpo-
ration of this amino acid into polypeptides, (2) the intermediary metabolic conversion of aspartate to other amino acids, and
(3) the biosynthesis of nucleic acid (pyrimidine) precursors.
The conversion of aspartate to N-carbamyl aspartate (Eq.
8–29) is the first step in the metabolic pathway toward pyrimidine synthesis. The enzymic reaction is inhibited by a later
product of the pyrimidine pathway, cytidine triphosphate
(CTP). It is believed that this inhibition is of consequence in
the regulation of the rate of pyrimidine synthesis, higher concentrations of the pyrimidine CTP being inhibitory to further
biosynthesis in the pyrimidine pathway. The CTP-sensitive
enzyme aspartate transcarbamylase exhibits kinetic behavior
characteristic of allosteric enzyme systems. The dependence of
reaction velocity on aspartate concentration and on CTP concentration are illustrated in Fig. 8–20. Note that in this case the
negative effector (CTP) accentuates the cooperativity of substrate interactions. This is in contrast to the behavior with the
positive effector of isocitrate dehydrogenase (AMP), in which
no AMP concentration-dependent change in the cooperativity
is observed (i.e., no change in the apparent order of reaction with
respect to isocitrate).

Aspartate transcarbamylase can be isolated in essentially pure
form. In aqueous solution the protein has an apparent molecular weight of about 300,000 (based on its hydrodynamic sedimentation and diffusion properties). Enzyme catalysis is affected by the addition of mercurial compounds (which form
complexes with sulfhydryl groups). In the presence of mercurials, all cooperative (allosteric) properties disappear; the reaction rate is substrate concentration-dependent as in a Michae-

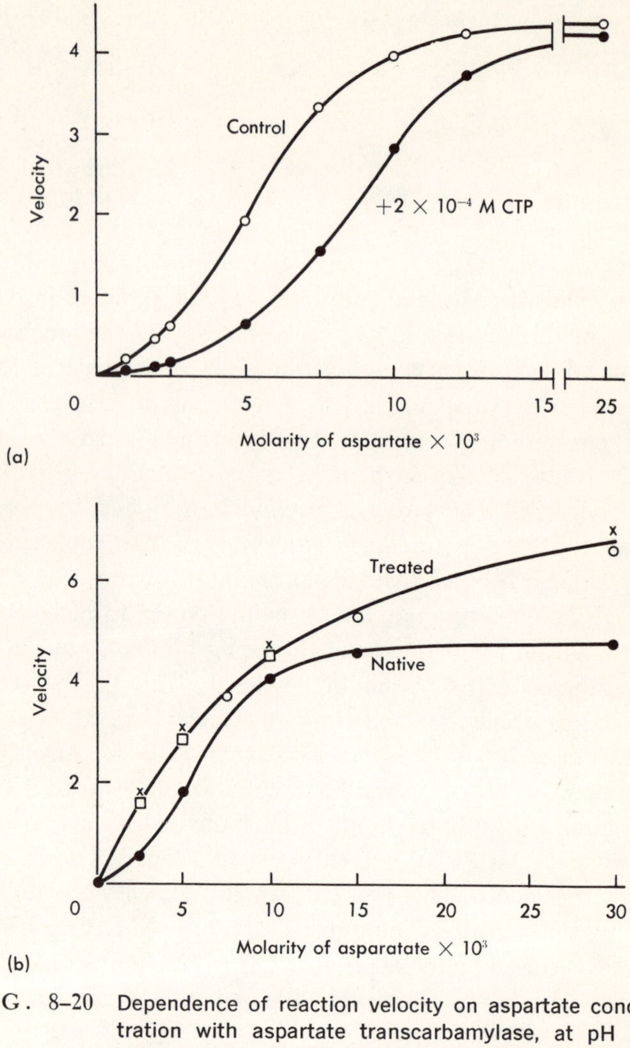

FIG. 8–20 Dependence of reaction velocity on aspartate concentration with aspartate transcarbamylase, at pH 7.0, under a variety of conditions. (a) Native oligomeric enzyme in the presence and absence of the negative effector, CTP. (b) In the presence of enzyme modified in the following ways: ●, native (untreated) enzyme; X, native enzyme in the presence of 10^{-6} *M* $Hg(NO_3)_2$; O, after heating native enzyme at 60°C for 4 min; □, heated enzyme in the presence of 2×10^{-4} *M* CTP. [From J. C. Gerhart and A. B. Pardee, *Cold Spring Harbor Symp. Quant. Biol.* **28,** 491 (1963).]

lis-Menten system, and is not affected by the presence of CTP. When enzyme preparations treated with *p*-hydroxymercuribenzoate are examined in the ultracentrifuge (Fig. 8–21), two distinct new protein components of lower molecular weight are observable. These components can be separated and the mercurial compound can be removed by dialysis. The higher molecular weight component thus isolated is catalytically active. When the dependence of reaction velocity on substrate concentration is examined with this component as catalyst, the cooperativity noted in Fig. 8–20 disappears, the concentration-rate profile resembling that predicted in a Michaelis-Menten (independent

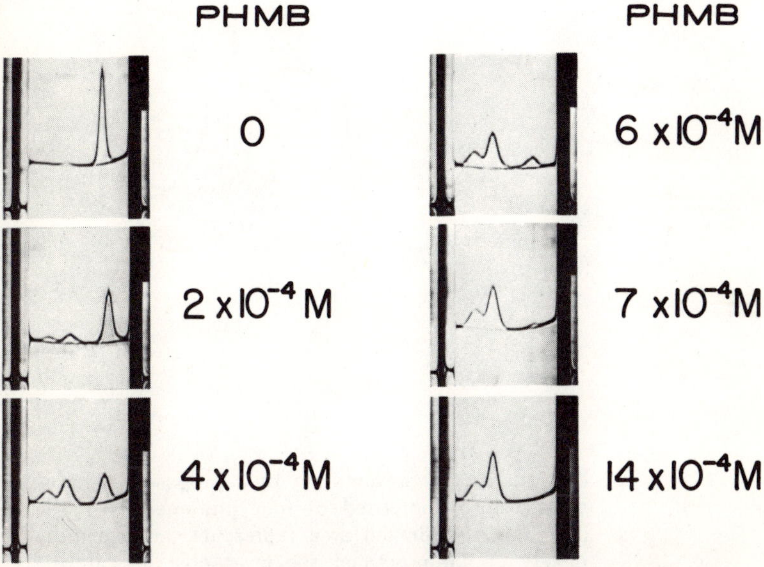

FIG. 8–21 Dissociation of aspartate transcarbamylase by mercurials. At pH 7.0 and varying concentrations of *p*-mercuribenzoate. Sedimentation patterns in the ultracentrifuge at 60,000 rpm about 40 min after treatment with the mercurial. The most dense protein will have a sedimentation peak closest to the right-hand edge of a photograph. Notice the appearance of progressively larger fractions of less dense protein with increasing concentration of the mercurial. [From J. C. Gerhart and H. K. Schachman, *Biochemistry* **4,** 1054 (1965).]

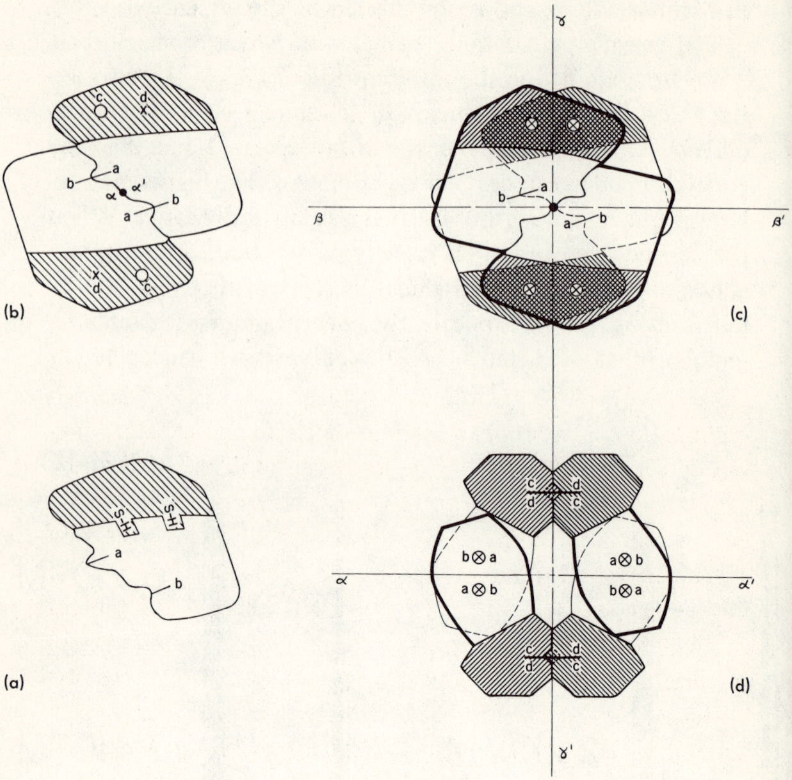

FIG. 8–22 Diagrammatic representation of a proposed oligomeric structure composed of four protomers. (a) The protomer: the shaded area represents the regulatory moiety of the protomer, the clear area its catalytic moiety consisting of one-half of the isolated catalytic subunit. The regulatory moiety and the catalytic moiety are linked through a heterologous association (H) which involves sulfhydryl groups contributed exclusively by the regulatory moiety. (b) Isologous association (IC) of 2 protomers through their catalytic moiety. The dimer obtained possesses a dyad axis of symmetry α–α' perpendicular to the plane of the figure. (c) The tetrameric molecule resulting from the isologous association (IR) of 2 such dimers through their regu-

site) model. Moreover, the reaction velocity is no longer inhibited by the former effector, CTP. The lower molecular weight component is catalytically inert. In a solution containing only this lower molecular weight protein component and CTP, the CTP has been shown to be specifically and firmly bound to the protein.

The two isolated protein components can be reincubated to restore the original catalytic properties of aspartate transcarbamylase. The reconstituted mixture again shows the cooperativity in the substrate-concentration-dependent reaction velocity, illustrated in Fig. 8–20, and is once again inhibited by CTP in a manner characteristic of allosteric effectors. By a more detailed examination of the cooperativity in substrate and effector binding, and from an analysis of the molecular weights and percentage compositions of the large and small components, the intact enzyme has been postulated to be composed of multiple substrate binding sites, and multiple allosteric effector sites. Indeed, under denaturing conditions, each of the two isolated protein components mentioned above can be dissociated into still smaller subunits. The arrangement of subunits into an orderly, organized regulatory enzyme structure is shown in Fig. 8–22.* The organized structure is now believed to be composed of six molecules each with two distinct types of subunit polypeptides. The larger subunits each contain a single substrate binding site, and the smaller subunits a single allosteric effector binding site.

* The oligomeric molecule of aspartate transcarbamylase was initially believed to be composed of four catalytic and four regulatory polypeptide subunits.

---

latory moieties. (d) Another view of the tetramer after clockwise rotation of the molecule by 90° around the γ–γ' axis. [From, J. P. Changeux, J. C. Gerhart, H. K. Schachman in *Regulation of Nucleic Acid and Protein Biosynthesis* (V. V. Konigsberger and L. Bosch, eds.), BBA Library, Vol. 10, p. 348, Elsevier, Amsterdam, 1967.]

8–6 ENZYMES OF KNOWN THREE-DIMENSIONAL STRUCTURE

After the determination of the three-dimensional structure of myoglobin and hemoglobin, the feasibility of determining the structures of crystalline enzymes became apparent. At the time of writing this text, the electron density maps for five enzymes, *carboxypeptidase*, *carbonic anhydrase*, *egg white lysozyme*, *α-chymotrypsin*, and *ribonuclease* have been determined to a resolution such that the conformation of the polypeptide chain can be followed over its entire length. It would be inappropriate to over-generalize at this time regarding the relationship between structure and function on the basis of these examples. Nevertheless, certain common features among these structures deserve mention:

(1) Each of these enzymes consists of a single polypeptide chain which is highly compact (folded) and nearly spherical in shape.

(2) In the native conformation of each enzyme there exists a single and quite distinct *cavity*, or *cleft*. This cavity is bounded, at least in part, by regions of the polypeptide involved in substrate binding and catalysis.

(3) Unlike myoglobin and hemoglobin, only a small fraction of the folded polypeptide chain is in the α-helical configuration.

Whether these results can be generalized to all enzymes remains to be determined. Let us examine in further detail the structures of the enzymes carbonic anhydrase and lysozyme.

CARBONIC ANHYDRASE

Carbonic Anhydrase catalyzes the reaction shown in Eq. 8–30.

$$CO_2 + H_2O \rightleftharpoons H_2CO_3 \qquad (8\text{–}30)$$

The dehydration of carbonic acid proceeds sufficiently slowly near pH neutrality, so as to require enzymic catalysis if the reaction is to be of physiological consequence. Native carbonic anhydrase contains one atom of Zn^{2+} per molecule of enzyme (the molecular weight of the enzyme is approximately 3.0×10^4 g). This atom of Zn^{2+} is essential for enzymic activity, and its occurrence happens to be a crystallographic convenience. The high electron density of the zinc atom will result in a prominent maximum in the electron density map, and hence provides for the facile location of a constituent of the active site. The conformation of the polypeptide chain and the location of the zinc atom, as derived from the electron density map, is shown schematically in Fig. 8–23. A more detailed view of the enzyme site region, to a resolution of 5 Å, is shown in Fig. 8–24. Notice the enzyme site *cavity* and the prominent location of the zinc atom within the cavity.

The zinc atom is known to be tightly bound to the enzyme site, presumably by chelation with specific amino acid side chains. Zinc is known to form tetrahedral complexes. The details of the site structure are sufficient to indicate that three of the four tetrahedral ligand contacts are supplied by the polypeptide-contained side chains. The fourth ligand binding position is unoccupied and points in the direction of the cavity. Apparently, it is accessible to the substrate which contributes the fourth ligand contact.

The resolution in Fig. 8–24 is to 5 Å. At this resolution individual atoms are not discernible, although, as it happens, the twists and turns of the polypeptide chain can be followed over its total length. If the amino acid sequence is known, some of the atomic positions may be presumed from the prior information we have regarding the configuration of a number of the larger amino acid side chains.

One interesting result derived from the structure determination is that the terminal carboxylate residue is a constituent of the active site (it is one of the ligands to the zinc atom). This should be compared with the results obtained with chymotrypsin and trypsin (Section 8–1) and with ribonuclease (Section

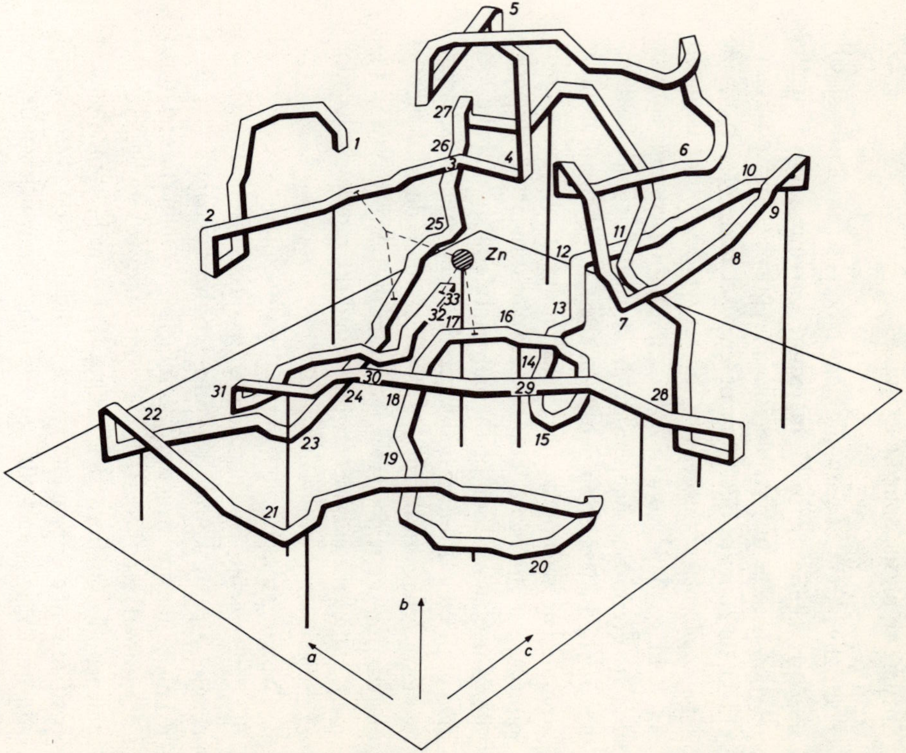

FIG. 8–23 The electron density map of carbonic anhydrase to a resolution of 5 Å. (Courtesy of Dr. B. Strandberg.)

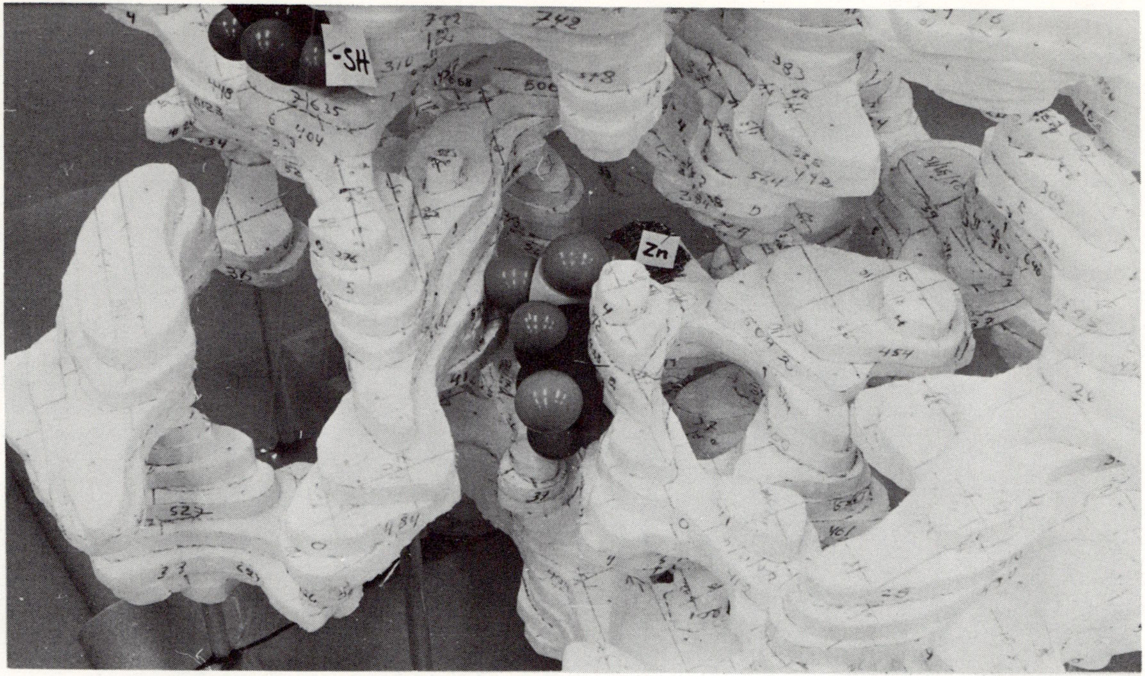

FIG. 8–24 A close-up view of the zinc-containing active site of carbonic anhydrase. (Courtesy of Dr. B. Strandberg.)

8–2), in which terminal amino acid residues have likewise been shown to be essential for catalysis. These results may relate to the experimental observation that no peptide bond cleavage procedure has ever been found which gives rise to an apprecia-bly smaller peptide with enzymic activity.

Carbonic anhydrase is rapidly and stoichiometrically inacti-vated by bromoacetate anion, and the inhibited enzyme has been shown to contain a single alkylated histidine residue (Eq. 8–31). Likewise, chemical modification studies indicate that

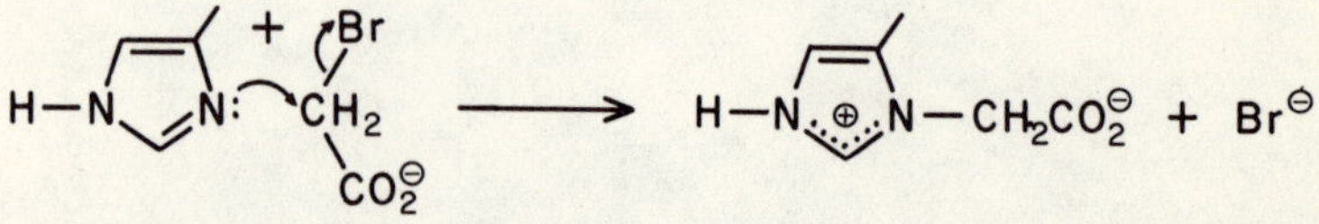

(8–31)

one methionine residue and the carboxyl terminal residue parti-cipate in the catalysis.

The turnover number for the dehydration of carbonic acid is unusually rapid (approx. 10^5 sec^{-1}). In addition to the hydration of carbon dioxide the enzyme catalyzes the hydration of acetal-dehyde (Eq. 8–32). Both reactions are catalyzed by electrophiles.

$$H_3C{-}C(H){=}O + H_2O \rightleftharpoons H_3C{-}C(H)(OH){-}OH \qquad (8\text{–}32)$$

Presumably, the zinc atom acts as an electrophilic catalyst, as described in Section 6–1.

EGG WHITE LYSOZYME

Egg white lysozyme is the first enzyme for which a complete three-dimensional structure determination has been carried out to atomic resolution. As in the case of myoglobin, a knowledge of the amino acid sequence was of aid in carrying out the complete structure determination. The primary sequence of egg white lysozyme is shown in Fig. 8–25. The molecule consists of a single polypeptide chain containing 129 amino acid residues, cross-linked by four disulfide bridges.

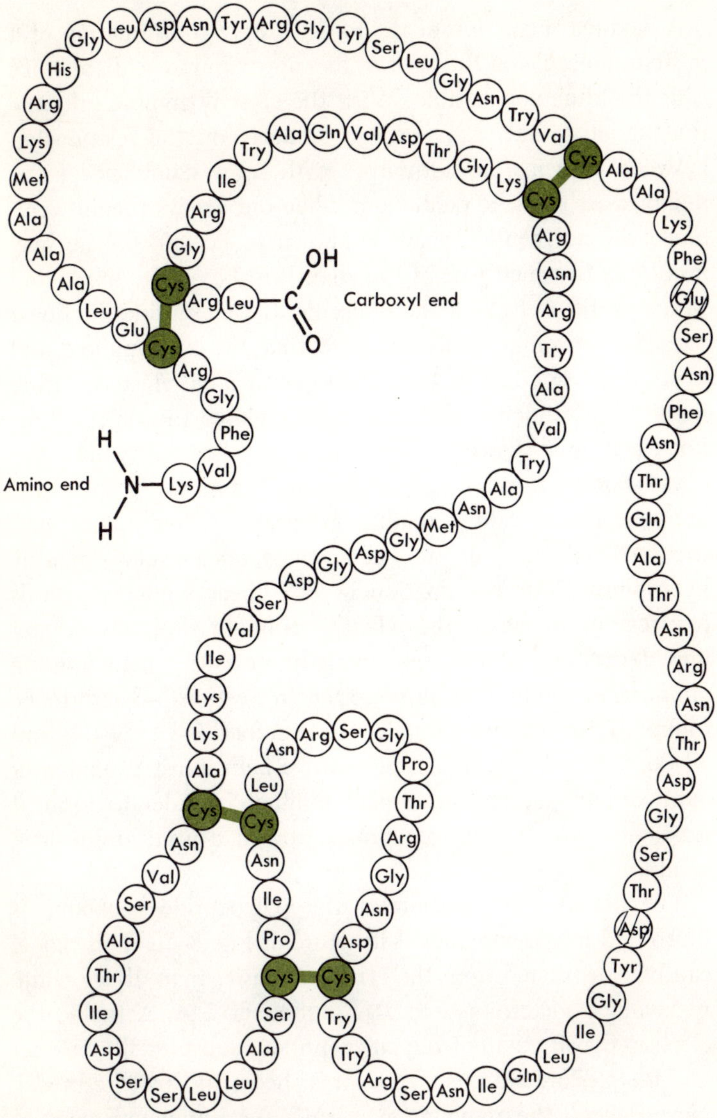

FIG. 8–25 The primary sequence of amino acid residues in egg white lysozyme.

A detailed prediction of the folded native conformation of a protein molecule on the basis of its known amino acid sequence is at this time not feasible. With the crystallographic information at hand, however, one can "reflect" on the relationship between sequence and structure with some confidence (it is much easier to make predictions when one knows the answer). Since the polypeptide chain is known to be folded, one may inquire as to which part of the chain will be on the inside and which on the outside of the molecule. If we examine the linear sequence of amino acids, we note that the hydrophobic and hydrophilic side chains tend to occur in large clusters, rather than to be randomly distributed throughout the polypeptide. For example, a sizeable proportion of the first thirty-nine residues, starting from the amino terminal end, are hydrophobic residues. The next forty residues are primarily hydrophilic, and the terminal third of the sequence is once again primarily hydrophobic. An examination of the linear sequence reveals few regions in which the α-helical configuration can be sustained over any considerable length. Potential helix-forming amino acid residues, as summarized in Section 3–3, are to be found in the regions bounded by residues 5–15, 24–34, and 88–96. Note that a stable region of α-helix must extend over approximately seven amino acid residues in order to form at least one pair of hydrogen bonds from a central amino acid residue.

The actual conformation of the polypeptide backbone is illustrated in the wire model structure of Fig. 8–26. This model can be constructed from the electron density map of egg white lysozyme as determined to a resolution of 2 Å. Actually, the electron density map of the polypeptide is substantially broader and more definitive in shape than is shown in this wire model. Nevertheless, the resolving power (2 Å) is not sufficient to permit us to recognize the individual atoms in the amino acid sequence. Recognition of the amino acid side chains and of their configurations is accomplished by combining the information on electron density with the known amino acid sequence in lysozyme. In this way it is possible to recognize from the relatively low resolution electron density map, the known con-

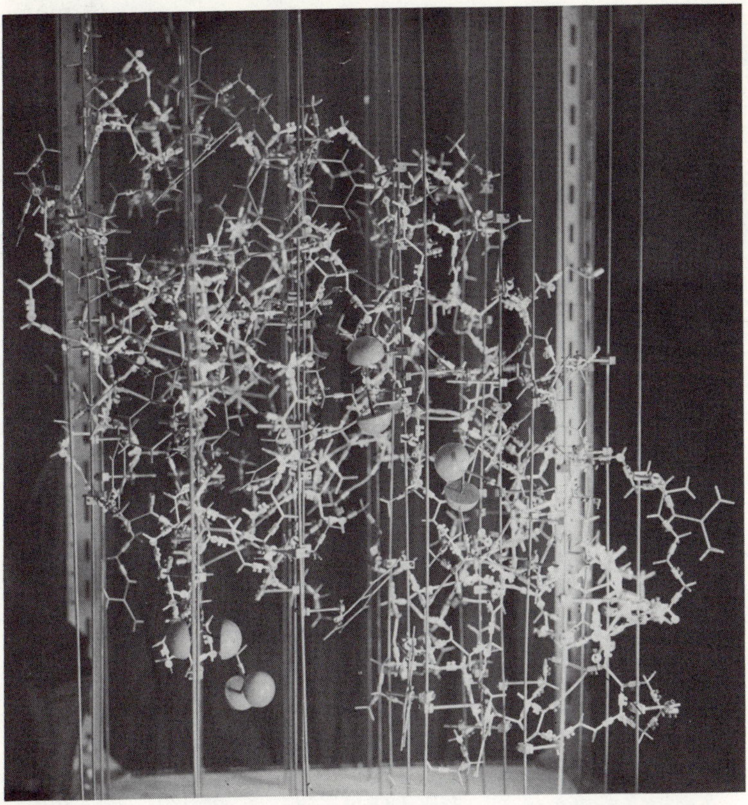

FIG. 8–26 A "wire" model of the conformation of the polypeptide chain in egg white lysozyme. (Courtesy of Dr. D. C. Phillips.)

figurations of the amino acid side chains. A three-dimensional molecular model of lysozyme has thus been constructed in which the locations of almost all of the atoms of the molecule are known with virtual certainty.

Let us now proceed to examine the results of this constructed model. One difficulty in such an examination is the problem of representing the three-dimensional array of so large a number of atoms. (The problem of spatial representation in two dimensions is less pronounced at lower resolution, as for example, in the structure of carbonic anhydrase discusssed above, since the details of atomic configuration are lacking.) The most obvious feature of the constructed lysozyme model is the single cleft in the structure, presumably the substrate binding site. How does this cleft arise within the folded conformation of the polypeptide?

There is now considerable evidence that in the biosynthesis of proteins, synthesis begins from the amino terminal end of the polypeptide chain.[2] It is not unreasonable to assume that the polypeptide starts folding as it is synthesized. The first 34 amino acids in the lysozyme sequence (starting from the amino terminal end) form two stretches of helix. The first (residues 5–15), is substantially in the configuration predicted by Pauling and Corey. The second helical stretch (residues 24–34) differs somewhat in specific hydrogen bonding patterns from that of the normal α-helix. These two helical residues contain primarily hydrophobic side chain "coats" and together form a large hydrophobic region of the molecule (Fig. 8–27a). These helical regions are connected via residues 35–40 to another type of regular polypeptide structure (also predicted by Pauling and Corey earlier), namely that of an "anti-parallel pleated sheet." In this type of structure the polypeptide is hydrogen bonded between one region and another via a "hairpin turn" of the chain, giving rise to intramolecular hydrogen bond pairs (Fig. 8–27b). Residues 41–45 and 50–54 form such a structure. The connecting residues (46–49) are so folded as to permit the "hairpin bend" required to connect the two sequences via hydrogen bonds. Most of the remainder of the sequence is folded in irregular (albeit highly specific) ways. Residues 88–96, however, form a third stretch of helix. The first 35 amino acid residues are folded into a hydrophobic wall. This wall is hinged to residues 41–54, which form a hydrophilic wall in such a way that the two regular structures together form the hydrophobic and hydrophilic wing structure shown in Fig. 8–27c. Into a part of this wing is fitted the third helical region (residues 88–96). The net result is the formation of the active site cleft. The remainder of the amino acid sequence is folded about this basic structure in a complex way, such that the cleft

[2] See, for example, Ingram, V. M., *The Biosynthesis of Macromolecules*, W. A. Benjamin, New York, 1965.

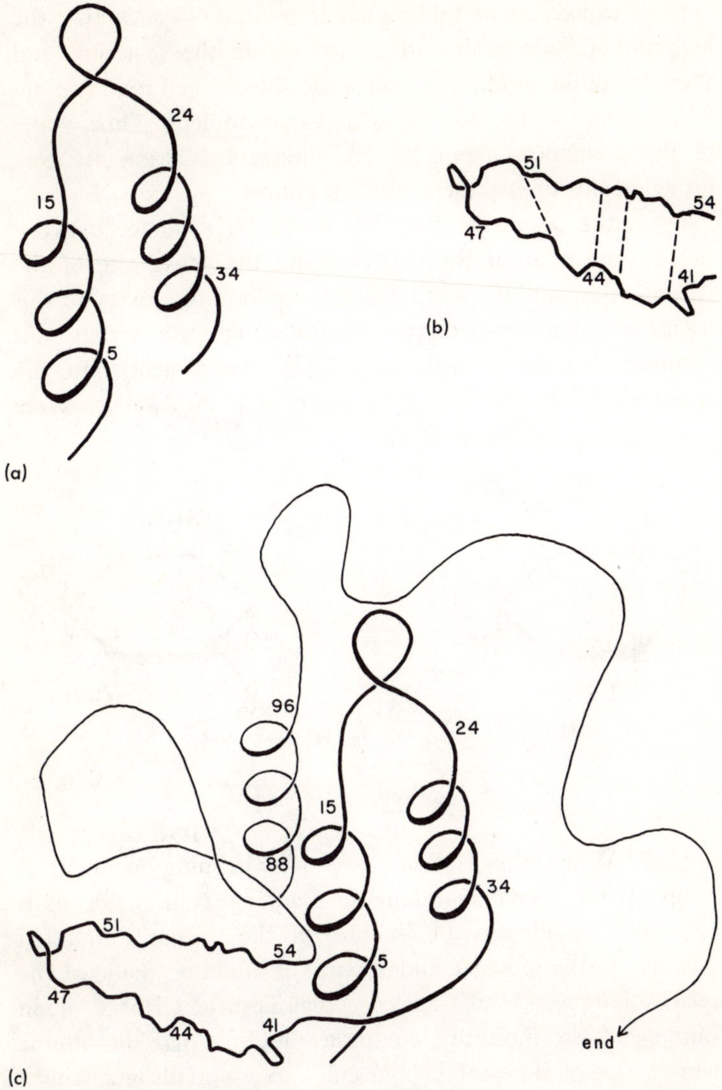

FIG. 8–27 The structural components of the lysozyme polypeptide defining the active site "cleft." (a) Two stretches of α-helix near the amino terminal end of the polypeptide. (b) The antiparallel pleated sheet structure. (c) A third stretch of helix (residues 88–96) interacts with both the hydrophobic helical wing [as in (a)] and the hydrophilic pleated-sheet structure [as in (b)] to form the cleft.

remains exposed. The folding of the molecule is such that the hydrophobic side chains are buried within the structure, and the hydrophilic residues (in particular the charged residues) are all accessible to the outside solvent environment. Thus, many of the predictions based on less direct observation are confirmed in this three-dimensional structure.

From the discussion thus far, the correspondence between the distinct cleft in the structure and the active site of the enzyme may only be presumed. An important discovery in this regard was that competitive inhibitors of enzymic activity, for example, N-acetylglucosamine (XVI) and N-acetylmuramic acid (XVII), can be incorporated indirectly into wet lysozyme

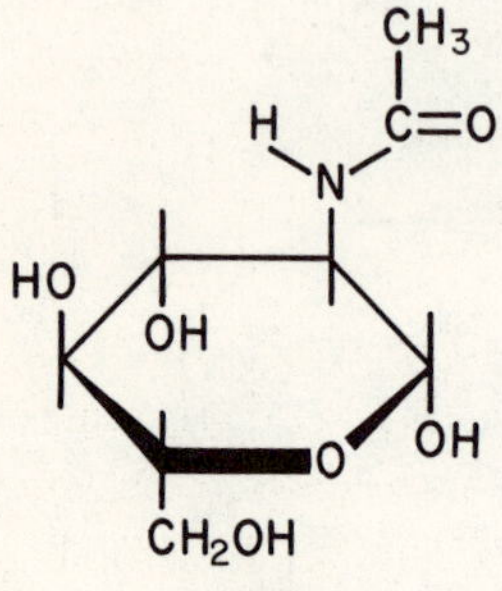

(XVI)

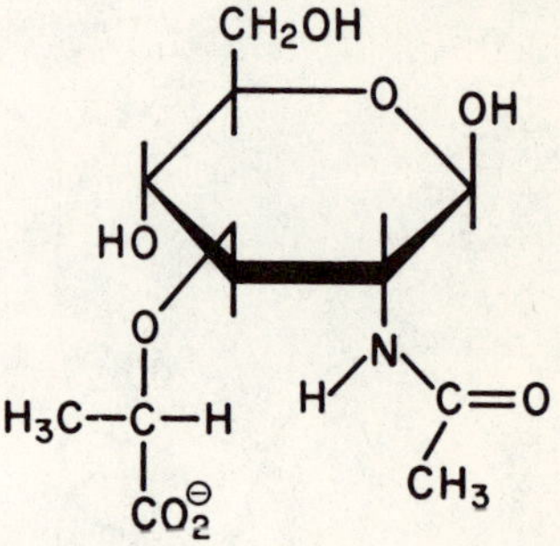

(XVII)

crystals. When this is done, the overwhelming majority of atoms in the crystal maintain their original coordinates, as is evidenced by identity of *location* of the various diffraction maxima, in the presence and absence of inhibitor. Some of the relative *intensities* are, however, significantly changed upon binding of the inhibitor; on the assumption that the atomic coordinates of the protein molecule are essentially unchanged (i.e., that the phases remain unchanged), the difference in structure between enzyme and enzyme inhibitor complex can be calculated by suitable mathematical procedures. When this type of analysis is carried out, the calculated "difference structure" gives the known structure of the inhibitor, confirming that the phase of the protein remains essentially unchanged in the E-I complex. Moreover, the precise location of the inhibitor within the enzyme molecule is thus determined. In this way, it has been established that the competitive inhibitor is bound to the enzyme within the cleft region. This type of analysis has been carried out for a number of amino sugars containing varying numbers of glucosyl residues. Such monomers, dimers, trimers, and so on, are of significance in relation to the configuration of true substrates of the enzyme. As more glucosyl residues are condensed in the proper configuration, the resultant carbohydrate becomes a better substrate for the enzyme (see Fig. 8–28). In the simple monomer (N-acetylglucosamine), a prominent mode of attachment of the inhibitor to the enzyme is via a hydrogen bond between the H-N–acetyl and an oxygen of the β-carboxylate residue of aspartate (residue 101). This hydrogen bond appears to be of crucial importance since the longer *oligomers,* which are bound deeper into the cleft, all contain this hydrogen bond between the first H-N acetyl and the protein. The rate of cleavage of oligomers of either N-acetylglucosamine or of repeating units of N-acetylglucosamine N-acetylmuramic acid (Fig. 8–28) increases rather abruptly when the oligomer contains 5 or more glucosyl residues (cleavage occurs at the linkage between rings D and E of Fig. 8–28). An examination of the dimensions of the cleft indicates that a maximum of six monomer units fits well into the cleft (Fig. 8–29). This is presumably the limit of substrate specificity of the site.

If the binding of substrate is as shown in Fig. 8–29, and cleavage occurs between rings D and E, it is of interest to survey the area surrounding the D-E linkage for potential catalytic centers. In this region two potentially catalytic amino

FIG. 8–28 A polysaccharide substrate for egg white lysozyme.

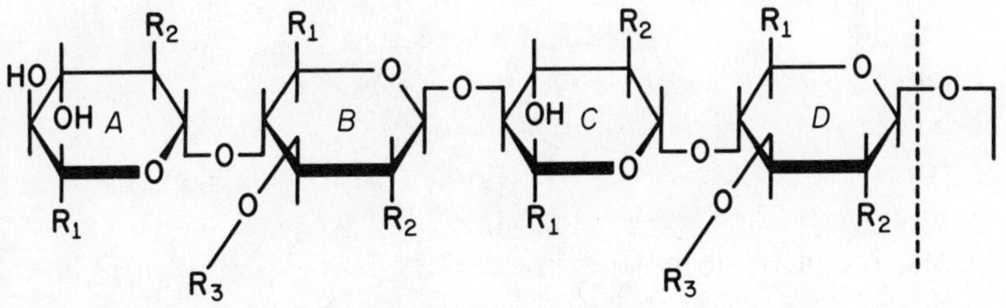

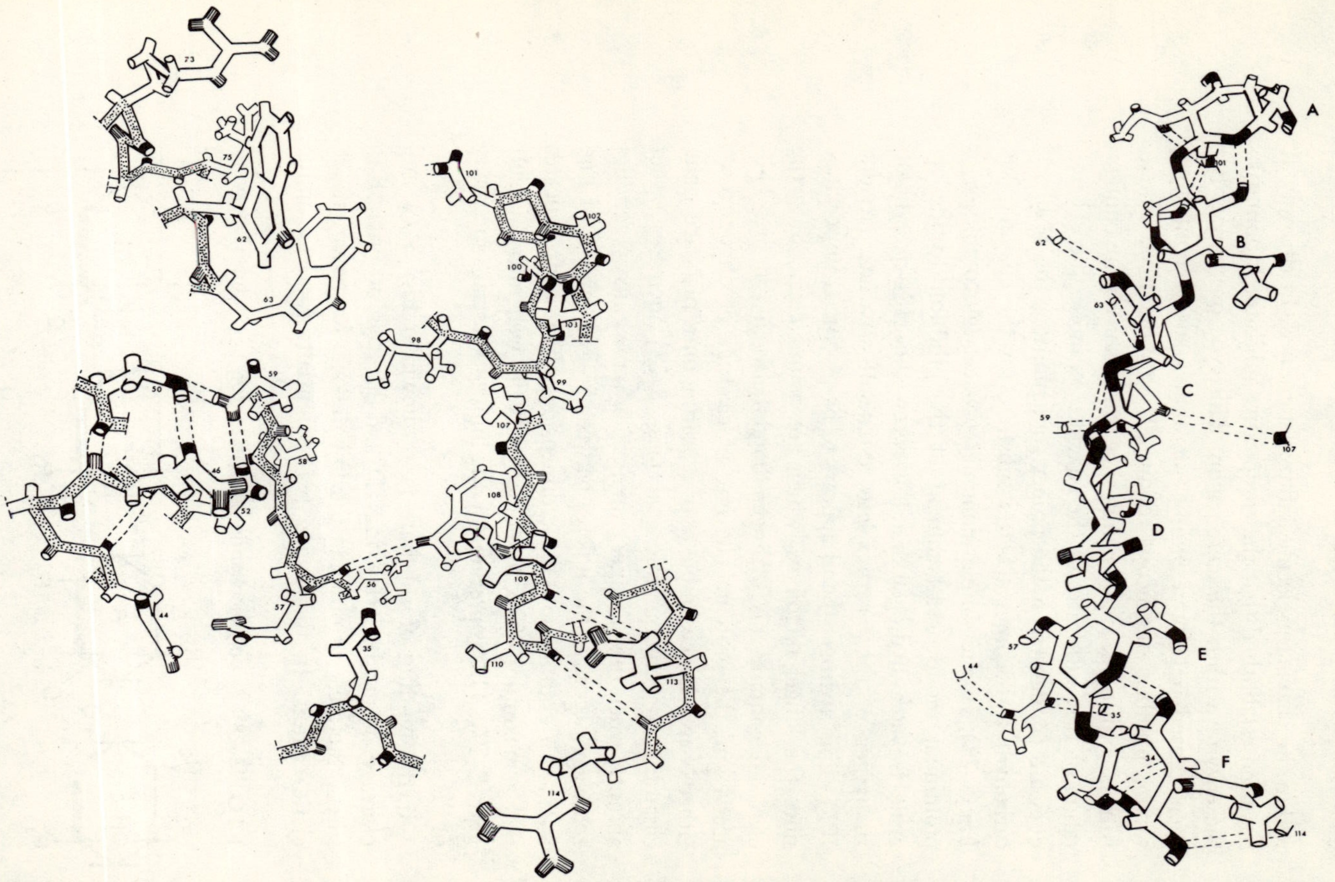

FIG. 8.29 Complementary structural relationship between substrate (right) and active site (left), to atomic dimensions

acid side chains are notable, namely the β and γ carboxylates of aspartate (residue 52) and glutamate (residue 35), respectively. Two alternative mechanisms, both involving acid-base catalysis by one carboxylic acid and one carboxylate ion may be proposed as the mechanism of lysozyme action, as summarized in Eq. 8-33. In one mechanism, general acid-base catalysis occurs via proton donation from glutamic acid 35 to the bridge oxygen of substrate, and is concerted with proton abstraction from water via aspartate 52, which facilitates attack by hydroxyl ion at position *C-1* of ring D. In the second mechanism, glutamic acid 35 again acts as a general acid catalyst, transferring a proton to the oxygen bridge. The formation of the positively charged oxygen bridge facilitates the rupture of the C-1 to oxygen linkage with the resultant formation of a C-1 carbonium ion in ring D. The transition state for the formation of this carbonium ion (a developing positive charge at C-1) should be stabilized by the proximity of the negatively charged carboxylate of aspartate 52.

The above discussion is indicative of the suggestive information concerning the mechanism of enzyme action which may be derived once we have a knowledge of the three-dimensional structure of enzyme proteins. Among the most dramatic results derived from the lysozyme structure is this speculative information concerning reaction mechanism. Undoubtedly, as more and more enzyme structures are determined, greater details of mechanism will become apparent. It should be noted that the suggestive mechanisms of lysozyme action are, *at this time*, based on crystallographic studies of the enzyme (monomer and trimer) inhibitor complexes—structures in which there is no interaction between "substrate" and the region of the polypeptide containing the two implicated catalytic centers (Glu-35 and Asp-52). It was noted earlier that the formation of *covalent* enzyme-substrate compounds can cause significant changes in the conformation of the protein in particular instances (perhaps generally). In contrast, these competitive inhibitors of lysozyme cause only minor changes in the polypeptide conformation—most notably, minor changes in the disposition of tryptophan residues. It

Mechanism I

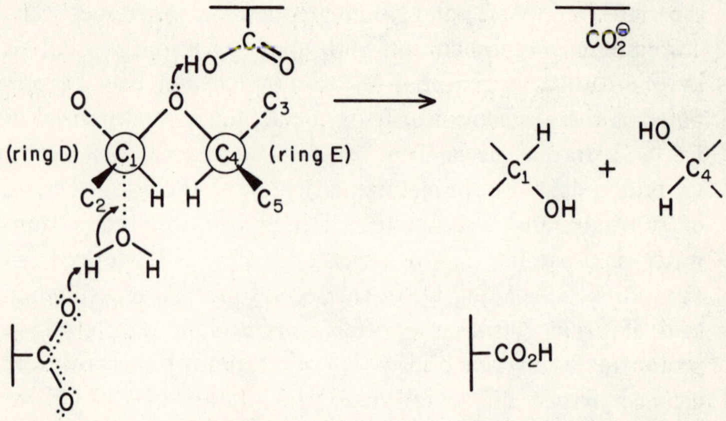

Mechanism 2

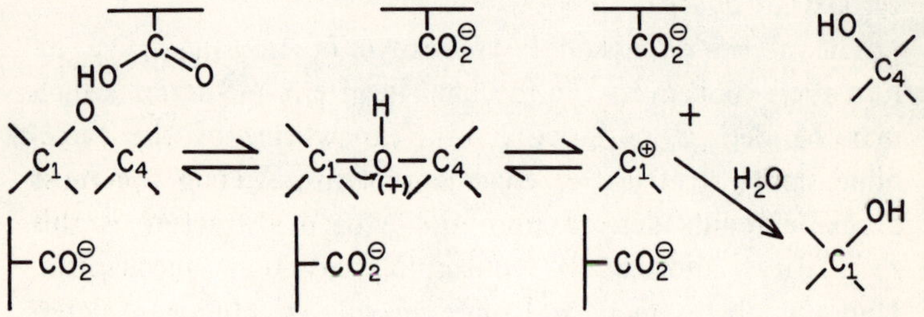

(8–33)

remains for the future to assess whether the relatively rigid conformation of the native enzyme (as is illustrated by the identity of phase of lysozyme and lysozyme-inhibitor complex crystals) contains all of the essentials for effective catalysis, or whether conformational changes in the protein molecule are of themselves a requirement for effective biological catalysis. It also remains for the future to establish the mechanisms of enzymic catalyses at the molecular level, and to define the general conditions (if such generalities exist) for effecting rapid catalysis in living systems.

All that has been discussed in these eight chapters represents (to the author) a suitable foreword. I have little doubt that the relevant *missing* material on the relationship between the structure and the function of enzymes at the molecular level will be available to the reader in the very near future.

REFERENCES

Despite the length of this chapter, the information presented on the various enzymes discussed is only fragmentary, and it often overlooks important contributions. To correct for some of the deficiencies in this chapter, a more detailed reference list is included. The references are arranged more or less in the order in which the subjects appear in the text. More comprehensive articles and reviews are distinguished by the length of the comments.

α-CHYMOTRYPSIN AND RELATED ENZYMES

Bender, M. L., "The Mechanism of α-Chymotrypsin Catalyzed Hydrolysis," *J. Am. Chem. Soc.* **84,** 2580 (1962).

Bender, M. L., and Kezdy, F., *J. Am. Chem. Soc.* **86,** 3704 (1964). These two articles admirably summarize the evidence for the formation of an acyl enzyme in α-chymotrypsin (and related enzyme) catalyzed reactions, and discuss and interpret the significance of the formation of intermediates to the mechanism of enzyme catalyzed reactions.

Important experimental observations pertaining to specialized subjects are noted below:

Primary Sequence

Hartley, B. S., "Amino Acid Sequence of Bovine Chymotrypsinogen A," *Nature 201,* 1284 (1964).

Walsh, K. A., and Neurath, H., "Trypsinogen and Chymotrypsinogen as Homologous Proteins," *Proc. Natl. Acad. Sci., U.S.* **52,** 884 (1964).

Neurath, H., "Protein Digesting Enzymes," *Scientific American,* December 1964. This article describes the molecular mechanism of the activation of zymogens.

pH Dependence and Catalytic Mechanism

Oppenheimer, H. L., Labouesse, B., and Hess, G. P., *J. Biol. Chem.* **241,** 2720 (1966).

Keizer, J., and Bernhard, S. A., *Biochemistry* **5**, 4127 (1966). These two papers consider the various weak acids involved in the catalytic mechanism.

Properties of Acyl Enzymes

Hartley, B. S., and Kilbey, B. A., *Biochem. J.* **56,** 288 (1954).

Caplow, M., and Jencks, W. P., *Biochemistry* **1,** 883 (1962).

Bender, M. L., Schonbaum, G. R., and Zerner, B., *J. Am. Chem. Soc.* **84,** 2540 (1962).

Bernhard, S. A., Lau, S. J., and Noller, H., *Biochemistry* **4,** 1108 (1965).

Inactivation by Irreversible Inhibitors

Wilson, I. B., Bergmann, F., and Nachmansohn, D., *J. Biol. Chem.* **186,** 781 (1950). This paper on the mechanism of inactivation of cholinesterase by diisopropylfluorophosphate contains the first suggestion of an acyl enzyme intermediate in the catalytic reaction of "active" serine enzymes with substrates.

Transient Reactions with Specific Substrates

Bernhard, S. A., and Gutfreund, H., *Proc. Natl. Acad. Sci., U.S.* **53,** 1238 (1965).

Barman, T. E., and Gutfreund, H., *Ibid.*, **53,** 1243 (1965).

Alkaline Phosphatase

Schwartz, J. H., Crestfield, A. M., and Lippman, L., "The Amino Acid Sequence of a Tetradecapeptide Containing the Reactive Serine in *E. coli* Alkaline Phosphatase," *Proc. Natl. Acad. Sci., U.S.* **49,** 722 (1963).

RIBONUCLEASE

Enzyme Models and Enzyme Structure

Anfinsen, C. B., "The Tertiary Structure of Ribonuclease," Brookhaven Symposium in Biology No. 15, p. 184 (1962). A good review of the information (through 1962) on the relationship between primary sequence and both conformation and catalysis.

Stein, W. H., and Moore, S. B., "The Chemical Structure of Proteins," *Scientific American*, February 1961. An excellent article on the methods utilized in the determination of primary structure, with particular emphasis on ribonuclease.

The Native Conformation

Richards, F. M., and Vithayathil, P. J., "Peptide-Protein Interactions in RNAase-S," Brookhaven Symposium in Biology No. 13, p. 115 (1960).

Anfinsen, C. B., Haber, E., Sela, M., and White, F. H., Jr., *Proc. Natl. Acad. Sci., U.S.* **47**, 1309 (1961).

The Catalytic Mechanism

Findlay, D., Herries, D. G., Nathias, H. P., and Rabin, B. R., *Biochem. J.* **85**, 152 (1962). A summary of evidence implicating general acid-base catalysis by two imidazole residues in the enzymic mechanism.

ENZYMES REQUIRING PYRIDOXAL PHOSPHATE

General Reference

Snell, E. E., Fasella, P. M., Braunstein, A., and Rossi Fanelli, A., (editors), *Chemical and Biological Aspects of Pyridoxal Catalysis*, Pergamon Press, London, 1963. A broad survey of the field is contained in this volume.

The Active Site Peptide

Hughes, R. C., Jenkins, W. T., and Fischer, E. H., *Proc. Natl. Acad. Sci., U.S.* **48**, 1615 (1962).

Equilibrium Observations of Enzyme-Pyridoxal Substrate Intermediates

Jenkins, W. T., and Sizer, I. W., *J. Biol. Chem.* **235**, 620 (1960).

Transient Observations of Intermediates

Hammes, G. G., and Fasella, P., *Biochemistry*, to be published.

Hammes, G. G., and Fasella, P., *J. Am. Chem. Soc.* **85**, 3929 (1963).

DEHYDROGENASES

General Reference

Boyer, P. D., Lardy, H., and Myrbäck, K. (editors), *The Enzymes*, 2nd ed., Vol. 7, Academic Press, New York, 1963. Excellent reviews, within this large volume, of particular relevance to the enzymes discussed in this chapter, are as follows: "Alcohol Dehydrogenases" by Sund, H., and Theorell, H.; "Lactate Dehydrogenase" by Schwert, G. W., and Winer, A. D.; "Glyceraldehyde 3-Phosphate Dehydrogenase" by Velick, S. F., and Furfine, C.

Stereospecificity of Hydrogen Transport in Nicotinamide Dehydrogenases

Vennesland, B., and Westheimer, F. H., in *The Mechanism of Enzyme Action* (McElroy, W. D., and Glass, B., editors),

Johns Hopkins, Baltimore, 1954. A fascinating review concerning the stereochemistry of hydrogen abstraction and addition between substrate and coenzyme and its implications for the enzymic mechanism.

Transient and Steady State Kinetics of Dehydrogenases

Alcohol Dehydrogenase:

Theorell, H., and Chance, B., *Acta Chem. Scand.* **5**, 1127 (1951).

Theorell, H., and McKinley McKee, J. S., *Acta Chem. Scand.* **15**, 1811 (1961).

Dalziel, K., *Acta Chem. Scand.* **1** (Suppl. 1), 27 (1963).

Malate Dehydrogenase:

Raval, D. N., and Wolfe, R. G., *Biochemistry* **2**, 220 (1963).

Czerlinski, G., and Schreck, G., *Biochemistry* **3**, 89 (1964).

Glyceraldehyde 3-Phosphate Dehydrogenase:

Furfine, C., and Velick, S. F., *J. Biol. Chem.* **240**, 844 (1965).

Fahien, L. A., *Ibid.* **241**, 4115 (1966).

Kirschner, K., Eigen, M., Bittman, R., and Voigt, B., *Proc. Natl. Acad. Sci., U.S.* **56**, 1661 (1966).

The Active Site of Glyceraldehyde 3-Phosphate Dehydrogenase

Krimsky, I., and Racker, E., *J. Biol. Chem.* **198**, 731 (1952).

Harris, I., Meriwether, B. P., and Park, J. H., *Nature* **198**, 154 (1963).

ALLOSTERIC ENZYMES

Models

Monod, J., Wyman, J., and Changeux, J.-P., *J. Mol. Biol.* **12**, 88 (1965).

Changeux, J.-P., *Cold Spring Harbor Symp. Quant. Biol.* **28**, 497 (1963).

Koshland, D. E., *Ibid.* **28**, 473 (1963).

The first two references deal with models involving an equilibrium between two conformational states of the enzyme protein with different affinity and (or) catalytic activity toward substrate, and with the influence of *effectors* on this equilibrium. The latter reference deals with substrate-induced changes in protein conformation. This latter model is favored in the interpretation of the allosteric behavior of isocitrate dehydrogenase.

Isocitrate Dehydrogenase

Atkinson, D. E., Hathaway, J. A., and Smith, E. C., *J. Biol. Chem.* **240,** 2682 (1965). A very detailed experimental steady-state analysis of the role of effectors.

Aspartate Transcarbamylase

Gerhart, J. C., and Pardee, A. B., *Cold Spring Harbor Symp. Quant. Biol.* **28,** 491 (1963). On the regulation of enzyme activity by protein subunit interactions.

Gerhart, J. C., and Schachman, H. K., *Biochemistry* **4,** 1054 (1965). On the molecular details of the subunit interactions.

ENZYMES OF KNOWN THREE-DIMENSIONAL STRUCTURE

Carboxypeptidase

Reeke, G. N., Hartouck, J. A., Ludwig, M. L., Quiocho, F. A., Steitz, T.A. and Lipscomb, W. N., *Proc. Natl. Acad. Sci., U.S.* **58,** 2220 (1967).

Carbonic Anhydrase

Fridborg, K., Kannan, K. K., Lilja, A., Lundin, J., Strandberg, B., Strandberg, R., Tilander, B., and Wirén, G., *J. Mol. Biol.* **25,** 505 (1967).

Lysozyme

Phillips, D. C., "The Three-Dimensional Structure of an Enzyme Molecule." *Scientific American,* November 1966. The first enzyme structure determination to atomic resolution is lucidly described in this well-illustrated article.

APPENDIX 8–I WIRE MODEL OF α-CHYMOTRYPSIN*

The wire model of mono-O-tosylated-(Ser-195)-α-chymotrypsin shown in Fig. I–1 was constructed from the electron density map at 2 Å resolution.

Note, relevant to the discussion in Section 8–1, that His-57 and the isoleucine terminal —NH_3^+ (B) are in proximity to the Ser-195 hydroxyl. This —NH_3^+ is in apparent contact with the —CO_2^- of Asp-194. [Figure from B. W. Mathews, P. B. Sigler, R. Henderson, and D. M. Blow, *Nature* **214,** 652 (1967).]

* The recently published material discussed in Appendixes 8–I and 8–II appeared too late to be included in the main body of the text.

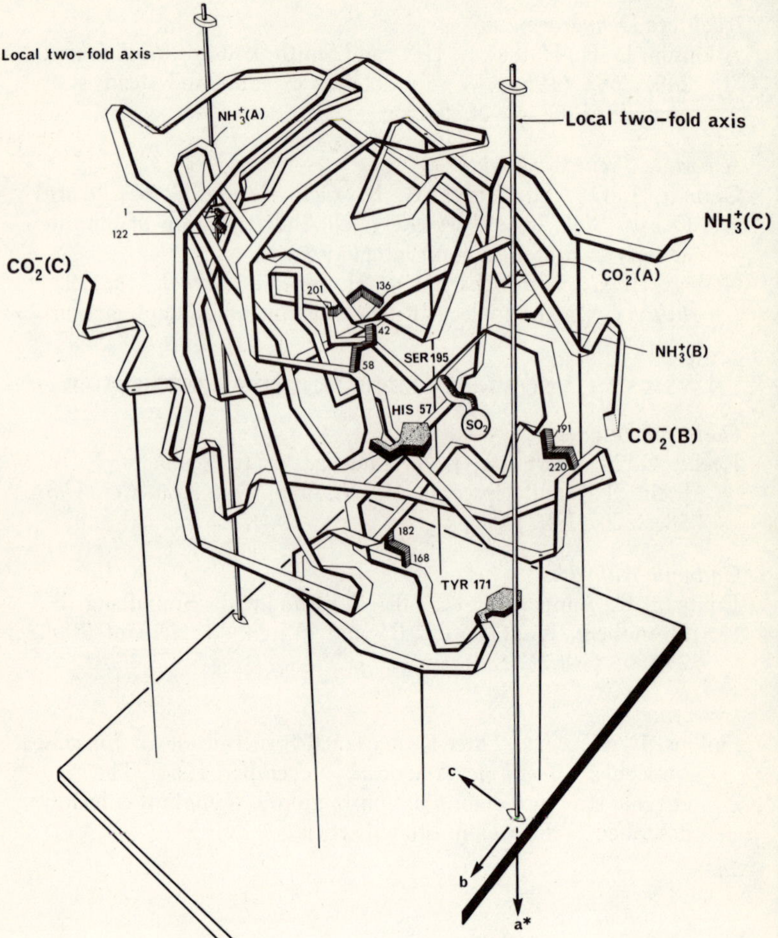

FIG. I–1.

APPENDIX 8–II SKELETAL MODEL OF RIBONUCLEASE-S

The skeletal model of ribonuclease-S, the recombined complex following specific subtilisin proteolysis at residue 20, is shown in Fig. II–1. The electron density map of RNase-S calculated from X-ray diffraction data on the protein and three heavy atom derivatives at 3.5 Å resolution is interpretable in terms of main

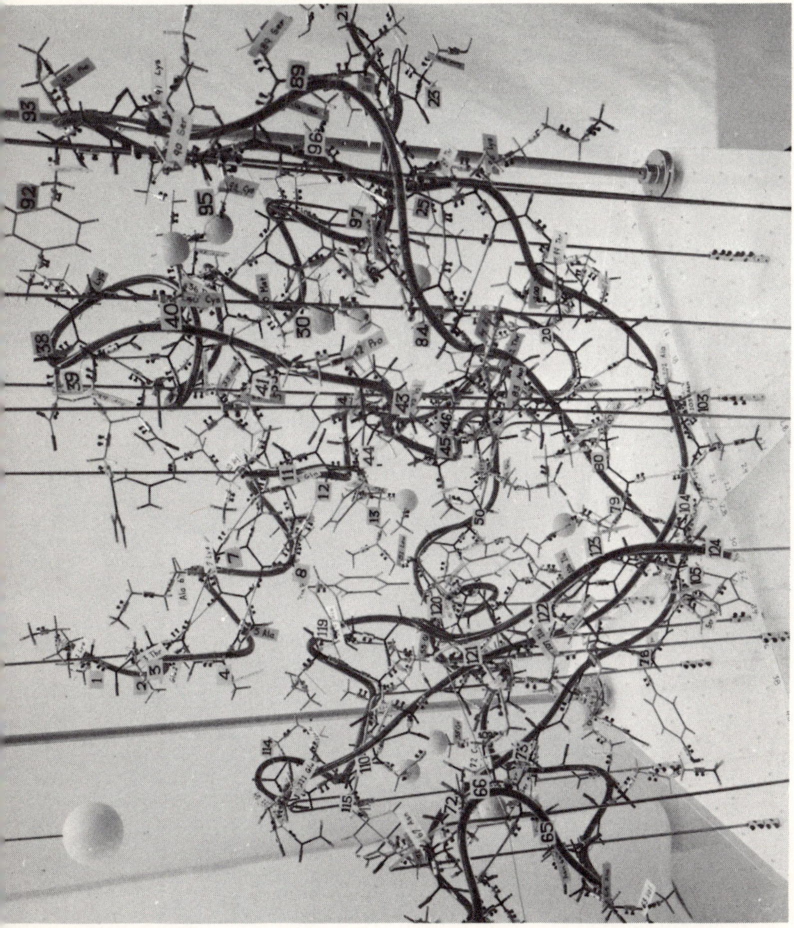

FIG. II-1.

chain and side chain conformation with the aid of pre-existing chemical sequence data and general stereochemical knowledge. Features of the structure include 15% helix, 15% hydrophobic core, and appreciable antiparallel β chain pairing. The configuration of the main chain and assignment of —S—S— bridges closely resembles the structure of RNase-A described by Kartha, Bellow, and Harker [*Nature* **213**, 862 (1967)], except where there is a chemical difference. The structure is also compatible with much of the relevant chemical literature: Note, in regard to the discussion in Section 8–2, that the two histidine residues (12 and 119) and the two lysine residues (7 and 41) implicated in the catalytic mechanism are in mutual proximity. [Figure from H. W. Wyckoff, K. D. Hardman, N. M. Allewell, T. Inagami, L. N. Johnson, and F. M. Richards, *J. Biol. Chem.* **242**, 3984 (1967).

INDEX